Encyclopaedia of Mathematical Sciences

Volume 63

Editor-in-Chief: R.V. Gamkrelidze

Yu. V. Egorov M. A. Shubin (Eds.)

Partial Differential Equations VI

Elliptic and Parabolic Operators

Springer-Verlag Berlin Heidelberg GmbH

Title of the Russian edition:
Itogi nauki i tekhniki, Sovremennye problemy matematiki,
Fundamental'nye napravleniya, Vol. 63,
Differentsial'nye uravneniya s chastnymi proizvodnymi 6
Publisher VINITI, Moscow 1990

Mathematics Subject Classification (1991):
35Jxx, 35J55, 35Kxx, 35Sxx, 58G03, 58G15

ISBN 978-3-642-08117-0 ISBN 978-3-662-09209-5 (eBook)
DOI 10.1007/978-3-662-09209-5

Cip data applied for

Originally published by Springer-Verlag Berlin Heidelberg New York in 1994
Softcover reprint of the hardcover 1st edition 1994

SPIN: 10020206 41/3140 - 5 4 3 2 1 0 - Printed on acid-free paper

List of Editors, Authors and Translators

Editor-in-Chief

R.V. Gamkrelidze, Russian Academy of Sciences, Steklov Mathematical Institute, ul. Vavilova 42, 117966 Moscow, Institute for Scientific Information (VINITI), ul. Usievicha 20a, 125219 Moscow, Russia, CIS

Consulting Editors

Yu. V. Egorov, Moscow State University, Leninskie Gory, 119899 Moscow, Russia, CIS

M. A. Shubin, Department of Mechanics and Mathematics, Moscow State University, Leninskie Gory, 119899 Moscow, Russia, CIS

Authors

M. S. Agranovich, Moscow Polytechnical Institute, Chair of Algebra and Analysis, Bolshoj Vuzovskij 3/12, 109028 Moscow, Russia, CIS

S. D. Ejdel'man, Kiev Institute of Radio-Engineering, 81 Mel'nikov Str., 252064 Kiev, Ukraina, CIS

S. Z. Levendorskij, Rostov Institute of National Economy, 69 Engels Str., 344709 Rostov-Don, Russia, CIS

B. Paneah, Department of Mathematics, Technion, 32000 Haifa, Israel

Translators

M. Capinski, Institute of Mathematics, Jagiellonian University, 30-059 Krakow, ul. Reymonta 4, Poland

R. Cooke, Department of Mathematics and Statistics, University of Vermont, 500 South Union Street, Burlington, VT 05401, USA

Contents

I. Elliptic Operators on Closed Manifolds

M. S. Agranovich

Translated from the Russian
by M. Capinski

Contents

Preface

0.1. The Scope of the Paper. This article is mainly devoted to the operators indicated in the title. More specifically, we consider elliptic differential and pseudodifferential operators with infinitely smooth symbols on infinitely smooth *closed* manifolds, i.e. compact manifolds without boundary. We also touch upon some variants of the theory of elliptic operators in $\mathbb{R}^n$. A separate article (Agranovich 1993) will be devoted to elliptic boundary problems for elliptic partial differential equations and systems.

We now list the main topics discussed in the article. First of all, we expound theorems on Fredholm property of elliptic operators, on smoothness of solutions of elliptic equations, and, in the case of ellipticity with a parameter, on their unique solvability. A parametrix for an elliptic operator A (and also for $A - \lambda I$) is constructed by means of the calculus of pseudodifferential operators in $\mathbb{R}^n$, which is first outlined in a simple case with uniform in x estimates of the symbols. As functional spaces we mainly use Sobolev L_2-spaces. We consider functions of elliptic operators and in more detail some simple functions and the properties of their kernels. This forms a foundation to discuss spectral properties of elliptic operators which we try to do in maximal generality, i.e., in general, without assuming selfadjointness. This requires presenting some notions and theorems of the theory of nonselfadjoint linear operators in abstract Hilbert space. This material is presented in separate sections and subsections and creates a background against which the specific features of elliptic operators can be better seen.

Obviously, the choice of topics has been determined by the personal interests of the author and by the article's length fixed in advance. Many interesting questions of the theory of elliptic equations were left out. This theory, as can be seen in any issue of any review journal, continues to get richer. The paper was written as an introductory one, and the author hopes that it can be read without much difficulty by any mathematician concerned in the subject.

We do not want to violate the logic of presentation, and so we again discuss some questions already covered in other volumes of the Encyclopaedia on partial differential equations (Egorov 1988), (Egorov and Shubin 1988b), and (Rozenblyum et al. 1989). As a rule, our method of presentation diverges from that of the cited books. Among the problems examined here that are not considered there at all, we would like to mention, first of all, the following ones: the structure of the kernels of polyhomogeneous pseudodifferential operators of nonpositive order, representation of pseudodifferential operators on a circle and a torus by the Fourier type series and complete symbolic calculus for these operators, calculus of pseudodifferential operators with a parameter, the relations between some simple functions of an elliptic operator, difference in structure between the parametrices of the resolvents of differential and pseudodifferential elliptic operators, complete asymptotics of the kernel and the trace of the resolvent, spectral theory on a circle, very weak perturbations of selfadjoint elliptic operators, elliptic operators far from being selfadjoint, some spectral problems for matrix operators elliptic in the sense of Douglis–Nirenberg, and examples of elliptic operators with infinite but not complete system of eigenfunctions.

0.2. Some Notation and Conventions. We use the standard multi-index notation adopted in the theory of partial differential equations. In particular,

$$\partial_x^\alpha = \frac{\partial^{|\alpha|}}{\partial(x^1)^{\alpha_1}\dots\partial(x^n)^{\alpha_n}}\,, \quad D_n^\alpha = (-\mathrm{i})^{|\alpha|}\partial_x^\alpha\,, \quad |\alpha| = \alpha_1 + \dots + \alpha_n\,,$$

$$\xi^\alpha = \xi_1^{\alpha_1}\dots\xi_n^{\alpha_n}\,, \quad \alpha! = \alpha_1!\dots\alpha_n!$$

($x = (x^1,\dots,x^n) \in \mathbb{R}^n$, $\xi = (\xi_1,\dots,\xi_n) \in \mathbb{R}^n$, $\alpha = (\alpha_1,\dots,\alpha_n) \in \mathbb{Z}_+^n$). The Fourier and the inverse Fourier transforms are defined by the formulas

$$v(\xi) = (Fu)(\xi) = \int \mathrm{e}^{-\mathrm{i}x\cdot\xi}u(x)dx\,, \quad u(x) = (F^{-1}v)(x) = \int \mathrm{e}^{\mathrm{i}x\cdot\xi}v(\xi)đ\xi\,.$$

Here and in what follows the integral without the limits is taken over the whole space;

$$x\cdot\xi = x^1\xi_1 + \dots + x^n\xi_n\,, \quad đ\xi = (2\pi)^{-n}d\xi \quad (x \in \mathbb{R}^n, \quad \xi \in \mathbb{R}_n)\,.$$

When considering a manifold M, we shall assume, unless otherwise stated, that it is n–dimensional, connected, closed, and of class C^∞. We also assume that it is equipped with a positive density dx of class C^∞. The local coordinates $(x^1,\dots,x^n)$ of a point x from a coordinate neighborhood O on M are the Euclidean coordinates of the image of this point in the corresponding open set $U \subset \mathbb{R}^n$. In addition, we assume, unless otherwise stated, that the local coordinates are *consistent with the density* dx. This means that the integral over O of a function $f(x)$ with support contained in O can be written as the Lebesgue integral $\int f(x)dx$ without an additional factor $\rho(x)$ under the integral sign. Such local coordinates can always be introduced, and this convention makes the study of the kernels of various operators much easier.

We assume that the reader is familiar with some elementary notions from

1) the theory of distributions and their Fourier transforms,
2) Sobolev L_2–spaces,
3) differential geometry, including cotangent bundle T^*M and some other simple notions,
4) linear operators in Hilbert spaces.

For references one can use for example

1) (Vladimirov 1979), (Egorov and Shubin 1988b), and (Shilov 1984);
2) (Egorov and Shubin 1988b), (Taylor 1981), and (Tréves 1980);
3) (Alekseevskij et al. 1988) and (Dubrovin et al. 1986);
4) (Gohberg and Krejn 1965), (Dunford and Schwartz 1958, 1963), and (Reed and Simon 1978).

Some of the notions will be recalled.

§1. Pseudodifferential and Elliptic Pseudodifferential Operators in $\mathbb{R}^n$

The references for §1 and §2 are: (Egorov 1984 and 1988), (Shubin 1978), (Friedrichs 1968), (Hörmander 1985a), (Kumano-go 1981), (Taylor 1981), and (Tréves 1980).

1.1. Definitions and Examples. Boundedness Theorems. Let $a(x,\xi)$ be a function from $C^\infty(\mathbb{R}^n \times \mathbb{R}_n)$. We write $a \in S^m = S^m(\mathbb{R}^n \times \mathbb{R}_n)$ $(m \in \mathbb{R})$ if for any α, $\beta \in \mathbb{Z}_+^n$

$$|\partial_x^\alpha \partial_\xi^\beta a(x,\xi)| \leq C_{\alpha\beta}(1+|\xi|)^{m-|\beta|}, \tag{1.1}$$

where[1] the constants $C_{\alpha\beta} = C_{\alpha\beta}(a)$ do not depend on (x,ξ). *A pseudodifferential operator* $A = a(x,D)$ *in* $\mathbb{R}^n$ *with symbol* $a(x,\xi) \in S^m$ is defined by the formula

$$Au(x) = \int e^{ix\cdot\xi} a(x,\xi)(Fu)(\xi) đ\xi \,. \tag{1.2}$$

It is sufficient to assume here that the function u belongs to the Schwartz space $S(\mathbb{R}^n)$ of infinitely smooth functions with finite norms $(N = 0, 1, 2, \ldots)$

$$\max_{|\alpha|\leq N; x} (1+|x|)^N |\partial^\alpha u(x)| \,. \tag{1.3}$$

The integral is absolutely convergent in this case, and it is easy to prove the following theorem.

Theorem 1.1.1. *The pseudodifferential operator given by* (1.2) *continuously maps* $S(\mathbb{R}^n)$ *into itself.*

The set of all pseudodifferential operators in $\mathbb{R}^n$ with symbols from S^m is denoted by $\Psi^m = \Psi^m(\mathbb{R}^n)$.

A simple and important example is a differential operator

$$A = \sum_{|\alpha|\leq m} a_\alpha(x) D^\alpha \tag{1.4}$$

$(m \in \mathbb{N})$ with coefficients $a_\alpha(x)$ belonging to the space $B^\infty(\mathbb{R}^n)$ of infinitely smooth functions with bounded derivatives of arbitrary order. In this case

$$a(x,\xi) = \sum_{|\alpha|\leq m} a_\alpha \xi^\alpha \,. \tag{1.5}$$

The operator of multiplication by a function $a(x) \in B^\infty(\mathbb{R}^n)$ is a pseudodifferential operator from Ψ^0 with symbol $a(x)$. For $m < -n$ any pseudodifferential operator from Ψ^m is an integral operator

[1] In what follows similar assumptions about the constants are always imposed and will not be repeated.

$$Au(x) = \int K(x, y)u(y)dy \tag{1.6}$$

with continuous kernel

$$K(x, y) = \int e^{i(x-y)\cdot\xi} a(x, \xi) đ\xi \,. \tag{1.7}$$

(We shall dwell on its properties in some detail at the beginning of Sect. 1.3.) This kernel becomes smoother as $m \to -\infty$. We put

$$S^{-\infty} = \bigcap_m S^m \,, \quad S^{\infty} = \bigcup_m S^m \,, \quad \Psi^{-\infty} = \bigcap_m \Psi^m \,, \quad \Psi^{\infty} = \bigcup_m \Psi^m \,. \tag{1.8}$$

It is easy to prove that the class $\Psi^{-\infty}$ coincides with the class of integral operators (1.6), where the kernels $K(x, y)$ are C^∞ and decrease rapidly together with all their derivatives as $|x - y| \to \infty$:

$$|\partial_x^\alpha \partial_y^\beta K(x, y)| \leq C_{\alpha\beta N}(1 + |x - y|)^{-N} \tag{1.9}$$

for any α, β, and N.

In Sect. 1.2 we describe a 'symbolic calculus' of pseudodifferential operators. It is constructed modulo addition of symbols from $S^{-\infty}$ and operators from $\Psi^{-\infty}$.

Operators from $\Psi^\infty(\mathbb{R})$ can be conveniently considered in the scale of *Sobolev spaces* $H_s(\mathbb{R}^n) = W_2^{(s)}(\mathbb{R}^n)$ ($s \in \mathbb{R}$). We recall that $H_s(\mathbb{R}^n)$ can be defined as the completion of $S(\mathbb{R}^n)$ with respect to the norm

$$\|u\|_s = \|u\|_{s,\mathbb{R}^n} = \left(\int |(Fu)(\xi)|^2 <\xi>^{2s} đ\xi \right)^{1/2} \,, \tag{1.10}$$

where $<\xi>= (1 + |\xi|^2)^{1/2}$. This norm equals $\|\Lambda^s u\|_0$, where Λ^s is the pseudodifferential operator from Ψ^s with the symbol $<\xi>^s$:

$$\Lambda^s = F^{-1} <\xi>^s F = F^{-1}(1 + |\xi|^2)^{s/2} F \,. \tag{1.11}$$

The space $H_0(\mathbb{R}^n)$ coincides with $L_2(\mathbb{R}^n)$.

Theorem 1.1.2. *If $A \in \Psi^m(\mathbb{R}^n)$, then we have the inequality*

$$\|Au\|_{s-m} \leq C_s \|u\|_s \quad (u \in S(\mathbb{R}^n)) \,, \tag{1.12}$$

and A can be extended to a continuous operator from $H_s(\mathbb{R}^n)$ to $H_{s-m}(\mathbb{R}^n)$ for any s.

The proof employs the composition theorem (see Sect. 1.2 below). We note that if the symbol does not depend on x, then the assertion is obvious.

An operator A that is bounded from $H_s(\mathbb{R}^n)$ to $H_{s-m}(\mathbb{R}^n)$ for all s is called an operator *of order m in $\mathbb{R}^n$*. If A has order m for any $m \in \mathbb{R}$ (or for negative m with arbitrarily large absolute value), then we say that A has *order* $-\infty$. So if A belongs to Ψ^m (to $\Psi^{-\infty}$), it is an operator of order m ($-\infty$, respectively) in $\mathbb{R}^n$. An example of an operator of order $-\infty$ which does not belong to $\Psi^{-\infty}$ is $T = F^{-1}a(\xi)F$, where $a(\xi)$ is a nonsmooth function with bounded support.

Remark. From now on until Sect. 3.1 we assume that the estimates (1.1) are uniform in x. This assumption is essential for Theorems 1.1.1 and 1.1.2. It also allows us to manage without introducing the so called properly supported pseudodifferential operators (see Sect. 3.1). We do not need more general operators to build the theory of elliptic pseudodifferential operators on a closed manifold. The notation Ψ^m in the case of more general estimates (see 3.1f) is used in (Hörmander 1985a) and (Trèves 1980). The alternative notation L^m is used in (Shubin 1978) and OPS^m in (Taylor 1981).

We assume that all functions are scalar until Sect. 3.2.

Pseudodifferential operators can be considered using L_p-Sobolev spaces and Hölder spaces C^α instead of H_s. The interested reader is referred to (Taylor 1981).

1.2. A Calculus of Pseudodifferential Operators

a. The pseudodifferential operator given by (1.2) can be written in the form of the 'iterated' integral (with integration first in y and next in ξ):

$$Au(x) = \int\int e^{i(x-y)\cdot\xi} a(x,\xi)u(y)dy đ\xi \,. \tag{1.13}$$

We consider an operator of more general form

$$Au(x) = \int\int e^{i(x-y)\cdot\xi} p(x,y,\xi)u(y)dy đ\xi \,. \tag{1.14}$$

Here p is called the *amplitude function* of A. We assume that p belongs to $S^m(\mathbb{R}^{2n} \times \mathbb{R}_n)$ for some $m \in \mathbb{R}$, i.e. $p \in C^\infty$ and

$$|\partial_x^\alpha \partial_y^\beta \partial_\xi^\gamma p(x,y,\xi)| \le C_{\alpha\beta\gamma}(1+|\xi|)^{m-|\gamma|} \tag{1.15}$$

for all α, β, and γ. Let $u \in S(\mathbb{R}^n)$. For $m < -n$ the 'double' integral (1.14) is absolutely convergent. For $m \ge -n$ it can be considered, for example, as the 'iterated' integral: it is easy to see that the inner integral in y decreases rapidly in absolute value as $|\xi| \to \infty$. There are also other methods of regularization of (1.14). For example, we can put

$$Au(x) = \lim_{\varepsilon \searrow 0} \int\int \theta_0(\varepsilon\xi) e^{i(x-y)\cdot\xi} p(x,y,\xi)u(y)dy đ\xi \,, \tag{1.16}$$

where $\theta_0(\xi)$ is a function from $C_0^\infty(\mathbb{R}_n)$ which is equal to 1 in a neighborhood of the origin. In particular, we can apply this to the integral (1.13).

The amplitude function of the operator (1.14) is not uniquely defined. It turns out that this operator can always be represented in the form (1.13):

Theorem 1.2.1. *The operator* (1.14) *with amplitude function from* S^m *is a pseudodifferential operator from* $\Psi^m(\mathbb{R}^n)$ *with symbol* $a(x,\xi)$ *having the following asymptotic expansion:*

$$a(x,\xi) \sim \sum \frac{1}{\alpha!} \partial_\xi^\alpha D_y^\alpha p(x,y,\xi)\,|_{y=x} \;. \tag{1.17}$$

We shall now explain the meaning of this and other asymptotic formulas below. Let $a_j(x,\xi)$ be functions from S_{m-j} $(j=0,1,\dots)$ and let $a(x,\xi) \in S^m$. We write

$$a(x,\xi) \sim \sum_j a_j(x,\xi) \tag{1.18}$$

and we call $a(x,\xi)$ *the asymptotic sum* of the formal series on the right if for any $N \in \mathbb{Z}_+$

$$a(x,\xi) - \sum_0^N a_j(x,\xi) \in S^{m-N-1} \,. \tag{1.19}$$

The proof of Theorem 1.2.1 is based on the realization of the following formal argument. The amplitude $p(x,y,\xi)$, as a function of y, can be expanded in powers of $y-x$ by Taylor's formula. Next, we can write $(y-x)^\alpha e^{i(x-y)\cdot\xi}$ in the form $(-1)^{|\alpha|} D_\xi^\alpha e^{i(x-y)\cdot\xi}$. Integration by parts with respect to ξ transfers D_ξ^α onto the coefficients of the expansion.

If an operator A is given in the form (1.14) with

$$p(x,y,\xi) = w\left(\frac{x+y}{2},\, \xi\right), \quad w(x,\xi) \in S^\infty(\mathbb{R}^n \times \mathbb{R}_n)\,,$$

then the function $w(x,\xi)$ is called the *Weyl symbol* of A.

b. Theorem 1.2.2. *For an arbitrary (formal) series consisting of functions* $a_j(x,\xi) \in S^{m-j}$ $(j=0,1,\dots)$ *there exists an asymptotic sum* $a(x,\xi) \in S^m$.

The function $a(x,\xi)$ can be constructed in the form

$$a(x,\xi) = \sum_{j=0}^\infty \theta\left(\frac{\xi}{h_j}\right) a_j(x,\xi)\,,$$

where $\theta(\xi)$ is a function from $B^\infty(\mathbb{R}_n)$, equal to 1 outside a neighborhood of the origin and 0 inside a smaller one, and h_j are positive numbers converging to ∞ sufficiently quickly. The asymptotic sum of the series is clearly defined modulo addition of a function from $S^{-\infty}$.

c. Let A be an operator acting in $S(\mathbb{R}^n)$ (this assumption is sufficient for our purposes). The operator A^*, also acting in $S(\mathbb{R}^n)$, is called *formally adjoint to* A if

$$(Au, v) = (u, A^*v) \quad \text{for} \quad u, v \in S(\mathbb{R}^n) . \tag{1.20}$$

Here and in what follows $(\cdot, \cdot)$ denotes the standard scalar product in $L_2(\mathbb{R}^n)$:

$$(u, v) = (u, v)_0 = (u, v)_{0,\mathbb{R}^n} = \int u(x)\overline{v(x)}dx . \tag{1.21}$$

The norm (1.10) introduced above coincides with $(u, u)^{1/2}$ for $s = 0$. If $A = A^*$, then A is called a *formally selfadjoint operator.* (The 'nonformal' role of A^* will be discussed in 2.3d below.) If A is written in the form (1.14) with $p = w(\frac{x+y}{2}, \xi)$, then A is formally selfadjoint if and only if the Weyl symbol w is real.

An operator A' acting in $S(\mathbb{R}^n)$ is called *formally transposed to A* if

$$<Au, v>=<u, A'v> \quad \text{for} \quad u, v \in S(\mathbb{R}^n) , \tag{1.22}$$

where

$$<u, v> = (u, \bar{v}) = \int u(x)v(x)dx . \tag{1.23}$$

Obviously, $A^*v = \overline{A'v}$. If A is a differential operator given by (1.4), then

$$A^* = \sum_{|\alpha|\leq m} D^\alpha[\overline{a_\alpha(x)}\cdot] , \quad A' = \sum_{|\alpha|\leq m} (-D)^\alpha[a_\alpha(x)\cdot] , \tag{1.24}$$

where the multiplication by coefficients precedes the differentiation. If A is a pseudodifferential operator given by (1.2), then, as is easy to see,

$$A^*v(x) = \int\int e^{i(x-y)\cdot\xi}\overline{a(y, \xi)}v(y)dy đ\xi , \tag{1.25}$$

$$A'v(x) = \int\int e^{i(x-y)\cdot\xi}a(y, -\xi)v(y)dy đ\xi . \tag{1.26}$$

Using Theorem 1.2.1, we obtain the following

Theorem 1.2.3. *Suppose that* $A = a(x, D) \in \Psi^m(\mathbb{R}^n)$. *Then* A^* *and* A' *belong to* $\Psi^m(\mathbb{R}^n)$, *and for their symbols we have*

$$a^{(*)}(x, \xi) \sim \sum \frac{1}{\alpha!}\partial_\xi^\alpha D_x^\alpha \overline{a(x, \xi)} \tag{1.27}$$

and

$$a^{(')}(x, \xi) \sim \sum \frac{1}{\alpha!}\partial_\xi^\alpha D_x^\alpha a(x, -\xi) . \tag{1.28}$$

If $a^{(')}(x, \xi)$ is the symbol of A', then the function

$$a_d(x, \xi) = a^{(')}(x, -\xi)$$

is called the *dual symbol* of A. Any pseudodifferential operator A can be expressed in terms of its dual symbol by the formula

$$Au(x) = \int\int e^{i(x-y)\cdot\xi} a_d(y,\xi)u(y)dy\text{đ}\xi \, . \tag{1.29}$$

For differential operators this expression is equivalent to the following one:

$$Au(x) = \sum_{|\alpha|\le m} D^\alpha[b_\alpha(x)u(x)] \, ,$$

and in this case $a_d(x,\xi) = \sum b_\alpha(x)\xi^\alpha$.

Theorem 1.2.3 allows us to define a pseudodifferential operator A on the Schwartz space $S'(\mathbb{R}^n)$ of distributions (linear continuous functionals on $S(\mathbb{R}^n)$) by the formula

$$<Af,\varphi> = <f, A'\varphi> \quad (\varphi \in S(\mathbb{R}^n)) \, , \tag{1.30}$$

where $<u,v>$ is the result of applying the distribution u to the test function v, which coincides with (1.23) if u is a regular distribution represented by the function $u(x)$ and the function $u(x)v(x)$ is in $L_1(\mathbb{R}^n)$ for all $v \in S(\mathbb{R}^n)$. We see at once that A is continuous in $S'(\mathbb{R}^n)$.

In particular, the expression $Ae^{ix\cdot\eta}$ makes sense for any fixed $\eta \in \mathbb{R}_n$. For the symbol $a(x,\eta)$ we obtain the formula

$$a(x,\eta) = e^{-ix\cdot\eta} A e^{ix\cdot\eta} \, . \tag{1.31}$$

The same formula is true for the symbol of any operator A that is well-defined on functions of the form $e^{ix\cdot\eta}$ and maps them into functions. If, in addition, A is continuous in $S'(\mathbb{R}^n)$ and the function (1.31) belongs to S^m, then A is a pseudodifferential operator from $\Psi^m(\mathbb{R}^n)$ with symbol (1.31).

d. We now state the *composition theorem* for pseudodifferential operators.

Theorem 1.2.4. *If A belongs to $\Psi^{m_1}(\mathbb{R}^n)$ and B to $\Psi^{m_2}(\mathbb{R}^n)$, then $C = AB$ is an operator from $\Psi^{m_1+m_2}(\mathbb{R}^n)$ with symbol*

$$c(x,\xi) \sim \sum_\alpha \frac{1}{\alpha!}\partial_\xi^\alpha a(x,\xi) D_x^\alpha b(x,\xi) \, . \tag{1.32}$$

For the proof we express B by means of the dual symbol and use Theorem 1.2.1.

If A is a differential operator, the sum on the right in (1.32) is finite and $\sim$ can be replaced by $=$. In the case of two differential operators the assertion can be verified directly (and it is sufficient to consider $A = a(x)D^\alpha$ and $B = b(x)D^\beta$). If $b(\xi)$ does not depend on x, the assertion follows at once from the definition of a pseudodifferential operator.

From (1.32) we can easily deduce that the commutator $[A,B] = AB - BA$ belongs to $\Psi^{m_1+m_2-1}(\mathbb{R}^n)$.

1.3. The Kernel of a Pseudodifferential Operator and the Pseudolocal Property. As we have already mentioned, for $m < -n$ an operator $A \in \Psi^m(\mathbb{R}^n)$ is an integral operator (1.6) with a continuous kernel $K(x, y)$. It can be easily verified that the kernel belongs to C^∞ and rapidly decreases in modulus as $|x-y| \to \infty$, together with the derivatives of an arbitrarily fixed order:

$$|x-y|^N |\partial_x^\alpha \partial_y^\beta K(x,y)| \leq C_{\alpha\beta N} \quad (|x-y| \geq C > 0) . \tag{1.33}$$

For $m \geq -n$ we write the pseudodifferential operator $A \in \Psi^m$ in the form $(A\Lambda^{-2l})\Lambda^{2l}$, where l is a positive integer and $m - 2l < -n$. As is clear from (1.11), the operator Λ^{2l} coincides with $(1-\Delta)^l$, where Δ is the Laplace operator. Therefore we obtain

$$Au(x) = \sum_{|\alpha| \leq 2l} \int K_\alpha(x,y) D^\alpha u(y) dy ,$$

where the properties of the kernels K_α are similar to the properties of K. Hence in the case $m \geq -n$ the Schwartz kernel of A (see for example (Egorov and Shubin 1988b)) clearly coincides for $x \neq y$ with an infinitely smooth function

$$\sum_{|\alpha| \leq 2l} (-1)^{|\alpha|} D_y^\alpha K_\alpha(x,y) , \tag{1.34}$$

which rapidly decreases in modulus as $|x-y| \to \infty$, together with all its derivatives. The Schwartz kernel of the differential operator (1.4) is 0 for $x \neq y$, i.e. its support lies on the diagonal of $\mathbb{R}^n \times \mathbb{R}^n$.

If A is a differential operator, then A has the *local property*

$$\operatorname{supp} Au \subset \operatorname{supp} u \quad (u \in S'(\mathbb{R}^n)) . \tag{1.35}$$

If A is a pseudodifferential operator but not a differential operator, then it does not have the local property, but as a result of what was said about the kernel $K(x,y)$, it has the *pseudolocal property*: for $u \in S'(\mathbb{R}^n)$

$$\operatorname{sing\,supp} Au \subset \operatorname{sing\,supp} u . \tag{1.36}$$

1.4. Change of Variables in Pseudodifferential Operators. Let $A = a(x, D)$ be a pseudodifferential operator from $\Psi^m(\mathbb{R}^n)$. Let Ω and $\widetilde{\Omega}$ be two domains in $\mathbb{R}^n$ with a diffeomorphism (of class C^∞)

$$\Omega \ni x \mapsto y = \varphi(x) \in \widetilde{\Omega} , \tag{1.37}$$

and let $x = \psi(y)$ be the inverse mapping. We take two functions $\alpha_j(x)$ ($j = 1, 2$) from $C_0^\infty(\mathbb{R}^n)$ with supports contained in Ω and such that $\alpha_2 \equiv 1$ in a neighborhood of the support of $\alpha_1(x)$. Define the operator $\widetilde{A}$ by

$$(\widetilde{A}v)(y) = \alpha_1(x) A[\alpha_2 u](x) \,|_{x=\psi(y)} , \tag{1.38}$$

where $u(x) = (v \circ \varphi)(x)$ for $x \in \Omega$.

Theorem 1.4.1. *The operator* $\widetilde{A}$ *is a pseudodifferential operator belonging to* $\Psi^m(\mathbb{R}^n)$ *with the symbol* $\widetilde{a}(y,\eta)$ *which has the following asymptotic expansion: for* $y=\varphi(x)$, $x\in\Omega$ *we have*

$$\widetilde{a}(\varphi(x),\eta)\sim\alpha_1(x)\sum_\gamma\frac{1}{\gamma!}D_z^\gamma e^{i\varphi_2(x,z)\cdot\eta}|_{z=x}\left(\partial_\xi^\gamma a\right)\left(x,\left(\frac{\partial\varphi(x)}{\partial x}\right)'\eta\right), \quad (1.39)$$

where $\partial\varphi/\partial x$ *is the Jacobi matrix, the dash denotes the transposed matrix, and*

$$\varphi_2(x,z)=\varphi(z)-\varphi(x)-\frac{\partial\varphi(x)}{\partial x}(z-x)\,. \quad (1.40)$$

This theorem permits us to define a pseudodifferential operator on a manifold (see Sect. 2.1 below).

1.5. Polyhomogeneous Pseudodifferential Operators. Principal and Subprincipal Symbols. In this section we consider a narrower class of pseudodifferential operators which is most important for us. We first introduce the class of homogeneous symbols. Let $b(x,\xi)$ be a function defined on $\mathbb{R}^n\times\dot{\mathbb{R}}_n$, where $\dot{\mathbb{R}}_n=\mathbb{R}_n\backslash\{0\}$. We write

$$b(x,\xi)\in S_{\mathrm{h}}^m=S_{\mathrm{h}}^m(\mathbb{R}^n\times\dot{\mathbb{R}}_n)\quad(m\in\mathbb{R})\,,$$

if $b(x,\xi)$ is positively homogeneous in ξ of degree m, i.e.

$$b(x,\tau\xi)=\tau^m b(x,\xi)\quad(\tau>0,\ \ \xi\neq0) \quad (1.41)$$

and if for any α and β we have

$$|\partial_x^\alpha\partial_\xi^\beta b(x,\xi)|\le C_{\alpha\beta}|\xi|^{m-|\beta|}\quad(\xi\neq0)\,. \quad (1.42)$$

Definition 1.5.1. Let A be a pseudodifferential operator from $\Psi^m(\mathbb{R}^n)$ with symbol $a(x,\xi)$. We say that A is a *polyhomogeneous*, or *classical*, pseudodifferential operator , and we write $A\in\Psi_{\mathrm{ph}}^m=\Psi_{\mathrm{ph}}^m(\mathbb{R}^n)$ if there is a sequence of homogeneous symbols $a_j(x,\xi)\in S_{\mathrm{h}}^{m-j}$ $(j=0,1,\ldots)$ such that

$$a(x,\xi)\sim\theta(\xi)a_0(x,\xi)+\theta(\xi)a_1(x,\xi)+\ldots \quad (1.43)$$

(in the same sense as in (1.18)) for some (and hence for any) function $\theta(\xi)\in B^\infty(\mathbb{R}_n)$ equal to 1 outside a neighborhood of the origin and equal to 0 in a smaller one. In such case we call $a(x,\xi)$ a *polyhomogeneous*, or *classical, symbol* and we write $a\in S_{\mathrm{ph}}^m=S_{\mathrm{ph}}^m(\mathbb{R}^n\times\dot{\mathbb{R}}_n)$. In place of (1.43) we shall write

$$a(x,\xi)\sim a_0(x,\xi)+a_1(x,\xi)+\ldots \quad (1.44)$$

assuming that this means that $a-\sum_0^N\theta a_j\in S^{m-N-1}$ for all N.

A differential operator (1.4) with the coefficients in $B^\infty(\mathbb{R}^n)$ is clearly a polyhomogeneous pseudodifferential operator belonging to Ψ_{ph}^m. The pseudodifferential operator (1.11) belongs to Ψ^s_{ph}, and the required asymptotic expansion of its symbol is obtained by means of the binomial series for $(|\xi|^{-2}+1)^{s/2}$. The function $<\xi> + <\xi>^{1/2} \ln <\xi>$ is an example of a symbol (from S^1) which is not polyhomogeneous.

In what follows we shall mainly consider polyhomogeneous pseudodifferential operators. More general operators from Ψ^∞ are convenient to build the calculus of pseudodifferential operators, and they appear in a natural way when we consider polyhomogeneous pseudodifferential operators depending on a parameter (see Sect. 4.2 below).

The function $a_0(x,\xi)$ (we assume that it does not identically vanish) is called the *principal symbol*[2] of the polyhomogeneous pseudodifferential operator A.

If A is a polyhomogeneous pseudodifferential operator with principal symbol a_0, then A^* and A' are also polyhomogeneous and their principal symbols are $\overline{a}_0$ and a_0, respectively. If A and B are polyhomogeneous pseudodifferential operators with principal symbols a_0 and b_0, then AB is polyhomogeneous and has the principal symbol $a_0 b_0$ (provided that the last product does not identically vanish), and the commutator $[A, B]$ has the principal symbol $i^{-1}\{a_0, b_0\}$ (if $\{a_0, b_0\} \not\equiv 0$), where $\{\cdot,\cdot\}$ is the *Poisson bracket*:

$$\{a_0, b_0\} = \sum_{\nu=1}^{n} (\partial_{\xi_\nu} a_0 \cdot \partial_{x^\nu} b_0 - \partial_{x^\nu} a_0 \cdot \partial_{\xi_\nu} b_0) \ . \tag{1.45}$$

All this can be easily derived from Theorems 1.2.3 and 1.2.4.

By Theorem 1.4.1, a change of variables (see Sect. 1.4) turns a polyhomogeneous pseudodifferential operator A into a polyhomogeneous one $\widetilde{A}$. Moreover, if $\alpha_1 \equiv 1$ in a subdomain $\Omega_1 \subset \Omega$, then for the corresponding y ($y = \varphi(x)$) the principal symbol of $\widetilde{A}$ is given by the formula

$$\widetilde{a}_0(y,\eta) = a_0(x,\xi) \ , \tag{1.46}$$

where on the right-hand side we have to make the substitution

$$x = \psi(y), \quad \xi = \left[\frac{\partial\psi(y)}{\partial y}\right]^{\prime -1} \eta \, . \tag{1.47}$$

The formula for $\widetilde{a}_1(y,\eta)$ is more complicated.

The *subprincipal symbol* of a polyhomogeneous pseudodifferential operator A is defined in the following way: if $w(x,\xi)$ is the Weyl symbol of A and $w(x,\xi) \sim w_0(x,\xi)+w_1(x,\xi)+\ldots$, then $(\operatorname{sub} A)(x,\xi) = w_1(x,\xi)$. (The function w_0 coincides with the principal symbol a_0.) This definition is equivalent to the following one:

---

[2] The notation $a \sim a_m + a_{m-1} + \ldots$ is often used in place of $a \sim a_0 + a_1 + \ldots$, and in this way the indices show the order of homogeneity. We shall always denote the principal symbol by the index 0.

$$(\operatorname{sub} A)(x,\xi) = a_1(x,\xi) - \frac{1}{2\mathrm{i}} \sum_{|\alpha|=1} \partial_\xi^\alpha D_x^\alpha a_0(x,\xi) \,. \tag{1.48}$$

The analogue of (1.46) for the subprincipal symbol has the form

$$(\operatorname{sub} \widetilde{A})(y,\eta) = (\operatorname{sub} A)(x,\xi) - \frac{1}{2\mathrm{i}} \sum_{|\alpha|=1} \partial_\xi^\alpha a_0(x,\xi) \partial_x^\alpha \ln |D(x)| \,, \tag{1.49}$$

where $D(x) = \det[\partial\varphi(x)/\partial x]$ and the substitution (1.47) is made on the right-hand side. In particular, if $|D(x)| \equiv 1$, then (1.49) becomes quite similar to (1.46):

$$(\operatorname{sub} \widetilde{A})(y,\eta) = (\operatorname{sub} A)(x,\xi) \,. \tag{1.50}$$

The subprincipal symbol of the product of polyhomogeneous pseudodifferential operators A and B with principal symbols a_0 and b_0 can be computed by the formula

$$\operatorname{sub}(AB) = \operatorname{sub} A \cdot b_0 + a_0 \operatorname{sub} B + (2\mathrm{i})^{-1}\{a_0, b_0\} \,. \tag{1.51}$$

We also note that if $A \in \Psi_{\mathrm{ph}}^m$, then

$$A - A^* \in \Psi_{\mathrm{ph}}^{m-1} \iff a_0 \equiv \bar{a}_0 \,; \tag{1.52}$$

$$A - A^* \in \Psi_{\mathrm{ph}}^{m-2} \iff a_0 \equiv \bar{a}_0 \,, \quad \operatorname{sub} A = \overline{\operatorname{sub} A} \,. \tag{1.53}$$

Proposition 1.5.2. *Let A be a pseudodifferential operator from Ψ_{ph}^m with principal symbol $a_0(x,\xi)$, and let $c(a_0) = \sup |a_0(x,\xi)|$ for $|\xi| = 1$ and $x \in \mathbb{R}^n$. Then for any $s \in \mathbb{R}$ and $\varepsilon > 0$ there is an operator T of order $-\infty$ such that*

$$\|(A-T)u\|_{s-m} \le [c(a_0) + \varepsilon] \, \|u\|_s \,. \tag{1.54}$$

1.6. Polyhomogeneous Pseudodifferential Operators of Negative Order as Integral Operators. For $m < 0$ a pseudodifferential operator $A \in \Psi_{\mathrm{ph}}^m$ always is an integral operator. We write it in the form

$$Au(x) = \int k(x, x-y) u(y) dy \,. \tag{1.55}$$

In this section we give a precise description of the properties of the kernels $k(x,z)$. Such a kernel is connected with the symbol $a(x,\xi)$ of A by the formula

$$a(x,\xi) = F_{z\to\xi} k(x,z). \tag{1.56}$$

Roughly speaking, we have the following situation. The kernel $k(x,z)$ is an infinitely smooth function for $z \neq 0$ which rapidly decreases in modulus, together with derivatives of all orders, as $|z| \to \infty$. For $z = 0$ this function has a weak (i.e. integrable) singularity characterized by the expansion of $k(x,z)$ in an asymptotic series with respect to kernels homogeneous in z in generalized sense (in the general case). To make the above more precise, we give two definitions.

Definition 1.6.1. Let $k_0(x,z)$ be a function from $C^\infty(\mathbb{R}^n \times \dot{\mathbb{R}}^n)$ and $\lambda > -n$. We say that $k_0 \in \mathcal{K}^\lambda_{\mathrm{gh}}$ if the following two conditions are satisfied.

1) If $\lambda \notin \mathbb{Z}_+$, then $k_0(x,z)$ is positively homogeneous in z of degree λ. If, however, $\lambda \in \mathbb{Z}_+$, then $k_0(x,z)$ is positively homogeneous in z of degree λ in the generalized sense:

$$k_0(x,z) = q_0(x,z) + p_0(x,z)\ln|z|\,, \tag{1.57}$$

where $q_0(x,z)$ is a function positively homogeneous in z of degree λ and $p_0(x,z)$ is a homogeneous polynomial in z of degree λ with coefficients from $B^\infty(\mathbb{R}^n)$.

2) In the former case ($\lambda \notin \mathbb{Z}$) we have

$$|\partial_x^\alpha \partial_z^\beta k_0(x,z)| \le C_{\alpha\beta}|z|^{\lambda-|\beta|} \quad (z \neq 0)\,, \tag{1.58}$$

and in the latter case we require similar inequalities to hold for $q_0(x,z)$.

Definition 1.6.2. Let $k(x,z)$ be a function from $C^\infty(\mathbb{R}^n \times \dot{\mathbb{R}}^n)$ ($\dot{\mathbb{R}}^n = \mathbb{R}^n\backslash\{0\}$). We call $k(x,z)$ a *polyhomogeneous kernel* and we write $k \in \mathcal{K}^\lambda_{\mathrm{ph}}$ if

1) for any α, β, and N

$$|\partial_x^\alpha \partial_z^\beta k(x,z)| \le C_{\alpha\beta N}|z|^{-N} \quad \text{for} \quad |z| \ge 1\,, \tag{1.59}$$

2) there is a sequence of kernels $k_j(x,z) \in \mathcal{K}^{\lambda+j}_{\mathrm{gh}}$, $j = 0,1,\ldots$, such that

$$k(x,z) \sim k_0(x,z) + \ldots + k_N(x,z) + \ldots \quad (z \to 0) \tag{1.60}$$

in the following sense: for any α, β, N, and for $0 < |z| \le h < 1$ we have

$$\left|\partial_x^\alpha \partial_z^\beta \left[k(x,z) - \sum_0^N k_j(x,z)\right]\right| \le C'_{\alpha\beta N}|z|^{\lambda+N+1-|\beta|}\,|\ln|z||\,. \tag{1.61}$$

For $\lambda \notin \mathbb{Z}_+$ the factor $|\ln|z||$ on the right-hand side should be omitted.

Theorem 1.6.3. *Let A be a polyhomogeneous pseudodifferential operator of order $m < 0$. Then A is an integral operator of the form* (1.55) *with a polyhomogeneous kernel $k(x,z) \in \mathcal{K}^\lambda_{\mathrm{gh}}$, where $\lambda = -n-m$. The converse is also true. Moreover, the terms a_j and k_j in the asymptotic expansions* (1.44) *and* (1.60) *of the symbol $a(x,\xi)$ of A and of the kernel $k(x,z)$ are connected by the formula*

$$a_j(x,\xi) = F_{z\to\xi}k_j(x,z) \quad \text{for} \quad \xi \neq 0\,, \tag{1.62}$$

where the Fourier transform and the equality (1.62) *(for fixed x) are understood in the sense of distributions.*

1.7. Polyhomogeneous Pseudodifferential Operators of Zero Order and Singular Integral Operators. A pseudodifferential operator A belongs to Ψ^0_{ph} if and only if it admits the representation

$$Au(x) = a_0(x)u(x) + \int k(x, x-y)u(y)dy \tag{1.63}$$

($u \in S(\mathbb{R}^n)$), where $a_0(x)$ is a function from $B^\infty(\mathbb{R}^n)$ and $k(x,z)$ is a kernel from $\mathcal{K}^{-n}_{\mathrm{ph}}$. The definition of this class is similar to the definition of the class $\mathcal{K}^{\lambda}_{\mathrm{ph}}$ for $\lambda > -n$ given in the previous section, but with the additional condition

$$\int_{|z|=1} k_0(x,z)dS_z = 0 , \tag{1.64}$$

where dS is the surface element of the unit sphere defined by

$$dz = r^{n-1}drdS, \quad \text{where} \quad r = |z| . \tag{1.65}$$

The integral in (1.63) is understood in the sense of the Cauchy principal value, i.e. as the limit

$$\lim_{\delta\to 0} \int_{|x-y|>\delta} k(x, x-y)u(y)dy .$$

Formula (1.56) is replaced by

$$a(x,\xi) = a_0(x) + F_{z\to\xi}k(x,z), \tag{1.66}$$

where in the computation of $F_{z\to\xi}k$ the integral is again understood in the sense of the principal value. Formula (1.62) is preserved for $j = 1, 2, \ldots$, and for $j = 0$ it is replaced by

$$a_0(x,\xi) = a_0(x) + F_{z\to\xi}k_0(x,z) \quad \text{for} \quad \xi \neq 0 \tag{1.67}$$

(in the sense of distributions, for a fixed x).

We note that the Schwartz kernel of the operator (1.63) has the form

$$a_0(x)\delta(x-y) + k(x, x-y), \tag{1.68}$$

where $\delta(z)$ is the Dirac delta-function. In the paper of Seeley (1965), which we have mainly followed in this and in the previous sections, the Schwartz kernels of polyhomogeneous pseudodifferential operators of positive order are also described.

Closely related to polyhomogeneous pseudodifferential operators are *singular integral operators*

$$A_0u(x) = a_0(x)u(x) + \int k_0(x, x-y)u(y)dy , \tag{1.69}$$

where the kernel $k_0(x,z)$ is positively homogeneous of degree $-n$ and satisfies condition (1.64). Here the integral is understood in the sense of the principal value at zero and at infinity, i.e. as the limit of the integrals over $\{y : \varepsilon < |x-y| < h\}$ as $\varepsilon \searrow 0$ and $h \nearrow \infty$. If we do not stint on the regularity hypotheses and assume that $a_0 \in B^\infty(\mathbb{R}^n)$ and $k_0 \in \mathcal{K}_{\text{gh}}^{-n}$, then the singular integral operator (1.69) is the sum of a pseudodifferential operator (1.63) of zero order (with $k(x,z) = \theta_0(z)k_0(x,z)$, where $\theta_0(z)$ is any function from $C_0^\infty(\mathbb{R}^n)$ equal to 1 in a neighborhood of the origin) and an operator of order $-\infty$. If, as is often done, we add $Tu(x)$ to the right-hand side of (1.69), where T is an arbitrary operator of order -1, then we obtain a class of operators broader than Ψ_{ph}^0.

1.8. Elliptic Operators in $\mathbb{R}^n$

Definition 1.8.1. A pseudodifferential operator A from Ψ_{ph}^m and its principal symbol $a_0(x,\xi)$ are called *elliptic at the point* $x \in \mathbb{R}^n$ if

$$a_0(x,\xi) \neq 0 \quad \text{for} \quad \xi \neq 0\,, \tag{1.70}$$

elliptic in Ω (where Ω is a subset of $\mathbb{R}^n$) if (1.70) holds for all $x \in \Omega$, and *uniformly elliptic in* Ω if

$$|a_0(x,\xi)| \geq c > 0 \quad \text{for} \quad x \in \Omega\,, \quad |\xi| = 1\,. \tag{1.71}$$

The uniform ellipticity without specifying the set Ω will denote the uniform ellipticity in $\mathbb{R}^n$.

For example, the pseudodifferential operator (1.11) is uniformly elliptic.

Definition 1.8.2. Let A be a pseudodifferential operator from Ψ^∞. A pseudodifferential operator $B \in \Psi^\infty$ is called a *parametrix for* A if

$$BA = I + T_1\,, \quad AB = I + T_2\,, \tag{1.72}$$

where I is the unit operator and T_1, $T_2 \in \Psi^{-\infty}$. If only the left (or right) of these two relations is satisfied, then we call B a *left* (or *right*, respectively) *parametrix for* A.

If A has a left parametrix B_1 and a right one B_2, then from the equality $(B_1A)B_2 = B_1(AB_2)$ it can be easily deduced that $B_1 - B_2 \in \Psi^{-\infty}$, so each of the pseudodifferential operators B_1 and B_2 is a (two-sided) parametrix for A.

Theorem 1.8.3. *Let A be a uniformly elliptic pseudodifferential operator from Ψ_{ph}^m. Then there exists a uniformly elliptic pseudodifferential operator $B \in \Psi_{\text{ph}}^{-m}$ which is a parametrix for A.*

We sketch the proof. Suppose that (1.44) and $b \sim b_0 + b_1 + \ldots$ are the asymptotic expansions of the symbols of A and B. By the composition theorem (Theorem 1.2.4) we have the asymptotic equality

$$\sum \frac{1}{\alpha!}\partial_\xi^\alpha[b_0 + b_1 + \ldots] \cdot D_\xi^\alpha[a_0 + a_1 + \ldots] \sim 1\,. \tag{1.73}$$

Comparing the terms of the same degree of homogeneity, we obtain the following chain of equalities, from which we can successively find $b_0, b_1, \ldots$:

$$b_0 a_0 = 1 \, , \tag{1.74$_0$}$$

$$b_0 a_1 + b_1 a_0 + \sum_{|\alpha|=1} \partial_\xi^\alpha b_0 \cdot D_x^\alpha a_0 = 0 \, , \tag{1.74$_1$}$$

and so on. We construct the symbol b having the expansion $\theta b_0 + \theta b_1 + \ldots$ (see Theorem 1.2.2), and so we obtain a pseudodifferential operator $B \in \Psi_{\mathrm{ph}}^{-m}$ which is a left parametrix for A. Similarly we construct a right parametrix, and by the remark preceding the theorem, each of them is a parametrix for A.

Simplicity of this argument is due to the fact that the technical difficulties have already been overcome within the calculus of pseudodifferential operators (in particular, we have in mind Theorem 1.2.4 on composition of pseudodifferential operators).

A parametrix is defined modulo addition of an operator from $\Psi^{-\infty}$. As a corollary we find that if a uniformly elliptic pseudodifferential operator $A \in \Psi_{\mathrm{ph}}^m$ has the inverse A^{-1} of order $-m$, then A^{-1} is a uniformly elliptic pseudodifferential operator from Ψ_{ph}^{-m}.

We give two important corollaries. They both can be easily verified using the left equality of (1.72).

Theorem 1.8.4 (on a priori estimate). *Let A be a uniformly elliptic pseudodifferential operator from Ψ_{ph}^{-m}. Then for any $s \in \mathbb{R}$ there exists a constant C_s independent of u such that*

$$\|u\|_s \leq C_s(\|Au\|_{s-m} + \|u\|_{s-1}) \quad (u \in H_s(\mathbb{R}^n)) \, . \tag{1.75}$$

The analogous estimate is true with $\|u\|_{s'}$ in place of $\|u\|_{s-1}$ for any $s' < s$.

Theorem 1.8.5 (on improved smoothness). *Let A be a uniformly elliptic pseudodifferential operator from Ψ_{ph}^{-m}. If $u \in H_s(\mathbb{R}^n)$ and $Au \in H_{s-m+h}(\mathbb{R}^n)$ where $h > 0$, then $u \in H_{s+h}(\mathbb{R}^n)$.*

In particular, if $Au \in H_\infty(\mathbb{R}^n) = \bigcap H_s(\mathbb{R}^n)$ (for example, if $Au = 0$), then $u \in H_\infty(\mathbb{R}^n)$. We recall that the functions from $H_\infty(\mathbb{R}^n)$ belong to $B^\infty(\mathbb{R}^n)$ (after correcting it, in general, on a set of measure zero[3]). This follows from the *embedding theorem* for Sobolev spaces: if $s > k + (n/2)$, then $H_s(\mathbb{R}^n)$ is continuously embedded in the space $B^{(k)}(\mathbb{R}^n)$ of functions $u(x)$ continuous and bounded together with the derivatives $\partial^\alpha u(x)$ of order $|\alpha| \leq k$. (The norm in $B^{(k)}(\mathbb{R}^n)$ is of course defined as the supremum of the moduli of the derivatives of orders $|\alpha| \leq k$.)

The last assertions can be given in a local form. In particular, we have

[3] We do not mention this in what follows.

Theorem 1.8.6. *Suppose that* $A \in \Psi^m_{\text{ph}}$ *for some* m, $u \in S'(\mathbb{R}^n)$, *and* $Au = f$. *Suppose that* Ω *is a ball in* $\mathbb{R}^n$, $f \in C^\infty$ *in* Ω, *and* A *is elliptic in* Ω. *Then* $u \in C^\infty$ *in* Ω.

Hence it follows that for the pseudodifferential operators that are elliptic in $\mathbb{R}^n$ we have the inclusion inverse to (1.36):

$$\operatorname{sing\,supp} Au \supset \operatorname{sing\,supp} u \tag{1.76}$$

($u \in S'(\mathbb{R}^n)$), so that for an elliptic operator A the left and the right-hand sides of (1.76) coincide. More precise result is that the wave fronts coincide (for the definition see, for example, (Egorov and Shubin 1988b)):

$$WF(u) = WF(Au)\,. \tag{1.77}$$

Returning to Theorem 1.8.4, we note that the converse assertion is also true and can be given in a local form:

Theorem 1.8.7. *Suppose that for a pseudodifferential operator* $A \in \Psi^m_{\text{ph}}$ *and all functions* $u \in H_s(\mathbb{R}^n)$ *(for some* s*) with supports contained in a sufficiently small neighborhood* O *of the point* x_0 *the inequality* (1.75) *holds with a constant independent of* u. *Then* A *is elliptic at* x_0. *If* (1.75) *holds for all functions* $u \in H_s(\mathbb{R}^n)$, *then* A *is uniformly elliptic.*

The theorem can be proved by contradiction, inserting suitable families of functions depending on a parameter into the a priori estimate.

Thus we see that the following assertions for $A \in \Psi^m_{\text{ph}}$ are equivalent:

1) A is uniformly elliptic.
2) A has a (uniformly elliptic) parametrix.
3) The a priori estimate (1.75) holds (for some, hence for any s).

§2. Pseudodifferential Operators and Elliptic Pseudodifferential Operators on a Closed Manifold

We recall that the assumptions about the manifold M and the local coordinates on M were given at the end of Sect. 0.2.

2.1. Pseudodifferential Operators on a Manifold M

Definition 2.1.1. Let A be an operator acting in $C^\infty(M)$. We call it a *pseudodifferential operator of class* $\Psi^m(M)$ ($m \in \mathbb{R}$) if the following conditions are satisfied:

1) Let φ, ψ be functions from $C^\infty(M)$ with disjoint supports. Then the composition $\varphi A(\psi\cdot)$ extends to an operator of order $-\infty$ on M (i.e. in the scale of Sobolev spaces $\{H_s(M)\}$).

2) Let Ω be a domain in M with the closure $\overline{\Omega}$ contained in a coordinate neighborhood O with the chart $O \to U \subset \mathbb{R}^n$ and local coordinates x. Then there is a pseudodifferential operator $A_\Omega \in \Psi^m(\mathbb{R}^n)$ such that if φ and ψ are functions from $C^\infty(M)$ with supports contained in Ω, then in local coordinates

$$\varphi A(\psi u)(x) = \varphi(x)(A_\Omega(\psi u))(x)\ . \tag{2.1}$$

Here and in what follows we make some natural identifications. Namely, the functions ψu and φ (with supports in Ω) on the right-hand side are transported to U and extended on $\mathbb{R}^n$ as functions that are equal 0 outside U. After the right-hand side is computed as a function on $\mathbb{R}^n$, it is restricted to U, transported to O, and extended on M as a function that is equal 0 outside O. In what follows similar identifications are always made and will not be mentioned explicitly. Regarding the space $H_s(M)$ we recall that it can be defined as the completion of $C^\infty(M)$ with respect to the norm

$$\|u\|_s = \|u\|_{s,M} = \left(\sum_{k=1}^{K} \|\varphi_k u\|_{s,\mathbb{R}^n}\right)^{1/2}\ . \tag{2.2}$$

Here and in what follows $\{\varphi_k\}_1^K$ is a *partition of unity* on M ($\sum \varphi_k(x) \equiv 1$), consisting of infinitely smooth functions, subordinated to the covering $\{O_k\}_1^K$ of the manifold M by the coordinate neighborhoods (i.e. $\operatorname{supp}\varphi_k \subset O_k$), and the norm $\|\varphi_k u\|_{s,\mathbb{R}^n}$ is computed in local coordinates. Various norms of the form (2.2) with the same s are *equivalent*, i.e. their ratio is contained between positive constants.

Let $\{\psi_k\}_1^K$ be another system of functions from $C^\infty(M)$ with $\operatorname{supp}\psi_k \subset O_k$ and $\psi_k \equiv 1$ in a neighborhood of the support of φ_k for each k. Let Ω_k be a subdomain of O_k such that $\operatorname{supp}\psi_k \subset \Omega_k$ and $\overline{\Omega}_k \subset O_k$, and let $A_k = A_{\Omega_k}$ be the corresponding pseudodifferential operator from $\Psi^m(\mathbb{R}^n)$. Then we have the following formulas for A:

$$A = \sum \varphi_k A_k(\psi_k \cdot) + T_1\ , \quad A = \sum \psi_k A_k(\varphi_k \cdot) + T_2\ , \tag{2.3}$$

where T_1 and T_2 are operators of order $-\infty$ on M. And conversely, if A is defined by any of the two formulas, then A is a pseudodifferential operator from $\Psi^m(M)$.

The topology in $C^\infty(M)$ is defined by a countable family of norms that are constructed in an obvious way using the partition of unity. From the theorems of Sect. 1.1 we have

Theorem 2.1.2. *Any pseudodifferential operator $A \in \Psi^m(M)$*

1) continuously acts in $C^\infty(M)$,
2) is an operator of order m on M, i.e. it continuously maps $H_s(M)$ to $H_{s-m}(M)$ for all $s \in \mathbb{R}$.

For $m \leq 0$ a pseudodifferential operator $A \in \Psi^m(M)$ is a bounded operator in any $H_s(M)$. For $m < 0$ it is a compact operator since the embedding $H_{s+|m|}(M) \subset H_s(M)$ is compact (because M is compact).

We put

$$\Psi^{-\infty}(M) = \bigcap \Psi^m(M), \quad \Psi^{\infty}(M) = \bigcup \Psi^m(M). \tag{2.4}$$

The class $\Psi^{-\infty}(M)$ coincides with the class of operators of order $-\infty$ on M (unlike the case of $\mathbb{R}^n$), which in turn coincides with the class of integral operators on M

$$Au(x) = \int_M K(x,y)u(y)dy \tag{2.5}$$

with infinitely smooth kernels $K(x,y)$.

Definition 2.1.3. If all A_Ω in Definition 2.1.1 are polyhomogeneous pseudodifferential operators, $A_\Omega \in \Psi^m_{\text{ph}}(\mathbb{R}^n)$, then we call A a *polyhomogeneous pseudodifferential operator on* M and write $A \in \Psi^m_{\text{ph}}(M)$.

This is equivalent to the fact that in (2.3) we have $A_k \in \Psi^m_{\text{ph}}(\mathbb{R}^n)$.

In particular, differential operators on M of order $m \in \mathbb{N}$ with coefficients (in any local representation) from C^∞ belong to $\Psi^m_{\text{ph}}(M)$. For $m < 0$ any pseudodifferential operator A from $\Psi^m_{\text{ph}}(M)$ is an integral operator (2.5) with infinitely smooth kernel for $x \neq y$. Its singularity for $x = y$ in local coordinates is described in Theorem 1.6.3.

The *principal symbol* $a_0(x,\xi)$ of a pseudodifferential operator $A \in \Psi^m_{\text{ph}}(M)$ is defined as a function that coincides locally in x with the principal symbol of the corresponding pseudodifferential operator A_Ω. By (1.46) and (1.47), the principal symbol is a function defined on the cotangent bundle $T^*M\backslash 0$. This function belongs to C^∞ and is positively homogeneous in ξ of degree m (in each fiber $T^*_x M\backslash 0$). The class of all such functions is denoted by $S^m_{\text{h}}(T^*M\backslash 0)$. We note that for a given function $a_0(x,\xi) \in S^m_{\text{h}}(T^*M\backslash 0)$ we can construct (with a great deal of arbitrariness) a pseudodifferential operator $A \in \Psi^m_{\text{ph}}(M)$ whose principal symbol is $a_0(x,\xi)$.

If, as we have agreed, we only use local coordinates consistent with the density dx, then (1.49) takes the form (1.50). This allows us to define a *subprincipal symbol* $\text{sub}A(x,\xi)$ of a polyhomogeneous pseudodifferential operator A on M as a function on $T^*M\backslash 0$. We note that there is also a possibility to consider pseudodifferential operators, instead of functions, on half-densities, and then $\text{sub}A$ is a function on $T^*M\backslash 0$ without any additional stipulation (see for example (Trèves 1980, Vol.2)).

The complete symbol a of a pseudodifferential operator A from $\Psi^m(M)$ or $\Psi^m_{\text{ph}}(M)$ and the terms of its asymptotic expansion in the latter case are in general defined only locally, and under a change of local coordinates they transform according to the formulas of Sect. 1.4.

We note that in the paper of Widom (1980) a complete symbol calculus of pseudodifferential operators on a manifold with a fixed (symmetric) connection was built. It is also explained how all formulas change when we pass from one connection to another. We shall not dwell on that. However, in Sect. 2.4 we shall discuss the complete and very simple symbolic calculus on a circle and a torus.

We now give the theorems of symbolic calculus that follow from Theorems 1.2.4 and 1.2.3. This calculus is, obviously, coarser than that in $\mathbb{R}^n$.

Theorem 2.1.4. *Suppose that $A \in \Psi^{m_1}(M)$ and $B \in \Psi^{m_2}(M)$. Then we have $AB \in \Psi^{m_1+m_2}(M)$. If, in addition, A and B are polyhomogeneous pseudodifferential operators with principal symbols a_0 and b_0, then AB is a polyhomogeneous pseudodifferential operator with the principal symbol $a_0 b_0$ (under the condition $a_0 b_0 \not\equiv 0$).*

The subprincipal symbol sub (AB) is expressed by the formula of the form (1.51). The principal symbol of the commutator $[A, B]$ is equal to $\mathrm{i}^{-1}\{a_0, b_0\}$ (if $\{a_0, b_0\} \not\equiv 0$).

Having a density dx, we define a scalar product on M in a natural way:

$$(u, v)_M = (u, v)_{0,M} = \int_M u(x)\overline{v(x)}dx \,. \tag{2.6}$$

The operator A^*, *formally adjoint to* A, and the operator A', *formally transposed to* A, are defined by

$$(Au, v)_M = (u, A^*v)_M \,, \qquad <Au, v>_M = <u, A'v>_M$$
$$(u, v \in C_0^\infty(M)) \,, \tag{2.7}$$

where $<u, v>_M = (u, \overline{v})_M$. If $A = A^*$, then we call A *formally selfadjoint*.

Theorem 2.1.5. *Let $A \in \Psi^m(M)$. Then A^*, $A' \in \Psi^m(M)$. Moreover, if A is a polyhomogeneous pseudodifferential operator with principal symbol $a_0(x, \xi)$, then A^* and A' are polyhomogeneous pseudodifferential operators with principal symbols $\overline{a_0(x, \xi)}$ and $a_0(x, -\xi)$. The analogues of the relations* (1.52), (1.53) *are also true.*

In particular, we can see that although A^* and A' depend on the choice of the density dx, their principal symbols do not. We also note that now we can define a pseudodifferential operator $A \in \Psi^m(M)$ on distributions from $\mathcal{E}'(M)$, the space of continuous linear functionals on $\mathcal{E}(M) = C^\infty(M)$, by the formula $<Af, \varphi>_M = <f, A'\varphi>_M$ $(\varphi \in C^\infty(M))$, and we automatically find that A is continuous in $\mathcal{E}'(M)$. We also note that the pseudolocal property is preserved: for $u \in \mathcal{E}'(M)$ the relation of the form (1.36) is true.

2.2. Elliptic Pseudodifferential Operators on M

Definition 2.2.1. A pseudodifferential operator $A \in \Psi^m_{\mathrm{ph}}(M)$ and its principal symbol $a_0(x, \xi)$ are called *elliptic at the point* x if $a_0(x, \xi) \neq 0$ on $T^*_x M \backslash 0$, *elliptic in* Ω (where Ω is a subset of M) if they are elliptic at x for all $x \in \Omega$, and *elliptic* if they are elliptic in M. (There is no need for the definition of uniform ellipticity on M since M is compact.)

An elliptic pseudodifferential operator A admits the representation (2.3) with uniformly elliptic A_k. If A is elliptic, then the same is true for A^* and A'.

Example 2.2.2. Suppose that we have a Riemannian metric on M, i.e. a covariant real (infinitely smooth) tensor field $\{g_{jk}(x)\}$ is given and the matrix $(g_{jk}(x))$ is symmetric and positive definite. The *Beltrami–Laplace operator* Δ is defined by the formula

$$\Delta u(x) = g(x)^{-1/2} \sum_{j,k=1}^{n} \partial_{x^j} \left[g(x)^{1/2} g^{jk} \partial_{x^k} u(x) \right] , \qquad (2.8)$$

where $g(x) = \det(g_{jk}(x))$ and $(g^{jk}(x))$ is the matrix inverse to $(g_{jk}(x))$. The principal symbol is equal to $\sum g^{jk}(x)\xi_j\xi_k$, and it is elliptic since the matrix is positive definite.

The Riemannian metric defines a density $\{g(x)^{1/2}dx\}$ on M (with our conventions on coordinates we have $g(x) \equiv 1$). It is natural to assume that the integral in (2.6) is taken with respect to this density, and then Δ is a formally selfadjoint differential operator.

Example 2.2.3. Let $\mathfrak{S} = M$ be a two-dimensional closed infinitely smooth surface in $\mathbb{R}^3$. The points from $\mathbb{R}^3$ will be denoted here by X, Y. We consider the operator

$$Au(X) = \frac{-1}{4\pi} \int_{\mathfrak{S}} \frac{\mathrm{e}^{-ik|X-Y|}}{|X-Y|} u(Y) d\mathfrak{S}_Y \qquad (2.9)$$

on $\mathfrak{S}$. Here $k > 0$ (for simplicity) and $d\mathfrak{S}_Y$ is the 'surface element' on $\mathfrak{S}$. For any $X \in \mathbb{R}^3 \setminus \mathfrak{S}$ the right-hand side of (2.9) is the potential of the single layer for the *Helmholtz equation*

$$\Delta u(X) + k^2 u(X) = 0 \qquad (2.10)$$

with, so called, *radiation condition* at infinity (see for example (Colton and Kress 1983)). The operator (2.9) is obtained by 'dropping' the point X on $\mathfrak{S}$. Using a local parametrization

$$X = X(x), \quad \text{or} \quad X^j = X^j(x^1, x^2) \quad (j = 1, 2, 3) , \qquad (2.11)$$

of the surface $\mathfrak{S}$, the Taylor expansions, and Theorem 1.6.3, we can show that A is an elliptic pseudodifferential operator from $\Psi_{\mathrm{ph}}^{-1}(\mathfrak{S})$. Without loss of generality we may assume that the vectors $\partial X(x)/\partial x^1$, $\partial X(x)/\partial x^2$ are orthogonal. Let $h_1(x)$, $h_2(x)$ denote their lengths ($h_1(x)h_2(x) \equiv 1$ by our convention on local coordinates). Then the principal symbol of the pseudodifferential operator (2.9) can be expressed by the formula

$$a_0(x,\xi) = -\frac{1}{2}\left[h_1^{-2}(x)\xi_1^2 + h_2^{-2}(x)\xi_2^2\right]^{-1/2} \tag{2.12}$$

(see (Agranovich 1977)). We note that the imaginary part of the kernel of A is the infinitely smooth function

$$\frac{1}{4\pi|X-Y|}\sin(k|X-Y|)\,,$$

therefore A differs from the selfadjoint operator $A_0 = \mathrm{Re}A = (A+A^*)/2$ by a term of order $-\infty$. In what follows we shall often return to this example. Its analogues can also be considered in other dimensions, see, in particular, Example 2.4.8 below.

For a pseudodifferential operator $A \in \Psi^\infty(M)$ we can define the *parametrix* literally in the same way as in the case of $\mathbb{R}^n$: it is a pseudodifferential operator $B \in \Psi^\infty(M)$ such that the relations of the form (1.72) hold, where T_1 and T_2 are operators of order $-\infty$ on M.

Theorem 2.2.4. *Let A be an elliptic pseudodifferential operator from the space $\Psi^m_{\mathrm{ph}}(M)$. Then there exists an elliptic pseudodifferential operator $B \in \Psi^m_{\mathrm{ph}}(M)$ which is a parametrix for A.*

If, for instance, we represent A by the left of the formulas (2.3) with $A_k = A_{\Omega_k}$, then we can construct a right parametrix B in the form

$$B = \sum \psi_k B_k(\varphi_k\cdot), \tag{2.13}$$

where B_k is a (uniformly elliptic) parametrix for A_k.

Theorem 2.2.4 yields the same corollaries as Theorem 1.8.3. First of all, if an elliptic pseudodifferential operator $A \in \Psi^m(M)$ has the inverse A^{-1} of order $-m$, then A^{-1} is an elliptic pseudodifferential operator from $\Psi^{-m}(M)$. Next, we have the theorems on the *a priori estimate* and *improved smoothness*:

Theorem 2.2.5. *Let A be an elliptic pseudodifferential operator from the space $\Psi^m_{\mathrm{ph}}(M)$. Then for any s there exists a constant C_s such that*

$$\|u\|_s \le C_s(\|Au\|_{s-m,M} + \|u\|_{s-1,M})\,. \tag{2.14}$$

The term $\|u\|_{s-1,M}$ on the right-hand side can be replaced by $\|u\|_{s',M}$ with any $s' < s$.

Theorem 2.2.6. *Let A be an elliptic pseudodifferential operator from the space $\Psi^m_{\mathrm{ph}}(M)$. Then if $u \in H_s(M)$ and $Au \in H_{s-m+h}(M)$, where $h > 0$, then $u \in H_{s+h}(M)$. In particular, $Au \in C^\infty(M)$ implies $u \in C^\infty(M)$, and hence $\mathrm{Ker}A = \{u : Au = 0\}$ lies in $C^\infty(M)$ and does not depend on s.*

The relations of the form (1.76) and (1.77) are also true, as well as the theorem inverse to Theorem 2.2.5. Therefore, for $A \in \Psi^m_{\mathrm{ph}}(M)$ the following three conditions are equivalent:

1) A is elliptic;
2) A has an (elliptic) parametrix;
3) the a priori estimate (2.14) holds (for some, and hence for any s).

All this is similar to what was said in Sect. 1.8. However, a crucial difference between the cases of $\mathbb{R}^n$ and M is concerned with the compactness of the embedding of $H_{s_1}(M)$ in $H_{s_2}(M)$ for $s_1 > s_2$.

2.3. Elliptic Operators as Fredholm Operators in Sobolev Spaces. In this section abstract definitions and theorems go by turns with their applications to elliptic operators.

a. Definition 2.3.1. Let X_1 and X_2 be two Banach spaces, and let A be a linear bounded operator from X_1 to X_2. We call A a *Fredholm operator* if its kernel $\mathrm{Ker}A$ is finite-dimensional, the range $\mathrm{Im}A$ is closed in X_2, and $\mathrm{Im}A$ has finite defect, i.e. $\mathrm{Coker}A = X_2/\mathrm{Im}A$ is finite-dimensional. The difference

$$\kappa(A) = \dim \mathrm{Ker}A - \dim \mathrm{Coker}A \tag{2.15}$$

is called the *index* of the Fredholm operator A.

(Sometimes the term 'Noether operator' is used instead of the term 'Fredholm operator', and then Fredholm operator denotes the Noether operator with zero index.)

A bounded operator $B : X_2 \longrightarrow X_1$ is called a *left* or *right parametrix* for the bounded operator $A : X_1 \longrightarrow X_2$ if

$$BA = I_1 + T_1\,, \quad AB = I_2 + T_2\,, \tag{2.16}$$

respectively, where I_j is the unit operator and T_j is a compact operator in X_j ($j = 1, 2$). The operator B is called (two-sided) *parametrix* if both relations are satisfied.

Theorem 2.3.2. *For a bounded operator $A : X_1 \longrightarrow X_2$ to be Fredholm it is necessary and sufficient that it have a parametrix. More precisely, the existence of a left parametrix is necessary and sufficient for* $\mathrm{Ker}A$ *to be finite-dimensional and* $\mathrm{Im}A$ *closed, and the existence of a right parametrix is necessary and sufficient for* $\mathrm{Im}A$ *to be closed and* $\mathrm{Coker}A$ *finite-dimensional.*

Any uniformly elliptic pseudodifferential operator from $\Psi^m_{\mathrm{ph}}(\mathbb{R}^n)$ is a bounded operator from $H_s(\mathbb{R}^n)$ to $H_{s-m}(\mathbb{R}^n)$, and any elliptic pseudodifferential operator from $\Psi^m_{\mathrm{ph}}(M)$ is a bounded operator from $H_s(M)$ to $H_{s-m}(M)$. In both cases A has a parametrix in the sense of the definitions in Sects. 1.8 and 2.2. But only in the latter case it is the parametrix in the sense of the definition given in the present section since the operators of negative order on M are, but on $\mathbb{R}^n$ in general are not, compact[4] operators in H_s. We obtain:

[4] For a pseudodifferential operator from $\Psi^0(\mathbb{R}^n)$ with symbol $a(x,\xi)$ to be compact it is sufficient that $a(x,\xi) \to 0$ as $|x| + |\xi| \to \infty$.

Theorem 2.3.3. *An elliptic pseudodifferential operator from $\Psi^m_{\mathrm{ph}}(M)$ is Fredholm as an operator from $H_s(M)$ to $H_{s-m}(M)$ for any s.*

b. We can now prove an abstract analogue of the a priori estimate (2.14). Let X_0 be a third Banach space such that $X_1 \subset X_0$ with compact embedding, and let $\|\cdot\|_j$ be the norm in X_j.

Theorem 2.3.4. *Let A be a bounded operator from X_1 to X_2. Then* Ker$A$ *is finite-dimensional and* ImA *is closed if and only if the following estimate holds:*

$$\|u\|_1 \leq C(\|Au\|_2 + \|u\|_0) \quad (u \in X_1)\,. \tag{2.17}$$

Thus the validity of such an estimate is equivalent to the existence of a left parametrix. In addition, if A is invertible, then the term $\|u\|_0$ on the right-hand side of (2.17) is redundant. The same remark is obviously true for the estimates (1.75) and (2.14).

We can also see that to the list of three equivalent conditions at the end of Sect. 2.2 we can add the forth one: A is a Fredholm operator from $H_s(M)$ to $H_{s-m}(M)$ (for some, and hence for any s).

The theorem on improved smoothness can also be given in an abstract form introducing two scales $X_{1,s}$, $X_{2,s}$ of Banach spaces and assuming that the operator T_1 in (2.16) has a 'negative order'. We shall not dwell on that.

c. An elliptic operator from $\Psi^m_{\mathrm{ph}}(M)$ of order $m > 0$ can be considered not only as a bounded operator from $H_s(M)$ to $H_{s-m}(M)$ or from $H_{s+m}(M)$ to $H_s(M)$, but also as an unbounded operator in $H_s(M)$ with the domain $H_{s+m}(M)$ dense in $H_s(M)$. Using Theorem 2.2.6 on improved smoothness, we can easily prove

Theorem 2.3.5. *Let A be an elliptic pseudodifferential operator from the space $\Psi^m_{\mathrm{ph}}(M)$, $m > 0$, and $s \in \mathbb{R}$. Then A as an operator in $H_s(M)$ with domain $D_A = H_{s+m}(M)$ is closed.*

The analogous assertion is true for uniformly elliptic operators of order $m > 0$ in $\mathbb{R}^n$.

Remark 2.3.6. The definition of a Fredholm operator given at the beginning of this section can be extended to unbounded operators. In particular, let X be a Banach space with the norm $\|\cdot\|$, and let A be an unbounded closed operator in X with dense domain D_A. It is called *Fredholm* if the kernel Ker$A$ is finite-dimensional and the range ImA is closed in X and has a finite-dimensional direct complement.

This definition can be put in accordance with the definition given above in the following way. We introduce a norm in $X_1 = D_A$ by

$$\|u\|_1 = \left(\|u\|^2 + \|Au\|^2\right)^{1/2},$$

i.e. we take the norm of the point of the graph of A. Since the graph is closed, X_1 is a Banach space. If we put $X_2 = X$, then the two definitions of Fredholm operator are equivalent.

If A is an elliptic operator of order $m > 0$ and $X = H_s(M)$, then the norm $(\|u\|_{s,M}^2 + \|Au\|_{s,M}^2)^{1/2}$ is equivalent to the norm in $H_{s+m}(M)$, i.e. the ratio of these norms is contained between positive constants, and so A is Fredholm in $H_s(M)$ in the sense of the new definition.

It is important to draw the attention of the reader to the fact that a pseudodifferential operator A of negative order m is *not* a Fredholm operator in $H_s(M)$: as a compact operator with finite-dimensional kernel it has the range in $H_s(M)$ which is not closed.

d. We turn to the explanation of the 'informal role' of the operator A^* formally adjoint to a pseudodifferential operator $A \in \Psi_{\text{ph}}^m(M)$; simultaneously we give some other useful results.

First of all, it is clear that if $m \leq 0$, then A and A^* are bounded operators in $H_0(M)$ and the first of the formulas (2.7) extends to $u, v \in H_0(M)$. Consequently, the operator A^* is adjoint to A in $H_0(M)$. In particular, if A is formally selfadjoint, then it is a selfadjoint operator in $H_0(M)$. The same is true for pseudodifferential operators of order $m \leq 0$ in $\mathbb{R}^n$. Ellipticity is not necessary here.

Next, if $m > 0$, then the operator A^* formally adjoint to an elliptic pseudodifferential operator $A \in \Psi_{\text{ph}}^m(M)$ is also elliptic of order m, and we can apply Theorem 2.3.5. The first of the formulas (2.7) extends to $u, v \in H_m(M)$. Using the theorem on improved smoothness again, it is easy to verify that the following theorem is true:

Theorem 2.3.7. *Let A be an elliptic pseudodifferential operator from the space $\Psi_{\text{ph}}^m(M)$, $m > 0$, and let A^* be the operator formally adjoint to A. Then A and A^*, as operators in $H_0(M)$ with domain $H_m(M)$, are adjoint operators in the Hilbert space $H_0(M)$. In particular, if A is formally selfadjoint, then it is selfadjoint in $H_0(M)$.*

The analogous assertion is true for a uniformly elliptic pseudodifferential operator of order $m > 0$ in $\mathbb{R}^n$.

It is well-known (and can be easily proved) that if H is a Hilbert space, A is a linear operator densely defined in H, and A^* is adjoint to A, then

$$H = \overline{\text{Im} A} \oplus \text{Ker} A^* \tag{2.18}$$

($\oplus$ is the orthogonal sum and the bar denotes the closure). In the case of elliptic operators this gives a description of ImA:

Theorem 2.3.8. *Let A be an elliptic pseudodifferential operator from the space $\Psi_{\text{ph}}^m(M)$, and let A^* be the operator formally adjoint to A. Then, if we consider A as an operator from $H_m(M)$ to $H_0(M)$, we have*

$$H_0(M) = \text{Im} A \oplus \text{Ker} A^* \, . \tag{2.19}$$

Consequently, the equation $Au = f$ with $f \in H_0(M)$ has a solution $u \in H_m(M)$ if and only if the following condition is satisfied:

$$(f, v)_M = 0 \quad \text{for all} \quad v \in \text{Ker}A^* \,. \tag{2.20}$$

Moreover, we have the following formula for the index:

$$\kappa(A) = \dim \text{Ker}A - \dim \text{Ker}A^*. \tag{2.21}$$

We can now construct on M analogues of the operator (1.11).

Theorem 2.3.9. *Let $m \in \mathbb{R}$ and let $a_0(x, \xi)$ be any real function from $S^m_{\text{h}}(T^*M\backslash 0)$ different from 0 everywhere. Then there exists an elliptic pseudodifferential operator $\Lambda_m \in \Psi^m_{\text{ph}}(M)$ selfadjoint in H_0, with the principal symbol $a_0(x, \xi)$, and such that it defines a continuous isomorphism of $H_m(M)$ onto $H_0(M)$.*

Indeed, let A be a pseudodifferential operator from $\Psi^m_{\text{ph}}(M)$ with the principal symbol a_0. The pseudodifferential operator $A_1 = \text{Re}A = (A + A^*)/2$ has the same principal symbol and is selfadjoint in $H_0(M)$. Let P be the orthogonal projection of $H_0(M)$ on $\text{Ker}A_1$ ($P = 0$ if $\text{Ker}A_1 = \{0\}$); it is an operator of order $-\infty$. We put $\Lambda_m = A_1 + P$.

Remark 2.3.10. Suppose that a pseudodifferential elliptic operator $A \in \Psi^m_{\text{ph}}(M)$ has the inverse A^{-1} if it is considered as an operator from $H_{s+m}(M)$ to $H_s(M)$ for some s. Then A^{-1} is an elliptic pseudodifferential operator from $\Psi^{-m}_{\text{ph}}(M)$ and is inverse to A for all s. Consequently, the pseudodifferential operator Λ_m (see Theorem 2.3.9) defines a continuous isomorphism of $H_{s+m}(M)$ onto $H_s(M)$ for any s.

Theorem 2.3.9 and the additional assertions given in Remark 2.3.10 can be improved after introducing the notion of the power of an elliptic operator (see Sect. 5.3 below).

e. We continue the discussion of the role played by the pseudodifferential operator A^* adjoint to a pseudodifferential operator A. As we shall see, Theorem 2.3.8 can be generalized to the case of $H_s(M)$ in place of $H_0(M)$. But first we have to give the corresponding abstract result.

Let again X_1 and X_2 be Banach spaces and $A : X_1 \longrightarrow X_2$ a bounded operator. Let X_j^* be the space dual to X_j ($j = 1, 2$), and let $(v, \varphi)_j$ be the form expressing the action of the functional $\varphi \in X_j^*$ on the element $v \in X_j$. This form is sesquilinear (linear in v and conjugate-linear in φ). The operator adjoint to A is defined by the equality $(Au, \varphi)_2 = (u, A^*\varphi)_1$, and it is a bounded operator from X_2^* to X_1^*.

Theorem 2.3.11. *Let A be a Fredholm operator from X_1 to X_2. Then A^* is a Fredholm operator from X_2^* to X_1^*. Moreover, we have*

$$\operatorname{Ker}A^* = \{\varphi \in X_2^* : (f,\varphi)_2 = 0 \text{ for all } f \in \operatorname{Im}A\}\,, \tag{2.22}$$

$$\operatorname{Im}A = \{f \in X_2 : (f,\varphi)_2 = 0 \text{ for all } \varphi \in \operatorname{Ker}A^*\}\,, \tag{2.23}$$

$$\kappa(A) = \dim\operatorname{Ker}A - \dim\operatorname{Ker}A^* = -\kappa(A^*)\,. \tag{2.24}$$

We turn to elliptic operators and Sobolev spaces. The space $H_s(M)$ is a Hilbert space, and its dual may be identified with $H_s(M)$. However, a different point of view is more convenient. The form $(u,v)_{0,M}$ can be extended from $u, v \in C^\infty(M)$ to $u \in H_s(M)$ and $v \in H_{-s}(M)$ by

$$(u,v)_{0,M} = (\Lambda_s u, \Lambda_s^{-1} v)_{0,M}, \tag{2.25}$$

and for fixed $v \in H_{-s}(M)$ it defines a continuous linear functional on $H_s(M)$ whose norm is equivalent to $\|v\|_{-s,M}$. It is easy to prove (using the general form of a continuous linear functional on a Hilbert space) that this is the general form of a continuous linear functional on $H_s(M)$. Thus $[H_s(M)]^*$ can be identified with $H_{-s}(M)$. (We see that here elliptic operators help us to complete the theory of Sobolev spaces.) Moreover, if a pseudodifferential operator $A \in \Psi^m_{\text{ph}}(M)$ is considered as an operator from $H_{s+m}(M)$ to $H_s(M)$, then the adjoint operator will coincide with the formally adjoint pseudodifferential operator A^* considered as an operator from $H_{-s}(M)$ to $H_{-s-m}(M)$. Together with Theorem 2.3.11 this gives the following result.

Theorem 2.3.12. *Let A be an elliptic pseudodifferential operator from the space $\Psi^m_{\text{ph}}(M)$. Then the equation $Au = f$ with $f \in H_s(M)$ has a solution $u \in H_{s+m}(M)$ if and only if the condition (2.20) is satisfied. For the index $\kappa(A)$ of A as an operator from $H_{s+m}(M)$ to $H_s(M)$ the formula (2.21) is preserved, so the index does not depend on s.*

f. We now discuss some properties of the index. We have the following abstract results.

Theorem 2.3.13. *Let X_j $(j = 1,2,3)$ be Banach spaces, and suppose that $A_1 : X_1 \longrightarrow X_2$, $A_2 : X_2 \longrightarrow X_3$ are Fredholm operators. Then $A_2A_1 : X_1 \longrightarrow X_3$ is a Fredholm operator with index $\kappa(A_2A_1) = \kappa(A_2) + \kappa(A_1)$.*

Theorem 2.3.14. *Let $A_1 : X_1 \longrightarrow X_2$ be a Fredholm operator. Then all bounded operators $B : X_1 \longrightarrow X_2$ that are sufficiently close in norm to A are Fredholm and have the same index as A.*

Consequently, if A_t is a function of $t \in [0,1]$ with values in the set of Fredholm operators from X_1 to X_2 and such that it is continuous in the operator norm, then the index $\kappa(A_t)$ does not depend on t.

These assertions applied to elliptic pseudodifferential operators allow us to establish the following results. If two elliptic pseudodifferential operators from $\Psi^m_{\text{ph}}(M)$ have the same principal symbol, then their indices coincide. Next, the indices of elliptic pseudodifferential operators $A^{(0)}$ and $A^{(1)}$ coincide if their principal symbols $a_0^{(0)}(x,\xi)$ and $a_0^{(1)}(x,\xi)$ can be connected by a homotopy $a_0^{(t)}(x,\xi)$ $(0 \le t \le 1)$ in the class of all elliptic symbols from $S_{\mathrm{h}}^m(T^*M\backslash 0)$. Moreover, it is sufficient to have a homotopy in the class of continuous nonzero functions $a_0(x,\xi)$ on $T^*M\backslash 0$ positively homogeneous of degree m in ξ, since such a homotopy can be approximated by a homotopy running over elliptic symbols from $S_{\mathrm{h}}^m(T^*M\backslash 0)$. We also note that the computation of the index for arbitrary m can be reduced to the computation of the index for $m = 0$ by the multiplication of the pseudodifferential operator by a selfadjoint elliptic pseudodifferential operator from $\Psi_{\mathrm{ph}}^{-m}(M)$ (a selfadjoint elliptic pseudodifferential operator clearly has zero index).

These and some other simple assertions concerned with the index that will be given in Sects. 5.5 and 5.6 are essential in computing the index of a general elliptic pseudodifferential operator on M (including matrix elliptic pseudodifferential operators and elliptic pseudodifferential operators acting in vector bundles) but do not solve the problem completely. The problem of computing the index of general elliptic operators was posed by Gel'fand (1960) and has drawn the attention of many mathematicians. It was solved for pseudodifferential operators on closed manifolds in the paper of Atiyah and Singer (1963). This question is discussed in detail in the survey (Fedosov 1990) where further references can be found. The Atiyah–Singer formula contains the following result as a relatively simple particular case: the index of a scalar elliptic pseudodifferential operator on a manifold of dimension $n \ge 2$ is equal to 0. In the case $n = 1$ the index of a scalar elliptic pseudodifferential operator can be nonzero, and in 2.4c we shall give a simple formula to compute it.

g. We conclude with the following remark. Let $A : X_1 \longrightarrow X_2$ be a Fredholm non-invertible operator with zero index. Then we can choose a finite-dimensional operator T such that the operator $A+T$ is invertible. In the case when A is an elliptic pseudodifferential operator, it is sufficient to fix the bases $f_1, \dots, f_l$ of Ker$A$ and $g_1, \dots, g_l$ of KerA^* and put

$$Tu = \sum_1^l (u, f_j)_M g_j \,. \tag{2.26}$$

This is clearly an operator of order $-\infty$.

2.4. Pseudodifferential Operators on a Circle and on a Torus

a. We denote by S the circle of radius 1 on the plane $\mathbb{R}^2$. The specific feature of S as a manifold is that it can be covered by coordinate neighborhoods $Q_{\alpha\beta}$ with the polar angle x as a local coordinate ($\alpha < x < \beta$, where $\beta - \alpha \le 2\pi$), and the transformation of one local coordinate into another in the intersection of two coordinate neighborhoods is a shift. We are going to use only such coordinate neighborhoods and local coordinates, and we shall identify functions on S with 2π-periodic functions on $\mathbb{R}$. The formula (1.39) on the change of variables in a pseudodifferential operator under the transformations $y = x + c$ ($c \in 2\pi\mathbb{Z}$) becomes much simpler. This allows us, as will be seen below, to build the complete symbolic calculus of pseudodifferential operators on S, analogous to the calculus of pseudodifferential operators on $\mathbb{R}$, with symbols $a(x,\xi)$ which are 2π-periodic in x and, in addition, with Fourier series in place of Fourier transforms.

The Fourier coefficients of a 2π-periodic function $u(x)$ with respect to the system $\{e^{inx}\}_{-\infty}^{\infty}$ will be denoted by $c_n(u)$:

$$c_n(u) = \frac{1}{2\pi}\int_{-\pi}^{\pi} u(x)e^{-inx}dx\,. \tag{2.27}$$

If $u(x)$ is a distribution from $\mathcal{E}'(S)$, then by $c_n(u)$ we understand the result of its action on the test function $(2\pi)^{-1}e^{-inx}$. In the space $H_r(S)$ ($r \in \mathbb{R}$) it is convenient to use the norm

$$\|u\|_{r,S} = \left[\sum_{-\infty}^{\infty}(1+|n|^2)^r|c_n(u)|^2\right]^{1/2}\,. \tag{2.28}$$

The operators of order $-\infty$ on S are integral operators

$$Tu(x) = \int_{-\pi}^{\pi} K(x,y)u(y)dy \tag{2.29}$$

with infinitely smooth kernels $K(x,y)$ which are 2π-periodic in x as well as in y.

Let A be an operator acting, say, in $C^\infty(S)$. Then we can associate with A a *discrete symbol*

$$a(x,n) = e^{-inx}A(e^{inx}) \tag{2.30}$$

defined on $S \times \mathbb{Z}$ (compare with (1.31)). If A is a continuous operator in $C^\infty(S)$, then for any function $u \in C^\infty(S)$ we have

$$Au(x) = \sum_{n=-\infty}^{\infty} e^{inx}a(x,n)c_n(u)\,. \tag{2.31}$$

Moreover, as is easy to prove, A has order $-\infty$ if and only if for any α and N

$$|\partial_x^\alpha a(x,n)| \le C_{\alpha N}(1+|n|)^{-N}\,, \tag{2.32}$$

where $C_{\alpha N}$ does not depend on x, n.

Theorem 2.4.1. *Let A be a pseudodifferential operator from the space $\Psi^m(S)$ (or from $\Psi^m_{\mathrm{ph}}(S)$). Then the function* (2.30) *is the restriction to $S \times \mathbb{Z}$ of a function $a(x,\xi)$ from $S^m(\mathbb{R}^1 \times \mathbb{R}_1)$ (from $S^m_{\mathrm{ph}}(\mathbb{R}^1 \times \mathbb{R}_1)$, respectively) 2π-periodic in x. Conversely, if $a(x,\xi)$ is a 2π-periodic in x function from $S^m(\mathbb{R}^1 \times \mathbb{R}_1)$ (or from $S^m_{\mathrm{ph}}(\mathbb{R}^1 \times \mathbb{R}_1)$), then* (2.31) *is a pseudodifferential operator from $\Psi^m(S)$ (from $\Psi^m_{\mathrm{ph}}(S)$ respectively) and* (2.30) *holds.*

This can be easily proved by any of the formulas (2.3) (with $K = 2$).

The continuation $a(x, \xi)$ of the function $a(x, n)$ is not unique, but the difference between any two is a 2π-periodic function from $S^{-\infty}(\mathbb{R}^1 \times \mathbb{R}_1)$. We shall call any such a continuation the (complete) *symbol* of a pseudodifferential operator (2.31).

The analogues of Theorems 1.2.3 and 1.2.4 can be proved in different ways (see (Agranovich 1984), (Elschner 1985)):

Theorem 2.4.2. 1) *Let A be a pseudodifferential operator from $\Psi^m(S)$ with symbol $a(x, \xi)$. Then A^* and A' are pseudodifferential operators from $\Psi^m(S)$ with symbols* (1.27) *and* (1.28) *respectively.*

2) *Let A and B be pseudodifferential operators from $\Psi^{m_1}(S)$ and $\Psi^{m_2}(S)$ with symbols $a(x, \xi)$ and $b(x, \xi)$. Then $C = AB$ is a pseudodifferential operator from $\Psi^{m_1+m_2}(S)$ with symbol* (1.32).

Here the spaces $\Psi^m(S)$ can be replaced by $\Psi^m_{\text{ph}}(S)$. All symbols are 2π-periodic in x.

The following global analogue of Theorem 1.4.1 is also true.

Theorem 2.4.3. *Suppose that a function $\varphi(x)$ from $C^\infty(\mathbb{R})$ with positive derivative satisfies the identity*

$$\varphi(x + 2\pi) \equiv \varphi(x) + 2\pi , \tag{2.33}$$

and set

$$Zu(x) = u[\varphi(x)] . \tag{2.34}$$

Then if A is a pseudodifferential operator from $\Psi^m(S)$ with symbol $a(x, \xi)$, then $\widetilde{A} = Z^{-1}AZ$ is a pseudodifferential operator from $\Psi^m(S)$ with symbol $\widetilde{a}(y, \eta)$ which for $y = \varphi(x)$ has the asymptotic expansion

$$\widetilde{a}[\varphi(x), \eta] \sim \sum_{\gamma \geq 0} \frac{1}{\gamma!} D_z^\gamma \left[e^{i\varphi_2(x,z)\cdot\eta} \right]_{z=x} \left(\partial_\xi^\gamma a \right) (x, \partial_x \varphi(x) \eta) , \tag{2.35}$$

where $\varphi_2(x, z) = \varphi(z) - \varphi(x) - (z - x)\partial\varphi(x)$.

Here the space $\Psi^m(S)$ can be also replaced by $\Psi^m_{\text{ph}}(S)$. The analogous theorem is true for $\varphi(x)$ with negative derivative and -2π in place of $+2\pi$ on the right-hand side of (2.33).

We now give a description of functions $a(x, n)$ that can be extended to symbols $a(x, \xi)$ from $\Psi^m(S)$ or $\Psi^m_{\text{ph}}(S)$. For a function $b(n)$ defined on $\mathbb{Z}$ we set

$$\Delta b(n) = b(n + 1) - b(n) , \quad \Delta^\beta b(n) = \Delta(\Delta^{\beta-1} b(n)) \quad (\beta \in \mathbb{N}) . \tag{2.36}$$

Theorem 2.4.4. *Let $a(x, n)$ be a function on $\mathbb{R} \times \mathbb{Z}$, 2π-periodic and infinitely smooth in x. Then:*

1) it has a 2π-periodic in x continuation $a(x, \xi) \in S^m(\mathbb{R}^1 \times \mathbb{R}_1)$ if and only if for all $\alpha, \beta \in \mathbb{Z}_+$ we have

$$|\partial_x^\alpha \Delta^\beta a(x,n)| \le C_{\alpha\beta}(1+|n|)^{m-\beta} . \tag{2.37}$$

2) For the function $a(x,n)$ *to have a* 2π*-periodic in* x *continuation* $a(x,\xi) \in S^m_{\text{ph}}(\mathbb{R}^1 \times \mathbb{R}_1)$ *it is necessary and sufficient that there exists a sequence of functions* $a_\nu(x,n)$ $(\nu = 0,1,\ldots)$ *infinitely smooth and* 2π*-periodic in* x*, positively homogeneous of degree* $m-\nu$ *in* n*, and such that for all* $\alpha, N \in \mathbb{Z}_+$, $n \neq 0$ *we have*

$$\left|\partial_x^\alpha \left[a(x,n) - \sum_{\nu=0}^{N-1} a_\nu(x,n)\right]\right| \le C_{\alpha N}|n|^{m-N}. \tag{2.38}$$

Moreover, then $a(x,\xi) \sim a_0(x,\xi) + \ldots$.

We also note that the theorems of the symbolic calculus can be formulated in terms of discrete symbols $a(x,n)$ using the difference operator Δ in place of ∂_ξ (see (Amosov 1988)).

b. We shall dwell on some examples.

Example 2.4.5. A differential operator

$$A = \sum_{\nu=0}^{m} a_\nu(x) D^{m-\nu} \tag{2.39}$$

with infinitely smooth 2π-periodic coefficients has the symbol $\sum a_\nu(x)\xi^{m-\nu}$ and the principal symbol $a_0(x)\xi^m$, so that $a_0 \neq 0$ is the ellipticity condition.

Example 2.4.6. We put

$$P_+u(x) = \sum_{n\ge 0} \mathrm{e}^{inx} c_n(u) , \quad P_-u(x) = \sum_{n<0} \mathrm{e}^{inx} c_n(u) . \tag{2.40}$$

Here we have pseudodifferential operators from $\Psi^0_{\text{ph}}(S)$ with discrete symbols $p_+(x,n) = 1$ for $n \ge 0$, $p_+(x,n) = 0$ for $n < 0$, and $p_-(x,n) = 1 - p_+(x,n)$, respectively. These operators are not elliptic, but their sum $P_+ + P_- = I$ and the difference $P_+ - P_-$ are elliptic. The operators P_+ and P_- appear in many questions of analysis. We recall that if we put

$$\mathrm{e}^{ix} = z , \quad u(x) = v(z) , \quad v_+(z) = \sum_{n\ge 0} c_n(u) z^n , \quad v_-(z) = \sum_{n<0} c_n(u) z^n ,$$

then under the condition $\sum |c_n(u)| < \infty$ the functions $v_+(z)$ and $v_-(z)$ continuously extend from the unit circle $\{z : |z| = 1\}$ to the unit disk $\{z : |z| \le 1\}$ and to the complement $\{z : |z| \ge 1\}$ of the open unit disk, respectively, the extensions being holomorphic for $|z| < 1$ and $|z| > 1$, and $v_-(\infty) = 0$. The operators $P_\pm$ are orthogonal projections in $H_0(S)$.

c. We give a formula for the index of an elliptic pseudodifferential operator on S.

Theorem 2.4.7. *Let A be an elliptic pseudodifferential operator on S with principal symbol $a_0(x,\xi)$. Then*

$$\kappa(A) = \frac{1}{2\pi}\left[\arg(a_0(x,-1)a_0^{-1}(x,+1))\right]_{-\pi}^{\pi}, \tag{2.41}$$

where $[\arg]_{-\pi}^{\pi}$ is the increment of a continuous branch of the argument when x runs from $-\pi$ to π.

To prove this formula we can reduce the general case to the case when $m = 0$ and

$$A = e^{il_- x}P_- + e^{il_+ x}P_+ , \quad \text{where} \quad l_\pm = \frac{1}{2\pi}[\arg a_0(x,\pm 1)]_{-\pi}^{\pi} ,$$

and in this case it can be directly verified. One can also use the Noether formula (see (Muskhelishvili 1962) or (Prößdorf 1987)) for the index of a singular integral operator on a closed curve.

d. All the results concerned with pseudodifferential operators on the circle are also valid for pseudodifferential operators on a one-dimensional closed compact C^∞ manifold or, which is the same, on a closed curve Γ of the class C^∞ in the Euclidean space, since such a curve is diffeomorphic to the circle. It is sufficient to take as a parameter the suitably normed length measured from a fixed point.

Example 2.4.8. Let $\Gamma \subset \mathbb{R}^2$ and let

$$Au(x) = \frac{i}{4}\int_\Gamma H_0^{(2)}(k|X-Y|)u(y)ds_Y . \tag{2.42}$$

Here $k > 0$, X, Y are the points of the plane $\mathbb{R}^2$ lying on Γ, and ds_Y is the length element of the curve. By $H_0^{(2)}$ we denote the zero order Hankel function of the second kind. For $X \notin \Gamma$ the right-hand side of (2.42) is the potential of a simple layer for the Helmholtz equation on the plane, and the operator A is obtained by dropping the point X on Γ. Evidently, this is the one-dimensional analogue of the pseudodifferential operator from Example 2.2.3. We introduce a parametrization of Γ

$$X = X(s) = (X_1(x), X_2(x)) , \quad \text{where} \quad x = 2\pi|\Gamma|^{-1}s , \tag{2.43}$$

s is the length of the curve Γ measured from a certain point, and $|\Gamma|$ is the length of the whole curve. We put $u(X(x)) = v(x)$ and rewrite the operator in the form

$$Av(x) = \int_{-\pi}^{\pi} k(x, x-y)v(y)dy , \tag{2.44}$$

where

$$k(x,y) = \frac{i}{8\pi}|\Gamma|\, H_0^{(2)}(k|X(x) - X(x-y)|) .$$

Using the well-known representation of the Hankel function in the form of the sum of a series, the Taylor expansions, and the results presented in Sect. 1.6, one can prove that (2.44) is a pseudodifferential operator from $\Psi_{\mathrm{ph}}^{-1}(S)$, and one can find first terms of the asymptotic expansion of its symbol. (In principle, any number of the terms of the expansion can be found, but the computation quickly becomes complicated.) Namely,

$$a(x,\xi) \sim \frac{\alpha_0}{|\xi|} + \frac{\alpha_2}{|\xi|^3} + \frac{\alpha_4(x)}{|\xi|^5} + \frac{\alpha_5(x)}{|\xi|^6} + \dots , \tag{2.45}$$

where

$$\alpha_0 = -\frac{|\Gamma|}{4\pi}, \quad \alpha_2 = -\frac{k^2|\Gamma|^3}{4(2\pi)^3}, \quad \alpha_4 = -\frac{k^2|\Gamma|^5}{4(2\pi)^5}\left[\frac{3}{4}k^2 + q^2(x)\right], \tag{2.46}$$

and $q(x)$ is the curvature (i.e. the modulus of the second derivative of $X(x)$ with respect to the arc length). In particular, A is elliptic. The imaginary part of its kernel belongs to C^∞ (since $\mathrm{i}H_0^{(2)}(\tau) = N_0(\tau) + \mathrm{i}J_0(\tau)$ and the Bessel function $J_0(\tau)$ even is an entire analytic function of τ^2), hence A differs by a term of order $-\infty$ from the selfadjoint pseudodifferential operator ReA.

e. All what was said so far in this section about complete symbolic calculus and the representation of pseudodifferential operators by Fourier series can be generalized to pseudodifferential operators on the n-dimensional torus $\mathbb{T}^n = \mathbb{R}^n/(2\pi\mathbb{Z})^n$ ($\mathbb{T}^1 = S$). The analogue of (2.31) has the form

$$Au(x) = \sum_{\gamma\in\mathbb{Z}^n} \mathrm{e}^{\mathrm{i}\gamma\cdot x} a(x,\gamma) c_\gamma(u) , \tag{2.47}$$

where

$$c_\gamma(u) = \frac{1}{(2\pi)^n}\int_{\mathbb{T}^n} u(x)\mathrm{e}^{-\mathrm{i}\gamma\cdot x}dx . \tag{2.48}$$

2.5. The Spectrum of a Linear Operator and the System of Its Root Vectors. In this section we recall some abstract definitions and assertions. Here and in what follows we denote by H an infinite-dimensional separable complex Hilbert space and by $(\cdot,\cdot)$ the scalar product in H.

a. Let A be a linear operator in H. We shall assume that A is either bounded and then defined on the whole H or unbounded and then closed and defined on a dense domain D_A. By A^* we denote the operator adjoint to A.

The point $\lambda \in \mathbb{C}$ belongs to the *resolvent set* $\rho(A)$ of A if the operator $A - \lambda I$ has bounded inverse

$$(A - \lambda I)^{-1} = R_A(\lambda) \tag{2.49}$$

called the *resolvent* of A at λ. The complement $\sigma(A)$ of $\rho(A)$ is called the *spectrum* of A; the spectrum is always closed. If A is bounded and has the norm $\|A\|$, then the spectrum $\sigma(A)$ lies inside the circle $\{\lambda : |\lambda| \leq \|A\|\}$. The spectra of the operators A and A^* are connected by the relation $\lambda \in \sigma(A) \Leftrightarrow \overline{\lambda} \in \sigma(A^*)$.

The point $\lambda \in \sigma(A)$ is called an *eigenvalue* of the operator A if $\operatorname{Ker}(A - \lambda I) \neq \{0\}$. In this case we have the *root lineal* $\mathfrak{N}_\lambda(A)$ consisting of 0 and *root vectors* (or *generalized eigenvectors*), that is, nonzero vectors f such that $(A - \lambda I)^k f = 0$ for some $k \in \mathbb{N}$. We shall assume that $k = k(f)$ is taken minimal. If $k(f) = 1$, then f is an *eigenvector*, and if $k(f) > 1$, we call it an *associated vector*. Eigenvectors and 0 form the *eigenspace* which coincides with $\operatorname{Ker}(A - \lambda I)$. The eigenspace, as well as $\mathfrak{N}_\lambda(A)$, is invariant with respect to A. The dimension $\dim \mathfrak{N}_\lambda(A)$ is called the *multiplicity* of the eigenvalue λ. If it is equal to 1, then we call λ a *simple* eigenvalue.

The eigenvalue λ is called *normal* if $\mathfrak{N}_\lambda(A)$ is finite-dimensional and

$$H = \mathfrak{N}_\lambda(A) \dot{+} \mathfrak{N}'_\lambda(A) \tag{2.50}$$

(the direct sum), where $\mathfrak{N}'_\lambda(A)$ is a subspace invariant with respect to A and the spectrum of the restriction of A to $\mathfrak{N}'_\lambda(A)$ does not contain λ. Automatically, in this case λ is an isolated point in $\sigma(A)$, $\mathfrak{N}_\lambda(A) = \operatorname{Ker}(A - \lambda I)^k$ for some finite $k = k(\lambda)$, and $\mathfrak{N}_\lambda(A)$ is closed, so that it is the *root subspace* of the operator A. The point $\overline{\lambda}$ is in this case a normal eigenvalue of A^*, and $\mathfrak{N}'_\lambda(A)$ coincides with the orthogonal complement to $\mathfrak{N}_{\overline{\lambda}}(A^*)$.

Let A be a bounded operator. The *essential spectrum* $\sigma_e(A)$ consists, by definition, of all λ such that $A - \lambda I$ is not a Fredholm operator in H. It is a closed subset of $\sigma(A)$ and may coincide with $\sigma(A)$. In each connected component of the complement of $\sigma_e(A)$ the index $\kappa(A - \lambda I)$ is constant. Moreover, if it is different from 0, then each point λ of this component belongs to $\sigma(A)$ and λ is an eigenvalue of A and/or $\overline{\lambda}$ is an eigenvalue of A^*. In the component with zero index either all λ are eigenvalues of A (and simultaneously, $\overline{\lambda}$ are eigenvalues of A^*) or we only have isolated points of $\sigma(A)$ which are normal eigenvalues of A.

If A is a compact operator, then $\sigma_e(A) = \{0\}$ and $\sigma(A)$ consists of 0 and a finite or countable set of normal eigenvalues with 0 the only possible condensation point. However, 0 does not have to be an eigenvalue. If there are no nonzero eigenvalues, then A is called a *Volterra operator*.

If A is compact and normal ($AA^* = A^*A$), then associated vectors do not exist and in H we have an orthonormal basis of eigenvectors of A. (Conversely, if there is such a basis, then A is a normal operator.)

Suppose now that A is an unbounded operator. Its spectrum $\sigma(A)$ may fill the whole plane. The essential spectrum $\sigma_e(A)$ is again defined as the set of all λ such that $A - \lambda I$ is not Fredholm (see Remark 2.3.6). The case most interesting for us is when $\rho(A)$ is nonempty and for some $\lambda \in \rho(A)$ the resolvent $R_A(\lambda)$ is compact. Then it is compact for all $\lambda \in \rho(A)$, and the spectrum $\sigma(A)$ consists of a finite number or countably many normal eigenvalues whose only condensation point may be infinity. In this case A is called the *operator with compact resolvent* or (less accurate but more common name) *with discrete spectrum* (the empty spectrum is not excluded). The operator A^* has discrete spectrum simultaneously with A. If in addition $0 \notin \sigma(A)$, then the eigenvalues of A are *characteristic numbers* of the compact operator A^{-1}. If T is a compact operator and 0 is not an eigenvalue of T nor of T^*, then putting $Af = g$ if $Tg = f$ we obtain a closed operator 'inverse to T' with dense domain D_A and discrete spectrum consisting of characteristic numbers of T.

b. We now give some definitions concerned with systems of vectors $\{f_j\}_1^\infty$ in H.

The system $\{f_j\}$ is called *complete* in H if finite linear combinations of these vectors are dense in H.

The system $\{f_j\}$ is called *minimal* if none of the vectors belongs to the closed linear hull of the remaining vectors of this system.

If $\{f_j\}$ is a minimal system, then there is a system $\{g_k\}_1^\infty$ biorthogonal to $\{f_j\}$:

$$(f_j, g_k) = \delta_{jk} \quad (j, k \in \mathbb{N}) . \tag{2.51}$$

As g_k we can take a vector that is proportional to the orthogonal projection of f_k on the orthogonal complement to the closed linear hull of the vectors f_j, $j \neq k$.

The system $\{f_j\}$ is minimal if there is a system $\{g_k\}$ biorthogonal to it. If $\{f_j\}$ is complete and minimal, then $\{g_k\}$ is unique and minimal but not necessarily complete.

A system $\{f_j\}$ is a *basis* of H if any vector f can be uniquely represented in the form

$$f = \sum_{j=1}^{\infty} c_j f_j, \tag{2.52}$$

where c_j are numerical coefficients.

Any basis is a minimal complete system, but converse is not true. If $\{f_j\}$ is a basis and $\{g_k\}$ is biorthogonal to $\{f_j\}$, then in (2.52) we have $c_j = (f, g_j)$. These numbers are called the *Fourier coefficients of the vector f with respect to the system* $\{f_j\}$, and the series (2.52) is called the *Fourier series of the vector f in this system.*

Such a series can be formally built if we have a system which is biorthogonal to $\{f_j\}$. But if $\{f_j\}$ is not a basis, this series may not converge, and if it converges, then not necessarily to f. If $\{g_k\}$ is not complete, then different f can have the same Fourier coefficients. If both systems $\{f_j\}$ and $\{g_k\}$ are complete, then we can seek the methods of summation to recover the vectors f from their formal Fourier series.

If $\{f_j\}$ is a basis, then $\{g_k\}$ is a basis too (the Banach theorem). Moreover, in this case

$$\sup \|f_j\| \cdot \|g_j\| < +\infty . \tag{2.53}$$

c. We return to operators. Let A be an operator with discrete spectrum. If $A = A^*$, then there are no associated vectors and in H there exists an orthonormal basis of eigenvectors of this operator. We now assume that $A \neq A^*$ and that all pairwise different eigenvalues of the operator A form an infinite sequence $\{\mu_j\}_1^\infty$ ordered in such a way that the moduli are not decreasing (we shall always keep this numbering unless otherwise stated). We shall write $\mathfrak{N}_j(A)$, $\mathfrak{N}_j(A^*)$ instead of $\mathfrak{N}_{\mu_j}(A)$, $\mathfrak{N}_{\overline{\mu_j}}(A^*)$. It is easy to prove that

1) $\mathfrak{N}_j(A) \perp \mathfrak{N}_k(A^*)$ for $j \neq k$;
2) $\dim \mathfrak{N}_j(A) = \dim \mathfrak{N}_j(A^*)$ for all j;
3) the matrix of the pairwise scalar products of the elements of the basis of $\mathfrak{N}_j(A)$ and the elements of the basis of $\mathfrak{N}_j(A^*)$ is non-degenerated for all j.

Let $\{f_j\}_1^\infty$ be the system composed from the basis $f_1, \ldots, f_{n_1}$ of $\mathfrak{N}_1(A)$, the basis $f_{n_1+1}, \ldots, f_{n_2}$ of $\mathfrak{N}_2(A)$, and so on. We shall call it the *system of root vectors* of the operator A. What was said above allows us to construct the system $\{g_j\}_1^\infty$ of root vectors of A^* biorthogonal to $\{f_j\}$ from the basis $g_1, \ldots, g_{n_1}$ of $\mathfrak{N}_1(A^*)$, the basis $g_{n_1+1}, \ldots, g_{n_2}$ of $\mathfrak{N}_2(A^*)$, and so on. From this it follows that the system of root vectors of an operator with discrete spectrum is always minimal. (This is also true in the case of finite spectrum, with which we are not concerned.) As regards completeness or, moreover, the basis property, they do not always hold, and the relevant examples of elliptic operators are given below in the next section.

Suppose now that the operator A is compact. If $A = A^*$, then, as we have already mentioned, in H there exists an orthonormal basis of eigenvectors of A. We assume that $A \neq A^*$ and that all pairwise different nonzero eigenvalues of A form an infinite sequence $\{\mu_j\}_1^\infty$ now ordered in such a way that the moduli are not increasing (we shall always keep this numbering in what follows unless otherwise stated). We again write $\mathfrak{N}_j(A)$, $\mathfrak{N}_j(A^*)$ instead of $\mathfrak{N}_{\mu_j}(A)$, $\mathfrak{N}_{\overline{\mu_j}}(A^*)$. Denote by $\mathfrak{N}(A)$ and $\mathfrak{N}(A^*)$ the linear closed hulls of the spaces $\mathfrak{N}_1(A), \ldots$ and $\mathfrak{N}_1(A^*), \ldots$, respectively. It is easy to show that if $\mathfrak{N}(A) \neq H$, then the orthogonal complement $[\mathfrak{N}(A)]^\perp$ to $\mathfrak{N}(A)$ is invariant with respect to A^* and the restriction of A^* to $[\mathfrak{N}(A)]^\perp$ is a Volterra operator. Conversely, if A^* is a Volterra operator on a subspace invariant with respect to A^*, then this subspace is contained in $[\mathfrak{N}(A)]^\perp$ so that $\mathfrak{N}(A) \neq H$.

We select the following relatively simple case.

Condition 2.5.1. If 0 is an eigenvalue of a compact operator A and/or of the operator A^*, then we assume that

1) the root subspaces $\mathfrak{N}_0(A)$ and $\mathfrak{N}_0(A^*)$ corresponding to this eigenvalue are finite-dimensional;
2) their dimensions are the same;
3) the matrix of the pairwise scalar products of the elements of the basis of $\mathfrak{N}_0(A)$ by the elements of the basis of $\mathfrak{N}_0(A^*)$ is non-degenerate.

Under this condition we can compose the system of root vectors $\{f_j\}_1^\infty$ from the basis $f_1, \dots, f_{n_0}$ of $\mathfrak{N}_0(A)$, the basis $f_{n_0+1}, \dots, f_{n_1}$ of $\mathfrak{N}_1(A)$, and so on. This system is minimal, and the system $\{g_j\}_1^\infty$ biorthogonal to it can be constructed from the basis $g_1, \dots, g_{n_0}$ of $\mathfrak{N}_0(A^*)$, the basis $g_{n_0+1}, \dots, g_{n_1}$ of $\mathfrak{N}_1(A^*)$, and so on.

It is easy to show that if Condition 2.5.1 is satisfied, then the operator A can be 'improved' by adding a finite-dimensional operator $\lambda_0 P$ (P is the projection on $\mathfrak{N}_0(A)$ and λ_0 is a nonzero number different from all $\lambda_1, \lambda_2, \dots$) such that $A + \lambda_0 P$ has the same root subspaces as A and the same spectrum except that the eigenvalue 0 is replaced by λ_0. This allows us to assume without loss of generality that 0 is neither an eigenvalue of A nor of A^* and to pass to the 'inverse' operator with discrete spectrum when considering the questions of completeness, asymptotics of eigenvalues, and so on, for operators A satisfying Condition 2.5.1.

We note that if A is a *dissipative* operator, i.e. $\mathrm{Im}A = (A - A^*)/2i \geq 0$, then A has no associated vectors corresponding to real eigenvalues and $\mathrm{Ker}A = \mathrm{Ker}A^*$. In this case Condition 2.5.1 reduces to the condition $\dim\mathrm{Ker}A < \infty$ (which is always satisfied for elliptic A).

In particular, the elliptic pseudodifferential operator A in Example 2.2.3 is dissipative. Let G be a bounded domain in $\mathbb{R}^3$ with the boundary $\mathfrak{S}$. Then $\mathrm{Ker}A \neq \{0\}$ precisely for k such that the interior Dirichlet problem

$$\Delta u + k^2 u = 0 \quad \text{in } G\,, \qquad u = 0 \quad \text{on } \mathfrak{S}$$

has nontrivial solutions. Moreover, $\dim\mathrm{Ker}A$ is equal to the dimension of the space of solutions of this problem. The relevant k's form a sequence converging to $+\infty$.

The same is true for the pseudodifferential operator of Example 2.4.8.

2.6. The Spectrum and the Root Functions of an Elliptic Operator on M. Examples

a. Let A be an elliptic operator from $\Psi^m_{\mathrm{ph}}(M)$. We shall discuss its spectral properties considering it as an operator in $H_s(M)$ (with domain $H_{s+m}(M)$ if $m > 0$). This is a separable Hilbert space, and for $s = 0$ the scalar product (2.6) is defined by the density given on M, whereas for $s \neq 0$ the scalar product can be defined in many ways and sometimes this may be conveniently done in a special manner (see (5.108) below).

For $m < 0$ the operator A is compact in $H_s(M)$. Its root functions belong to $C^\infty(M)$ by Theorem 2.2.6 on improved smoothness (if $\lambda \neq 0$, then the leading part of the operator $(A - \lambda I)^k$ is $(-\lambda)^k I$ so that this operator is elliptic). Hence it follows that the spectrum and the root subspaces do not depend on s. Condition 2.5.1, as can be proved, reduces to the relations

$$\kappa(A) = 0 \quad \text{and} \quad \dim\mathfrak{N}_0(A) < \infty\,, \tag{2.54}$$

and if they are satisfied, then this is so for all s at once.

For $m > 0$ the operator A is unbounded and its spectrum $\sigma(A)$ either fills the whole plane for all s or is discrete, and then it is again independent of s and so are the root subspaces lying in $C^\infty(M)$. The essential spectrum is always empty.

For $m = 0$ the operator A is bounded in $H_s(M)$ and its essential spectrum $\sigma_e(A)$ coincides with the set of values of the principal symbol and therefore does not depend on s. If $\lambda \in \sigma(A)\backslash\sigma_e(A)$, then λ is an eigenvalue of A and/or $\overline{\lambda}$ is an eigenvalue of A^*. The corresponding root functions belong to $C^\infty(M)$. Therefore, the spectrum $\sigma(A)$ is independent of s also for $m = 0$.

b. Now we shall give a number of examples. In particular, we want to show that various possibilities mentioned above in this section can be met with when elliptic operators are considered.

Example 2.6.2. Any elliptic formally selfadjoint pseudodifferential operator of order $m > 0$ on M can be now considered as a selfadjoint operator in $H_0(M)$ with domain $H_m(M)$. Its spectrum does not contain nonreal points and hence is discrete. This in particular refers to the Beltrami–Laplace operator Δ on a Riemannian manifold M (see Example 2.2.2).

If $M = S^n$ is the unit sphere in $\mathbb{R}^{n+1}$ with Riemannian metric induced by the standard Riemannian metric of $\mathbb{R}^{n+1}$, then the eigenfunctions of the operator Δ are called *spherical functions* (for $n = 1$ they are ordinary trigonometrical functions). Pairwise different eigenvalues are equal to $-k(k+n-1)$ ($k = 0, 1, \dots$). For $n = 1$ they are double eigenvalues (except 0 which is a simple eigenvalue), and for $n > 1$ they have increasing multiplicities or order k^{n-1} (see for example (Shubin 1978, §22)).

Example 2.6.3. Let us consider a differential operator $A = e^{i\beta\cdot x}p(D)$ on the torus $\mathbb{T}^n = \mathbb{R}^n/(2\pi\mathbb{Z})^n$, where $p(\xi)$ is polynomial of degree m with the leading term $p_0(\xi)$ and $0 \neq \beta \in \mathbb{Z}^n$. We impose the ellipticity condition $p_0(\xi) \neq 0$ ($\xi \in \mathbb{R}^n$). It is easy to prove that if $p(\gamma) \neq 0$ for all $\gamma \in \mathbb{Z}^n$, then the spectrum $\sigma(A)$ is empty, and if $p(\gamma_0) = 0$ for some $\gamma_0 \in \mathbb{Z}^n$, then $\sigma(A) = \mathbb{C}$ and each λ is an eigenvalue of A. In the former case the elliptic pseudodifferential operator A^{-1} of order $-m$ is a Volterra operator (in any of the spaces $H_s(\mathbb{T}^n)$).

Example 2.6.4. Let us consider a differential operator of the first order on the circle $\mathbb{T}^1 = S$

$$A = a_0(x)D + a_1(x)\ . \tag{2.55}$$

Here $a_0(x)$ and $a_1(x)$ are complex-valued 2π-periodic functions from $C^\infty(\mathbb{R})$. Suppose that the ellipticity condition $a_0(x) \neq 0$ is satisfied. We put

$$\alpha = \frac{1}{2\pi}\int_{-\pi}^{\pi} a_0^{-1}(t)dt\ , \qquad \beta = \alpha^{-1} \quad \text{if} \quad \alpha \neq 0\ ,$$

$$\gamma = \frac{1}{2\pi}\int_{-\pi}^{\pi} a_1(t)a_0^{-1}(t)dt\ . \tag{2.56}$$

An elementary calculation shows that

1) if $\alpha = 0$ and $\gamma \notin \mathbb{Z}$, then the spectrum $\sigma(A)$ is empty so that A^{-1} is a Volterra operator;
2) if $\alpha = 0$ and $\gamma \in \mathbb{Z}$, then the whole complex plane consists of eigenvalues of A;
3) if $\alpha \neq 0$, then $\sigma(A)$ consists of the eigenvalues $\lambda_k = \beta(k+\gamma)$ $(k \in \mathbb{Z})$ and they all are simple eigenvalues.

In what follows we shall assume that $\alpha \neq 0$. Modulo multiplication by a positive constant, the eigenfunction corresponding to the eigenvalue λ_k is of the form

$$\varphi_k(x) = g^k(x)h(x)\,, \quad \text{where} \quad g(x) = \exp\left(\mathrm{i}\beta \int_{-\pi}^{x} a_0^{-1}(t)dt\right) \tag{2.57}$$

and $h(x)$ is a certain everywhere nonzero function.

We consider the curve Γ defined by the equation $z = g(\tau)$ $(-\pi \leq \tau \leq \pi)$ on the complex plane. This is a closed curve, and it does not pass through the origin.

Theorem 2.6.5 (Agranovich–Markus 1989). *The system $\{\varphi_k\}_{-\infty}^{\infty}$ of eigenfunctions is complete in $H_0(S)$ precisely when Γ does not intersect itself. If this condition is not satisfied, then the defect of the system is infinite. In the case when the system $\{\varphi_k\}$ is complete, it is a basis of $H_0(S)$ if and only if* $\arg a_0(x) \equiv \text{const}$.

The same is true in any of the $H_r(S)$.

In particular, if we put

$$a_0(x) = (1 + \mathrm{i}b\mathrm{e}^{\mathrm{i}x})^{-1} \qquad (b \in \mathbb{R}, \quad b \neq \pm 1)\,, \tag{2.58}$$

then the condition $\alpha \neq 0$ is satisfied and the completeness holds if $|b| < 1$ and does not hold if $|b| > 1$. If we put

$$a_0(x) = (1 - \cos 2x + \mathrm{i}\cos x)^{-1}\,, \tag{2.59}$$

then the system $\{\varphi_k\}$ is complete. It is not a basis in the latter case nor is in the former if $b \neq 0$.

Using the separation of variables, we can obtain examples of systems of eigenfunctions on the two-dimensional torus $\mathbb{T}^2$ that are not complete:

Example 2.6.6 (Agranovich and Markus 1989). We consider the differential operator

$$A = a_0(x)(D_x + \mathrm{i}D_y) \tag{2.60}$$

on $\mathbb{T}^2$. Here we write (x, y) instead of (x^1, x^2). Let $a_0(x)$ be a 2π-periodic function from $C^\infty(\mathbb{R})$ different from 0 everywhere and satisfying the condition $\alpha \neq 0$ (see (2.56)). Then $\sigma(A)$ consists of eigenvalues $\lambda_{kl} = k - \mathrm{i}l$ $(k, l \in \mathbb{Z})$ which are all simple, and the corresponding eigenfunctions are

$$u_{kl}(x,y) = \exp\left[\mathrm{i}(k - \mathrm{i}l)\int_{-\pi}^{x} a_0^{-1}(t)dt + lx + \mathrm{i}ly\right] . \tag{2.61}$$

This system of functions is complete in $H_0(\mathbb{T}^2)$ if and only if the system of eigenfunctions of the differential operator $a_0(x)D_x$ is complete in $H_0(S)$. In the case when the system $\{u_{kl}\}$ is not complete, its defect is infinite. In the case when it is complete, it is not a basis (by any numbering of the functions u_{kl} into a sequence with a single index) if and only if the system of eigenfunctions of $a_0(x)D_x$ is not a basis of $H_0(S)$. The same is true in $H_s(\mathbb{T}^2)$.

Of course, Theorem 2.6.5 gives a very special result. We shall be able to give much more general theorems on completeness of root functions of elliptic pseudodifferential operators and on summability of the corresponding Fourier series in §6 employing the tools introduced there. However, in these theorems the conditions are of a different character. We shall return to Theorem 2.6.5 at the end of §6.

We now give two examples of pseudodifferential operators of zero order.

Example 2.6.7. Let A be an elliptic pseudodifferential operator from the space $\Psi^0_{\mathrm{ph}}(\mathbb{T}^n)$ on the torus $\mathbb{T}^n$, and suppose that the symbol $a(\xi)$ does not depend on x. Then A is a normal operator in $H_0(\mathbb{T}^n)$ with the eigenvalues $a(\gamma)$, and the eigenfunctions $\{e^{\mathrm{i}\gamma\cdot x}\}$ ($\gamma \in \mathbb{Z}^n$) form an orthonormal basis. (In general, this is true for a pseudodifferential operator on $\mathbb{T}^n$ of arbitrary order with symbol independent of x.) If we fix a nonzero $\gamma_0 \in \mathbb{Z}^n$, then the eigenvalues $a(t\gamma_0)$, where $t > 0$ and $t\gamma_0 \in \mathbb{Z}^n$, condense at the point $a_0(\gamma_0)$ of the essential spectrum as $t \to +\infty$.

Example 2.6.8. Let $a(x)$ be a complex-valued function from $C^\infty(M)$, everywhere different from 0. The operator of multiplication by $a(x)$ is an elliptic pseudodifferential operator of zero order, and $a(x)$ is its symbol and principal symbol. The spectrum and the essential spectrum coincide with the set of the values of $a(x)$. If one of the values λ is taken on a set of positive measure and $\chi(x)$ in the characteristic function of any measurable subset of it of positive measure, then $\chi(x)$ is an eigenfunction with the eigenvalue λ. It is clear that in general $\chi \notin C^\infty(M)$.

Concluding, we shall show that the second of the relations in (2.54) does not follow from the first one.

Example 2.6.9. On the circle S we put

$$Au(x) = \sum_{n\neq 0} e^{\mathrm{i}(n+1)x}\frac{1}{(1+n^2)}c_n(u) . \tag{2.62}$$

This is an elliptic pseudodifferential operator of order $m = -2$ with zero index (see (2.41)). The eigenfunction 1 and the infinite chain of associated functions $e^{-\mathrm{i}x}, e^{-2\mathrm{i}x}, \ldots$ correspond to the eigenvalue 0.

§3. Some Variants and Generalizations

Throughout § 3 we only show the first step of possible generalizations; the combinations of the generalizations are not considered.

3.1. Other Classes of Scalar Pseudodifferential Operators in $\mathbb{R}^n$ and on M

a. Generalizing Definition 1.5.1 of polyhomogeneous pseudodifferential operators in $\mathbb{R}^n$, we can assume that for the symbol $a(x,\xi)$ we have a decreasing sequence $\{m_j\}_0^\infty$ of real numbers such that $m_0 = m$, $m_j \searrow -\infty$ as $j \to \infty$, and the functions $a_j(x,\xi)$ in the asymptotic expansion (1.44) are positively homogeneous of degree m_j (instead of $m - j$). The expansion (1.44) is here understood in an obvious sense: for any α, β, N

$$\left| \partial_x^\alpha \partial_\xi^\beta \left[a(x,\xi) - \sum_{j=0}^{N} \theta(\xi) a_j(x,\xi) \right] \right| \le C_{\alpha\beta}(1+|\xi|)^{m_{M+1}-|\beta|} . \tag{3.1}$$

Such symbols and the corresponding pseudodifferential operators are also called *polyhomogeneous* or *classical.* The simplest situation, where they necessarily appear, is when A is a pseudodifferential operator of noninteger order $m > 0$, polyhomogeneous in the former sense, and λ is a parameter. Then $A - \lambda I$ and the resolvent $R_A(\lambda)$ for $\lambda \in \rho(A)$ are polyhomogeneous in the new sense. The definitions of ellipticity in $\mathbb{R}^n$ and on M are preserved, and all what was said above about elliptic pseudodifferential operators remains valid almost without any change.

b. For a pseudodifferential operator from Ψ^m with symbol $a(x,\xi)$ we can also introduce the definition of *ellipticity.* The condition of *uniform ellipticity* in $\mathbb{R}^n$ is that for some constants $C_1 > 0$ and C_2 we have

$$|a(x,\xi)| \ge C_1 |\xi|^m \quad \text{for} \quad |\xi| \ge C_2 \tag{3.2}$$

and all x. To define ellipticity on M, the analogous condition is imposed on the local complete symbol in a neighborhood of an arbitrary point $x_0 \in M$. If $a_1 \in S^m$ and $a - a_1 \in S^{m_1}$ with $m_1 < m$, then we call a_1 a *principal symbol.* It is easy to prove that the main results from Sects. 2.2 and 2.3 remain valid.

c. Let us consider a pseudodifferential operator in $\mathbb{R}^n$ of the form

$$A = F_{\xi\to x}^{-1} \theta(\xi) a_0(x,\xi) F + T , \tag{3.3}$$

where $a_0(x,\xi)$ is a homogeneous symbol from $S_{\mathrm{h}}^m(\mathbb{R}^n \times \dot{\mathbb{R}}^n)$. The function $\theta(\xi)$ is the same as in (1.43), and T is an arbitrary operator of order not greater than $m-1$ or at least smaller than m (cf. (Dynin 1961) or (Agranovich 1965)). For the operators of the form (3.3) we can construct a coarse calculus on the level of principal symbols, and it is preserved when we pass to a manifold. The main results of Sects. 2.2 and 2.3 remain valid. We construct a coarse parametrix B for an elliptic operator A, i.e. we require that $BA - I$ and $I - AB$ be operators of negative order (not of order $-\infty$), but this is sufficient to obtain the theorems on Fredholm property and improved smoothness.

d. Returning to the definition of polyhomogeneous pseudodifferential operators given in Sect. 1.5, we mention that the symbols $a_j(x,\xi)$ can be assumed to have complex degrees of homogeneity:

$$a_j(x,t\xi) = t^{m-j} a_j(x,\xi) \quad (\xi \neq 0,\ t > 0)\,, \tag{3.4}$$

where $m \in \mathbb{C}$. Such a generalization, used in (Seeley 1967a), is essential when considering complex powers of operators that are polyhomogeneous and elliptic in the former sense. The necessary modifications of the above argument is that A is now a bounded operator from H_s to $H_{s-\mathrm{Re}\,m}$, and an elliptic operator on M is now a Fredholm operator from $H_s(M)$ to $H_{s-\mathrm{Re}\,m}(M)$.

e. In the Seeley's paper (1963) singular integral operators on $\mathbb{R}^n$ of the form $A = A_0 + T$ are considered. Here A_0 has symbol $a_0(x,\xi) \in S^0_{\mathrm{h}}$, and it is assumed that all derivatives $\partial_x^\alpha a_0(x,\xi)$ have limits as $|x| \to \infty$ uniformly in ξ, $|\xi| = 1$. It is assumed that the component T is a compact operator in $L_p(\mathbb{R}^n)$ for all p, $1 < p < \infty$. It is shown that A is Fredholm in $L_p(\mathbb{R}^n)$ if and only if it is elliptic. The space $\mathbb{R}^n$ is in fact replaced here by its one-point compactification $\mathbb{R}^n \cup \{\infty\}$. A similar point of view is taken in some other papers (see the bibliographical notes in (Egorov and Shubin 1988b, Sect. 1.6)).

f. The calculus of pseudodifferential operators can be constructed if instead of $\mathbb{R}^n$ we consider a domain $\Omega \subset \mathbb{R}^n$, which does not have to be bounded, and if the estimates of the symbols are not assumed to be uniform in x. The symbol $a(x,\xi)$ is given on $\Omega \times \mathbb{R}_n$ and is subjected to the following inequalities:

$$|\partial_x^\alpha \partial_\xi^\beta a(x,\xi)| \leq C_{\alpha\beta}(K)(1+|\xi|)^{m-|\beta|} \quad (x \in K) \tag{3.5}$$

for any compact set $K \subset \Omega$ and for any α, β. For example, $a(x,\xi)$ may be a polynomial in ξ with coefficients from $C^\infty(\Omega)$. If $A = a(x,D)$ is given in the form (1.14), then the condition analogous to (3.5) is imposed on the amplitude function.

Such a pseudodifferential operator acts from $C_0^\infty(\Omega)$ to $C^\infty(\Omega)$ and also from $H_{s,\mathrm{comp}}(\Omega)$ (the restrictions to Ω of functions from $H_s(\mathbb{R}^n)$ with compact supports contained in Ω) to $H_{s-m,\mathrm{loc}}(\Omega)$ (the restrictions to Ω of functions locally from $H_{s-m}(\mathbb{R}^n)$). Theorems 1.2.1-3 and 1.4.1 can be easily generalized to the new class of pseudodifferential operators. Using the transposed operator, a pseudodifferential operator A can be extended to the space $\mathcal{E}'(\Omega)$ of distributions with compact supports (continuous linear functionals over $\mathcal{E}(\Omega) = C^\infty(\Omega)$) mapping it into the space $\mathcal{D}'(\Omega)$ of all distributions in Ω. In all situations discussed here, A is a continuous operator in a sense which can be easily made precise.

It is a little bit more complicated to obtain a theorem on composition (see Theorem 1.2.4), since A is not defined on all Bu (even if $u \in C_0^\infty(\Omega)$). To overcome this difficulty, the notion of properly supported pseudodifferential operator is introduced. Let $K(x, y)$ be the Schwartz kernel of A. We denote by π_1, π_2 the projections of $(x, y) \in \Omega \times \Omega$ on x and y respectively. The operator A is called *properly supported* if the inverse image in supp$K$ of any compact set contained in $\pi_1(\text{supp}K)$ or $\pi_2(\text{supp}K)$ is compact.

Any differential operator is properly supported, since the support of its Schwartz kernel lies on the diagonal of $\Omega \times \Omega$. Any pseudodifferential operator A can be written as the sum of a properly supported pseudodifferential operator and an integral operator with infinitely smooth kernel, by multiplying the amplitude by φ and $1 - \varphi$, where φ is a suitable cutting off function equal to 1 in a neighborhood of the diagonal of $\Omega \times \Omega$.

If A is a properly supported pseudodifferential operator, then A' and A^* are also properly supported. A properly supported operator acts continuously in $C_0^\infty(\Omega)$, $C^\infty(\Omega)$, $\mathcal{E}'(\Omega)$, and $\mathcal{D}'(\Omega)$. The composition theorem can be generalized to the case when A is properly supported. In addition, if A and B are properly supported pseudodifferential operators, then so is AB. As a result, we obtain a calculus where pseudodifferential operators are considered modulo addition of integral operators with infinitely smooth kernels.

For elliptic pseudodifferential operators (it does not make sense any more to consider the uniform ellipticity) we can construct a parametrix (which also is elliptic) and prove the theorem on improved smoothness. In particular, if $u \in \mathcal{E}'(\Omega)$ and $Au \in C^\infty(O)$ in a subdomain $O \subset \Omega$ (i.e. the restriction of the function Au to O is C^∞), then $u \in C^\infty(O)$.

g. In the papers of Boutet de Monvel, Krée, and Tréves (see (Tréves 1980) and the references therein) analytic pseudodifferential operators are considered. We give some definitions from (Tréves 1980).

We first recall that a complex-valued function $f(x)$ defined in a domain $\Omega \subset \mathbb{R}^n$ is called *real-analytic* if each point $x_0 \in \Omega$ has a neighborhood where $f(x)$ has the Taylor expansion in a convergent power series in $x^j - x_0^j$. The space of all such functions will be denoted be $\mathcal{A}(\Omega)$.

We say that a linear operator T from $\mathcal{E}'(\Omega)$ to $\mathcal{D}'(\Omega)$ is *analytically smoothing* if its image is contained in $\mathcal{A}(\Omega)$.

A function $p(x, y, \xi)$ from $C^\infty(\Omega \times \Omega \times \mathbb{R}_n)$ is called a *pseudoanalytic amplitude function* if there is a neighborhood Ω^c of Ω in $\mathbb{C}^n$ such that p can be continued to a function $p(z, w, \xi)$ from $C^\infty(\Omega^c \times \Omega^c \times \mathbb{R}_n)$ holomorphic in (z, w) and satisfying the following estimates: for any compact set $K \subset \Omega^c \times \Omega^c$ there are constants $C = C(K)$ and $R_0 = R_0(K)$ such that

$$|D_\xi^\alpha p(z, w, \xi)| \leq C^{|\alpha|+1} \alpha! \, |\xi|^{m-|\alpha|} \tag{3.6}$$

for $(z, w) \in K$ and $|\xi| \geq R_0 \max(|\alpha|, 1)$.

A pseudodifferential operator A in Ω is called *analytic* if for any subdomain Ω_1 with $\overline{\Omega_1} \subset \Omega$ there is an analytically smoothing operator T_1 such that $A - T_1$ considered in Ω_1 can be represented in the form (1.14) with pseudoanalytic amplitude function.

For an elliptic analytic pseudodifferential operator A we can show, constructing the corresponding parametrix, that if $u \in \mathcal{E}'(\Omega)$ and $Au \in \mathcal{A}(O)$ in a subdomain $O \subset \Omega$, then $u \in \mathcal{A}(O)$. For elliptic differential operators with analytic coefficients such a result has been known since long, in the case of matrix operators it is due to Petrovskij (1939).

Finally, we can consider analytic pseudodifferential operators on real-analytic manifolds. For example, if $\mathfrak{S}$ is a sphere, then (2.9) is analytic (Friedlander 1980). (By the way, it is normal in $L_2(\mathfrak{S})$ in this case (Ramm 1973).)

h. A detailed discussion of elliptic pseudodifferential operators on non-compact manifolds goes beyond the frames of this survey. But we would like to mention that if one not only wishes to have the theorem on improved smoothness but also the theorems on Fredholm property and/or unique solvability in, say, $\mathbb{R}^n$, then usually some conditions on the behavior of the functions at infinity are assumed. For example, when equations of the Helmholtz type are considered, the radiation condition is imposed (for this condition and its generalizations see, for example, (Egorov and Shubin 1988b, §7, Chap. II)). We shall briefly dwell on a class of differential and pseudodifferential operators in $\mathbb{R}^n$ considered in the books of Shubin (1978) and Helffer (1984).

In (Shubin 1978) the classes G^m of pseudodifferential operators A in $\mathbb{R}^n$ are introduced. They have symbols $a(x, \xi)$ satisfying the conditions

$$|\partial_x^\alpha \partial_\xi^\beta a(x, \xi)| \le C_{\alpha\beta}(1 + |x| + |\xi|)^{m-|\alpha|-|\beta|} \quad (x \in K) \tag{3.7}$$

for any α, β. The calculus of such (and more general) pseudodifferential operators is constructed. They act continuously in $S(\mathbb{R}^n)$, for $m \le 0$ they extend to bounded operators in $L_2(\mathbb{R}^n)$, and for $m < 0$ they are compact operators in $L_2(\mathbb{R}^n)$. Let $\Lambda^{(s)}$ be a pseudodifferential operator with symbol $(1+|x|^2+|\xi|^2)^{s/2}$. We denote by $Q_s(\mathbb{R}^n)$ the space of all $u \in S'(\mathbb{R}^n)$ such that $\Lambda^{(s)}u \in L_2(\mathbb{R}^n)$. Each $A \in G^m$ acts continuously from $Q^s(\mathbb{R}^n)$ to $Q^{s-m}(\mathbb{R}^n)$ for any $s \in \mathbb{R}$. The following ellipticity condition guarantees that A is Fredholm as an operator from $Q^s(\mathbb{R}^n)$ to $Q^{s-m}(\mathbb{R}^n)$: there exist positive constants C, C', and C'' such that for $|x| + |\xi| \ge C$ we have

$$C'(|x| + |\xi|)^m \le |a(x, \xi)| \le C''(|x| + |\xi|)^m . \tag{3.8}$$

A formally selfadjoint A of that form can be given, for example, in the Weyl form with a real Weyl symbol. If in addition A is elliptic and $m > 0$, then A can be considered as a closed operator in $L_2(\mathbb{R}^n)$ with domain $Q^m(\mathbb{R}^n)$. This is a selfadjoint operator with discrete spectrum whose eigenfunctions belong to $S(\mathbb{R}^n)$.

In (Helffer 1984) 'classical' pseudodifferential operators were selected from G^m with symbols admitting the asymptotic expansion $a \sim a_0 + a_1 + \ldots$ for $|x| + |\xi| \to \infty$, where the functions $a_j(x, \xi)$ are positively homogeneous in (x, ξ) of degree $m - j$. The function $a_0(x, \xi)$ is called the *principal symbol.* A classical operator from G^m is called *globally elliptic* if there exists a positive constant C' such that

$$|a_0(x, \xi)| \geq C'(|x| + |\xi|)^m \quad \text{for} \quad |x| + |\xi| \neq 0\,. \tag{3.9}$$

From this the ellipticity in the sense of (3.8) obviously follows. A model example is the differential operator

$$Au(x) = \frac{1}{2}[-\Delta u(x) + |x|^2 u(x)]\,, \tag{3.10}$$

corresponding to the harmonic oscillator in quantum mechanics.

In those books an advanced spectral theory of selfadjoint pseudodifferential operators of such a form is presented. We shall mention some of the results in the bibliographical notes to § 6.

i. On a circle S we consider the operator

$$T = P_+ A P_+\,, \tag{3.11}$$

where A is a polyhomogeneous pseudodifferential operator of order m on S and P_+ is the projection defined in (2.40). Clearly, T is a pseudodifferential operator of order m. By the composition theorem its complete symbol is equal to 0 for $\xi < 0$ and coincides with the complete symbol $a(x, \xi)$ of A for $\xi > 0$ modulo addition of a function from $S^{-\infty}$.

We put

$$H_r^+(S) = P_+ H_r(S)\,. \tag{3.12}$$

It is a subspace of $H_r(S)$ consisting of all

$$u(x) = \sum_{n \geq 0} e^{inx} c_n(u)$$

such that

$$\sum_{n \geq 0} (1 + |n|^2)^r\, |c_n(u)|^2 < \infty\,. \tag{3.13}$$

We define a norm in $H_r^+(S)$ as the square root of the left-hand side of (3.13). The spaces $H_r^+(S)$ form a scale with properties analogous to that of the scale of Sobolev spaces $H_r(S)$.

The operator T can be considered as a bounded operator from $H_{r+m}^+(S)$ to $H_r^+(S)$, or as an operator in $H_r^+(S)$ unbounded for $m > 0$ and compact for $m < 0$. It is then called a *Toeplitz operator* on S (more precisely, a *generalized*

Toeplitz operator, since the ordinary Toeplitz operator is obtained for $m = 0$ and the operator A of multiplication by a function, see, for example (Prößdorf 1987)). The *symbol* and the *principal symbol* of T are the symbol $a(x,\xi)$ and the principal symbol $a_0(x,\xi)$ of A for $\xi > 0$. The operator T is called *elliptic* if $a_0(x,\xi) \neq 0$ for $\xi > 0$. For such a T there exists a parametrix P_+BP_+, where B is a pseudodifferential operator of order $-m$, and its principal symbol is equal to $a_0^{-1}(x,\xi)$ for $\xi > 0$. Consequently, the elliptic Toeplitz operator (3.11) is a Fredholm operator from $H^+_{r+m}(S)$ to $H^+_r(S)$. For $m > 0$ it is a closed operator in $H^+_r(S)$ with domain $H^+_{r+m}(S)$. If the resolvent set $\rho(T)$ is nonempty, then T has discrete spectrum.

There is a series of papers devoted to multidimensional generalized Toeplitz operators, see the book of Boutet de Monvel and Guillemin (1981) and the references therein. In particular, such operators can be considered on the boundary of a strictly pseudoconvex domain in $\mathbb{C}^n$, but they do not have to be pseudodifferential. In (Boutet de Monvel and Guillemin 1981) an advanced spectral theory of selfadjoint elliptic Toeplitz operators is presented. We shall mention it again in the remarks and bibliographical notes to § 6.

3.2. Matrix Pseudodifferential Operators and Pseudodifferential Operators in Sections of Bundles. Elliptic Complexes

a. The functions, on which operators act, were assumed scalar–valued so far, but now we shall admit vector–valued functions. The simplest matrix pseudodifferential operator in $\mathbb{R}^n$ has the form

$$Au(x) = \begin{pmatrix} A_{11} & \dots & A_{1p} \\ \vdots & \ddots & \vdots \\ A_{p1} & \dots & A_{pp} \end{pmatrix} \begin{pmatrix} u_1(x) \\ \vdots \\ u_p(x) \end{pmatrix}, \tag{3.14}$$

where A_{ij} are pseudodifferential operators from $\Psi^m(\mathbb{R}^n)$ or $\Psi^m_{\text{ph}}(\mathbb{R}^n)$ with the same m. In such cases we shall write $A \in \Psi^m(\mathbb{R}^n)$ or $A \in \Psi^m_{\text{ph}}(\mathbb{R}^n)$, respectively. The symbol $a(x,\xi)$ of A is a matrix composed of the symbols of A_{ij}. If $A \in \Psi^m_{\text{ph}}$, then the principal symbol $a_0(x,\xi)$ of A is composed of the principal symbols of A_{ij}, but we are not assuming now that all these scalar symbols are not identically zero. Only the matrix a_0 has to be not identically zero. For M in place of $\mathbb{R}^n$ we have analogous conventions. The spaces $S(\mathbb{R}^n)$, $H_s(\mathbb{R}^n)$, $C^\infty(M)$, and $H_s(M)$ consist now of vector–valued functions, and the scalar products in $H_0(\mathbb{R}^n)$ and $H_0(M)$ are defined by (1.21) and (2.6), where by $u\overline{v}$ we understand the sum $\sum u_j\overline{v}_j$.

We need to make the following evident modification in (1.27): $\overline{a}$ is replaced by the Hermitian adjoint matrix a^*. If A belongs to Ψ^m_{ph} and has the principal symbol a_0, then the formally adjoint A^* has the principal symbol a_0^*. The condition $A^* = A$ is equivalent to the Weyl symbol of this operator being Hermitian. The commutator of A and B of orders m_1 and m_2 has order $m_1 + m_2 - 1$ only in the case when the principal symbols a_0 and b_0 commute. In (1.28) the matrix a is replaced by the transposed matrix a'.

A matrix pseudodifferential operator A from $\Psi^m_{\mathrm{ph}}(\mathbb{R}^n)$ and its principal symbol $a_0(x,\xi)$ are called *elliptic* at the point x if

$$\det a_0(x,\xi) \neq 0 \quad (\xi \neq 0)\,. \tag{3.15}$$

The condition of uniform ellipticity in $\Omega \subset \mathbb{R}^n$ has the form

$$|\det a_0(x,\xi)| \geq c > 0 \quad (x \in \Omega,\ |\xi| = 1)\,. \tag{3.16}$$

Ellipticity of a matrix pseudodifferential operator on M is defined in an obvious way.

All what was said in § 1 and § 2 remains valid with natural modifications. In particular, for $m = 0$ the essential spectrum of an elliptic pseudodifferential operator on M consists of eigenvalues of its principal symbol on $T^*M\backslash 0$.

b. We now assume that we are given two sets of real numbers

$$\{l_1, \ldots, l_p\} \quad \text{and} \quad \{m_1, \ldots, m_p\} \tag{3.17}$$

and that in (3.14) we have $A_{ij} \in \Psi^{l_j+m_k}$ or $\Psi^{l_j+m_k}_{\mathrm{ph}}$. In this case we say that A is a matrix pseudodifferential operator *of type* $\{l_j, m_k\}$. In the case considered above $l_1 = \ldots = l_p = 0$, $m_1 = \ldots = m_p = m$ and we could speak about operators of type $\{0, m\}$, but in such a case we omit the remark about the type or speak about operators of ordinary, or standard, type. If A is a differential operator, then the numbers l_j, m_k are integers and $A_{ik} = 0$ if $l_j + m_k < 0$ (see (Douglis and Nirenberg 1955)).

The principal symbol $a_0(x,\xi)$ of a polyhomogeneous matrix pseudodifferential operator A of type $\{l_j, m_k\}$ consists of the principal symbols of A_{jk}. They are positively homogeneous in ξ of degree l_j+m_k and are not necessarily all different from 0. If $\det a_0(x,\xi) \neq 0$ for $\xi \neq 0$, then A is called *elliptic at the point x in the sense of Douglis–Nirenberg* (in more detail: of type $\{l_j, m_k\}$ elliptic at the point x in the sense of Douglis–Nirenberg). In an apparent way we can define the uniform ellipticity in the sense of Douglis–Nirenberg in $\Omega \subset \mathbb{R}^n$ and ellipticity in the sense of Douglis–Nirenberg on M. Ellipticity in the sense of definition (3.15) ($l_j = 0$, $m_j = m$) will be sometimes called ordinary, or standard, ellipticity. For $l_1 = \ldots = l_p = 0$ and arbitrary m_j we may speak about *ellipticity in the sense of Petrovskij* (see (Petrovskij 1939)).

Differential operators elliptic in the sense of Douglis–Nirenberg appear when we reduce a scalar elliptic equation to the system of the first order. For example, if we put $u_1 = u$, $u_2 = \partial_1 u$, and $u_3 = \partial_2 u$, then the Laplace equation $(\partial_1^2 + \partial_2^2)u = 0$ reduces to the system

$$\begin{pmatrix} 0 & \partial_1 & \partial_2 \\ \partial_1 & -1 & 0 \\ \partial_2 & 0 & -1 \end{pmatrix} \begin{pmatrix} u_1 \\ u_2 \\ u_3 \end{pmatrix} = 0\,, \tag{3.18}$$

that is not elliptic in the ordinary sense but is elliptic in the sense of Douglis–Nirenberg with $l_1 = 0$, $l_2 = l_3 = -1$, $m_1 = 2$, and $m_2 = m_3 = 1$. Important examples of systems elliptic in the sense of Douglis–Nirenberg can be found in hydrodynamics and the theory of elasticity, but usually with some boundary conditions. We also mention that the matrix operators of type $\{l_j, m_k\}$ arise in a natural way from scalar differential and pseudodifferential operators that depend on the parameter λ polynomially if we make the linearization with respect to λ (see 4.3c below).

We put

$$\mathcal{H}_s^{(1)}(\mathbb{R}^n) = H_{s+m_1}(\mathbb{R}^n) \times \ldots \times H_{s+m_p}(\mathbb{R}^n) , \tag{3.19}$$

$$\mathcal{H}_s^{(2)}(\mathbb{R}^n) = H_{s-l_1}(\mathbb{R}^n) \times \ldots \times H_{s-l_p}(\mathbb{R}^n) . \tag{3.20}$$

In a similar way we define the spaces $\mathcal{H}_s^{(1)}(M)$ and $\mathcal{H}_s^{(2)}(M)$. A pseudodifferential operator A of type $\{l_j, m_k\}$ acts in a bounded way from $\mathcal{H}_s^{(1)}$ to $\mathcal{H}_s^{(2)}$. The main result in the case of a closed manifold is the following

Theorem 3.2.1. *A pseudodifferential operator of type $\{l_j, m_k\}$ on M is elliptic if and only if it is Fredholm as an operator from $\mathcal{H}_s^{(1)}(M)$ to $\mathcal{H}_s^{(2)}(M)$.*

The corresponding a priori estimate for the equation $Au = f$ has the form

$$\sum_{k=1}^{p} \|u_k\|_{s+m_k,M} \leq C_s \left\{ \sum_{j=1}^{p} \|f_j\|_{s-l_j,M} + \sum_{k=1}^{p} \|u_k\|_{s+m_k-1,M} \right\} . \tag{3.21}$$

In the case of uniqueness the last sum over k is omitted. A parametrix B for A is a pseudodifferential operator of type $\{-m_j, -l_k\}$ elliptic in the sense of Douglis–Nirenberg.

To derive Theorem 3.2.1 from the analogous result for matrix pseudodifferential operators elliptic in the ordinary sense, it is sufficient to introduce, using the scalar elliptic invertible operators Λ_s of order s (see Theorem 2.3.9), the pseudodifferential operator

$$\widetilde{A} = \operatorname{diag}(\Lambda_{-l_1}, \ldots, \Lambda_{-l_p}) \cdot A \cdot \operatorname{diag}(\Lambda_{-m_1}, \ldots, \Lambda_{-m_p}) \tag{3.22}$$

of order 0, elliptic in the ordinary sense. The equation $Au = f$ is equivalent to the equation $\widetilde{A}v = g$, where

$$v_k = \Lambda_{m_k} u_k \;, \quad g_j = \Lambda_{l_j} f_j \quad (j, k = 1, \ldots, p) . \tag{3.23}$$

Remark 3.2.2. For the purposes of spectral theory the transformation (3.23) is inconvenient. If A is elliptic in the sense of Douglis–Nirenberg with

$$l_1 + m_1 = \ldots = l_p + m_p = m \tag{3.24}$$

(the elements of the main diagonal in the principal symbol have the same order), then the equation $Au - \lambda u = f$ can be easily transformed into $\widetilde{A}v - \lambda v = g$, where $\widetilde{A}$ has order m and is elliptic in the ordinary sense, by the substitution

$u_k = \Lambda_{m_k}^{-1} v_k$, $f_j = \Lambda_{m_j}^{-1} g_j$ and the multiplication of the original equation by $\mathrm{diag}(\Lambda_{m_1}, \ldots, \Lambda_{m_p})$. The operators A and $\widetilde{A}$ are similar and have essentially the same spectral properties.

However, condition (3.24) is far from being always satisfied, and in general the investigation of spectral properties of operators elliptic in the sense of Douglis–Nirenberg constitutes a separate problem.

Rejecting (3.24) in general, we assume that A is of type $\{l_j, m_k\}$, elliptic in the sense of Douglis–Nirenberg, and we assume that $l_j + m_j > 0$ for all j. Then A can be considered as an unbounded operator in $\mathcal{H}_s^{(2)}(M)$ with domain $\mathcal{H}_s^{(1)}(M)$. This operator is closed, and its spectrum is independent of s. If the resolvent set $\rho(A)$ is nonempty, then A has discrete spectrum.

We give the statement of the 'diagonalization theorem' from the paper of Kozhevnikov (1981) which concerns such pseudodifferential operators. We put $\mu_j = l_j + m_j$ and assume, without loss of generality, that

$$\mu_1 = \ldots \mu_{k_1} > \mu_{k_1+1} = \ldots = \mu_{k_2} > \ldots > \mu_{k_{r-1}+1} = \ldots \mu_{k_r} ,$$

where $\mu_{k_r} = p$. Next, we put $i_1 = k_1$, $i_2 = k_2 - k_1, \ldots, i_r = k_r - k_{r-1}$.

Theorem 3.2.3. *Suppose that the minors of the principal symbol $a_0(x, \xi)$ that lie on the intersection of the first k columns and the first k rows for $k = k_1, \ldots, k_r$ are nonzero everywhere on $T^*M \backslash 0$. Then there is an invertible pseudodifferential operator T such that*

$$B = T^{-1} A T \tag{3.25}$$

has the block-diagonal structure modulo addition of an operator of order $-\infty$:

$$B = \mathrm{diag}(B_{11}, \ldots, B_{rr}) + C , \quad \mathrm{ord}\, C = -\infty . \tag{3.26}$$

The block B_{ll} has dimension $i_l \times i_l$ and consists of pseudodifferential operators of order μ_{k_l}.

The theorem also describes the structure of T.

c. As a further generalization we consider pseudodifferential operators acting in sections of a p-dimensional complex vector bundle E of class C^∞ over M. The definitions of the vector bundle, its section, etc. are given for example in (Alekseevskij et al. 1988), (Dubrovin et al. 1985), or (Hörmander 1983a). We briefly recall that E is locally built as the direct product $O \times \mathbb{C}^p$ of a coordinate neighborhood O in M by the fiber $\mathbb{C}^p$. These direct products are glued together in a certain way. In more detail, the elements of E can be represented over O as pairs (x, w), where x is the image of a point of O in a local chart and w is a column vector from $\mathbb{C}^p$. If the intersection $O_\alpha \cap O_\beta$ of two coordinate neighborhoods is nonempty, then we have a *transition function* $\mathcal{T}_{\beta\alpha}(x^{(\alpha)})$ which is a nondegenerate complex $p \times p$-matrix, infinitely smoothly depending on the image $x^{(\alpha)}$ of a point of the intersection $O_\alpha \cap O_\beta$ in the chart U_α. If x^α, x^β are images of the same point in two charts U_α, U_β, then we identify $(x^{(\alpha)}, w^{(\alpha)})$ with $(x^{(\beta)}, w^{(\beta)})$, where $w^{(\beta)} = \mathcal{T}_{\alpha\beta}(x^{(\alpha)})w^{(\alpha)}$. The transition functions are subjected to the following conditions whose sense can be easily made precise:

$$\mathcal{T}_{\beta\alpha} = \mathcal{T}_{\alpha\beta}^{-1}, \qquad \mathcal{T}_{\gamma\alpha} = \mathcal{T}_{\gamma\beta}\mathcal{T}_{\beta\alpha} .$$

If all the $\mathcal{T}_{\alpha\beta}$ are unit matrices, then $E = M \times \mathbb{C}^p$ and such a bundle is called *trivial.* A *section* of a bundle is a mapping $M \to E$ such that the image of each point lies in the fiber over it. A section of the trivial p-dimensional bundle is a vector-valued function $u(x)$ on M with values in $\mathbb{C}^p$.

Any tensor field on M is a section of some vector bundle over M. In particular, differential forms of the first order, written locally in the form

$$u_1(x)dx^1 + \ldots + u_n(x)dx^n$$

with complex coefficients $u_j(x)$, are sections of the complexification of T^*M; we denote it by Λ^1. The differential forms of the k-th order $(1 \leq k \leq n)$ with the local form

$$\sum_{1 \leq i_1 < \ldots < i_k \leq n} u_{i_1 \ldots i_k}(x) dx^{i_1} \wedge \ldots \wedge dx^{i_k}$$

are sections of the k-th exterior power of the bundle Λ^1 which we denote by Λ^k.

We can easily define the space $C^\infty(M, E)$ of infinitely smooth sections of the bundle E.

In the fibers of the bundle E we can introduce a scalar product in an invariant way (i.e. in a way independent of the choice of local coordinates), and this can be done so that the scalar product $u(x) \cdot \overline{v(x)}$ of smooth sections will depend smoothly on $x \in M$. The bundle E is then called *Hermitian.* Integrating the scalar product of sections in a fiber with respect to a density given on M, we obtain a scalar product $(u, v)_M$ of sections over M. This allows us to introduce the space $L_2(M, E)$ of square integrable sections as the completion of the space $C^\infty(M, E)$ in the corresponding norm.

The Sobolev norms of sections can be defined by formulas of the type (2.2). For $s = 0$ we obtain a norm which is equivalent to the norm mentioned in the previous paragraph. The Sobolev spaces of sections are denoted by $H_s(M, E)$.

So far we have considered pseudodifferential operators and elliptic pseudodifferential operators acting in sections of trivial bundles. In general case the definition of a pseudodifferential operator as an operator acting in $C^\infty(M, E)$ is analogous to that given in Sect. 2.1. The principal symbol written in local coordinates undergoes the similarity transformation when they are changed. To explain the invariant sense of the principal symbol we consider the bundle π^*E over T^*M which is a 'lifting' of the bundle E from M to T^*M induced by the projection $\pi : T^*M \to M$ $(\pi(x, \xi) = x)$. The bundle π^*E is p-dimensional, and the principal symbol $a_0(x,\xi)$ is a homomorphism of this bundle over $T^*M\backslash 0$, i.e. it is an invariantly defined linear transformation of a fiber over $T^*M\backslash 0$ smoothly depending on (x,ξ). Consequently, the ellipticity condition also has the invariant sense: ellipticity means that the principal symbol is an isomorphism at all points of $T^*M\backslash 0$.

It can be proved that an elliptic pseudodifferential operator of order m is Fredholm as an operator from $H_s(M,E)$ to $H_{s-m}(M,E)$. Suppose that the bundle is Hermitian, and for simplicity let $s=m$. Then

$$L_2(M,E) = \mathrm{Im}A \oplus \mathrm{Ker}A^* , \tag{3.27}$$

where A^* is the elliptic pseudodifferential operator formally adjoint to A.

d. As mentioned in (Hörmander 1985a), in geometric problems usually arise elliptic complexes instead of elliptic operators. We briefly dwell on this notion.

Suppose we are given complex vector bundles $E_0,\dots,E_N$ of class C^∞ over M and polyhomogeneous pseudodifferential operators

$$A_l : C^\infty(M,E_l) \longrightarrow C^\infty(M,E_{l+1}) \quad (l=0,\dots,N-1)$$

of the same order. If $A_{l+1}A_l = 0$ for all l (i.e. $\mathrm{Ker}A_{l+1} \supset \mathrm{Im}A_l$), then we say that a *complex* is given

$$0 \longrightarrow C^\infty(M,E_0) \xrightarrow{A_0} C^\infty(M,E_1) \xrightarrow{A_1} \dots \xrightarrow{A_{N-1}} C^\infty(M,E_N) \longrightarrow 0 . \tag{3.28}$$

It is convenient to denote the outermost mappings by A_{-1} and A_N. A_l are operators belonging to $\varphi^{ml}_{\mathrm{ph}}(M)$ whose principal symbols $a_0^{(l)}(x,\xi)$ are at first scalar-valued ($l=0,\dots,N-1$). For each $(x,\xi)\in T^*M\backslash 0$ it is a linear transformation of the fiber $E_{l,(x,\xi)}$ over (x,ξ) in π^*E_l to the fiber $E_{l+1,(x,\xi)}$ over (x,ξ) in π^*E_{l+1}. We consider the sequence

$$0 \longrightarrow E_{0,(x,\xi)} \xrightarrow{a_0^{(0)}(x,\xi)} E_{1,(x,\xi)} \xrightarrow{a_0^{(1)}(x,\xi)} \dots \xrightarrow{a_0^{(N-1)}(x,\xi)} E_{N,(x,\xi)} \longrightarrow 0 .$$

If it is *exact* for all $(x,\xi)\in T^*M\backslash 0$, i.e. the image of each mapping is equal to the kernel of the next one, then the complex (3.28) is called *elliptic*.

We restrict ourselves to two examples.

1) The ellipticity of the complex

$$0 \longrightarrow C^\infty(M,E) \xrightarrow{A} C^\infty(M,E) \longrightarrow 0 \tag{3.29}$$

is equivalent, as is easy to see, to the ellipticity of the pseudodifferential operator A acting in $C^\infty(M,E)$.

2) The complex

$$0 \longrightarrow C^\infty(M,\Lambda^0) \xrightarrow{d_0} C^\infty(M,\Lambda^1) \xrightarrow{d_1} \dots \xrightarrow{d_{n-1}} C^\infty(M,\Lambda^n) \longrightarrow 0 , \tag{3.30}$$

where $C^\infty(M,\Lambda^0) = C^\infty(M)$ and d_l is the operator of exterior differentiation, is called (complexificated) *de Rham complex*. It is easy to prove that it is elliptic.

Suppose that all bundles E_l are Hermitian. Then in a natural way we can define a pseudodifferential operator

$$A_l^* : C^\infty(M, E_{l+1}) \to C^\infty(M, E_l) ,$$

(formally) adjoint to A_l. The ellipticity of the complex (3.28) is equivalent to the ellipticity of the selfadjoint pseudodifferential operators

$$\Delta_l = A_{l-1}A_{l-1}^* + A_l^* A_l \tag{3.31}$$

($l = 0, \ldots, N$) acting in $C^\infty(M, E_l)$. We shall assume that the complex is elliptic. Then $\mathrm{Ker}\Delta_l$ are finite-dimensional. Obviously,

$$\mathrm{Ker}\Delta_l = \{u \in C^\infty(M, E_l) : \; A_l u = 0, \; A_{l-1}^* u = 0\} .$$

It is easy to prove that the quotient spaces

$$H^l = \mathrm{Ker}A_l / \mathrm{Im}A_{l-1} \tag{3.32}$$

are finite-dimensional and have the same dimensions as $\mathrm{Ker}\Delta_l$. More precisely, in each equivalence class U of the quotient space (3.32) there is exactly one representative $u \in \mathrm{Ker}\Delta_l$, and the correspondence $U \leftrightarrow u$ is an isomorphism between H^l and $\mathrm{Ker}\Delta_l$.

The spaces (3.32) are called *cohomology spaces* of the elliptic complex (3.28), and the alternating sum

$$\sum_{l=0}^{N} (-1)^l \dim H_l \tag{3.33}$$

is called the *Euler characteristic* of this complex.

In example 1) $\mathrm{Ker}\Delta_0 = \mathrm{Ker}A$, $\mathrm{Ker}\Delta_1 = \mathrm{Ker}A^*$, and (3.33) is the index of the elliptic operator A.

In example 2) Δ_l is the Laplace operator on differential forms of the l-th degree, and the elements of its kernel are called *harmonic forms.* What was said about the isomorphism of the spaces H^l and $\mathrm{Ker}\Delta_l$, is the content of one of the main assertions of the Hodge theorem on differential forms. By the de Rham theorem H^l are in this case the cohomology groups $H^l(M, \mathbb{C})$ of the manifold M, and (3.33) is its Euler characteristic. For details see, for example, (Alekseevskij et al. 1988), (Dubrovin et al. 1986).

The reader can find further information on elliptic complexes and their applications in (Atiyah and Bott 1967, 1968), (Gilkey 1974), (Hörmander 1985), and the survey (Fedosov 1990).

§4. Pseudodifferential Operators with a Parameter

4.1. Pseudodifferential Operators Depending on a Parameter Polynomially. Ellipticity with Parameter. We consider a pseudodifferential operator

$$A(\lambda) = \sum_{l=0}^{p} \lambda^{p-l} A_l \, , \tag{4.1}$$

where A_l are pseudodifferential operators from $\Psi_{\mathrm{ph}}^{ml}(M)$ with principal symbols $a_{l0}(x,\xi)$, scalar at first ($l = 1, \ldots, p$, $m > 0$), and A_0 is the operator of multiplication by the function $a_0(x) \in C^\infty(M)$. Let $\mathcal{L}$ be a closed angle in the complex plane with the apex at the origin.

Definition 4.1.1. A pseudodifferential operator (4.1) is called *elliptic with parameter in* $\mathcal{L}$ if

$$\sum_{l=0}^{p} \lambda^{p-l} a_{l0}(x,\xi) \neq 0 \quad \text{for} \quad (x,\xi) \in T^*M \, , \quad \lambda \in \mathcal{L} \, , \quad (\xi,\lambda) \neq 0 \, , \tag{4.2}$$

where $a_{00}(x,\xi) \equiv a_0(x)$ and the functions $a_1(x,\xi), \ldots$ are assumed to be equal to 0 for $\xi = 0$.

The ordinary ellipticity for a fixed λ clearly follows from the ellipticity with parameter in $\mathcal{L}$.

We note that from the ellipticity with parameter in the angle

$$\mathcal{L}_{\alpha\beta} = \{\lambda : \alpha \leq \arg\lambda \leq \beta\} \cup \{0\} \tag{4.3}$$

there follows the ellipticity in a slightly bigger angle $\mathcal{L}_{\alpha-\varepsilon,\beta+\varepsilon}$ ($\varepsilon > 0$). The set of all rays starting from the origin along which $A(\lambda)$ is elliptic with parameter is open.

We pick out the case

$$A(\lambda) = A - \lambda I \, , \tag{4.4}$$

where A is a pseudodifferential operator from $\Psi_{\mathrm{ph}}^{m}(M)$. The ellipticity with parameter in $\mathcal{L}$ in this case simply means that $a_0(x,\xi) \notin \mathcal{L}$ for $\xi \neq 0$, where $a_0(x,\xi)$ is the principal symbol of A. If this condition is satisfied, then A will be also called *elliptic with parameter in* $\mathcal{L}$.

For example, if $a_0(x,\xi)$ is a positive function or a real function nonzero for $\xi \neq 0$, then A is elliptic with parameter in any angle $\mathcal{L}$ having no common points with the positive semi-axis $\overline{\mathbb{R}}_+$ or, respectively, the real axis $\mathbb{R}$, except its apex which is at the origin. Therefore, the existence of an angle of ellipticity with parameter can be considered as a sign of some remote resemblance of A to a selfadjoint operator. If A is elliptic with parameter in the closed left half-plane $\{\lambda : \mathrm{Re}\lambda \leq 0\}$, then the operator $\partial_t + A$ is called *parabolic*. We also note that if A is elliptic with parameter along some ray, then the index $\kappa(A)$ equals 0, since there is a homotopy of A into a selfadjoint elliptic operator. One more remark: if A is a differential operator, $n \geq 2$, and if a ray of ellipticity with parameter exists, then m is even (since $\dot{\mathbb{R}}_n$ is connected and for odd m the function $a_0(x,\xi)$ takes at $\xi \neq 0$ and $-\xi$ opposite values).

We shall see that in the presence of ellipticity with parameter one can obtain a number of essential advancements in many problems of general and spectral theory but first of all the following result:

Theorem 4.1.2. *Suppose that a pseudodifferential operator $A(\lambda)$ is elliptic with parameter in $\mathcal{L}$. Then for $\lambda \in \mathcal{L}$ sufficiently large in modulus there exists the operator $[A(\lambda)]^{-1}$ and it is of order $-mp$. Moreover, a solution of the equation $A(\lambda)u = f$ satisfies for any s the estimate*

$$\|u\|_{s+mp,M} + |\lambda|^p \|u\|_{s,M} \leq C_s \|f\|_{s,M} , \tag{4.5}$$

where C_s does not depend on f, λ.

In the next section we shall outline the proof for $A(\lambda) = A - \lambda I$, mainly following the papers of Seeley (1967a, 1969a) (the general case will be considered in 4.3c). It is based on the construction of a parametrix $B^{(N)}(\lambda)$ for $A - \lambda I$. We can show that

$$(A - \lambda I)B^{(N)}(\lambda) = I + T_N(\lambda) , \tag{4.6}$$

where $T_N(\lambda)$ is an operator with small norm in Sobolev spaces for $\lambda \in \mathcal{L}$ sufficiently large in modulus. Consequently, $I + T_N(\lambda)$ is invertible for such λ and the operator $A(\lambda)$ has the right inverse

$$R(\lambda) = B^{(N)}(\lambda)[I + T_N(\lambda)]^{-1} . \tag{4.7}$$

It is also the left inverse since $\kappa(A - \lambda I) = \kappa(A) = 0$. The number N characterizes the accuracy of the construction of the parametrix. For the proof of the theorem small accuracy is sufficient, but the construction with N arbitrarily large is necessary in the study of functions of A and their kernels (see § 5).

Remark 4.1.3. The condition of ellipticity with parameter is necessary for the estimate (4.5) to be valid for $\lambda \in \mathcal{L}$ large in modulus.

4.2. A Parametrix for $A - \lambda I$

a. We first consider a pseudodifferential operator $A \in \Psi^m_{\mathrm{ph}}$ in $\mathbb{R}^n$, $m > 0$. For simplicity we assume that the symbol $a(x,\xi)$ and the terms $a_0(x,\xi), \ldots$ of its asymptotic expansion do not depend on x for $|x| \geq c$ with sufficiently large c. We call A *elliptic with parameter in* $\mathcal{L}$ if $a_0(x,\xi) \notin \mathcal{L}$ for $\xi \neq 0$.

We begin with the case when A is a differential operator with coefficients from $B^\infty(\mathbb{R}^n)$ constant for large $|x|$ and the symbol of A is of the form

$$a(x,\xi) = \sum_{j=0}^{m} a_j(x,\xi) , \tag{4.8}$$

where a_m does not depend on ξ. For $\lambda \in \mathcal{L}$ we construct functions $b_k(x,\xi,\lambda)$ rational in λ ($k = 0, 1, \ldots$) and such that

$$\left[\sum_{j=0}^{m} a_j(x,\xi) - \lambda\right] \circ \sum_{k=0}^{\infty} b_k(x,\xi,\lambda) = 1 , \tag{4.9}$$

where $\circ$ is the formal composition of the symbols corresponding to the multiplication of the respective operators in the frames of the calculus in S^∞ for fixed λ. Here for each k the function $b_k(x,\xi,\lambda)$ is positively homogeneous in $(\xi, \lambda^{1/m})$ of degree $-m-k$:

$$b_k(x,\tau\xi,\tau^m\lambda) = \tau^{-m-k} b_k(x,\xi,\lambda) \quad (\tau > 0,\ \lambda \in \mathcal{L},\ (\xi,\lambda) \neq 0) . \tag{4.10}$$

Comparing the symbols of the same degrees of homogeneity on both sides of (4.9), we obtain the following chain of relations:

$$(a_0 - \lambda) b_0 = 1 , \tag{4.11$_0$}$$

$$(a_0 - \lambda) b_k + \sum_{\substack{j+l+|\alpha|=k \\ l<k}} \frac{1}{\alpha!} \partial_\xi^\alpha a_j \cdot D_x^\alpha b_l = 0 \quad (k = 1, 2, \ldots) . \tag{4.11$_k$}$$

Hence the functions $b_0, b_1, \ldots$ can be successively computed (for $\lambda \in \mathcal{L}$, $(\xi,\lambda) \neq 0$):

$$b_0(x,\xi,\lambda) = \frac{1}{a_0(x,\xi) - \lambda} ; \tag{4.12$_0$}$$

$$b_1(x,\xi,\lambda) = \frac{-a_1(x,\xi)}{[a_0(x,\xi)-\lambda]^2} + \frac{-\mathrm{i}\sum \partial_{\xi_\nu} a_0(x,\xi) \cdot \partial_{x^\nu} a_0(x,\xi)}{[a_0(x,\xi)-\lambda]^3} \tag{4.12$_1$}$$

etc. For any $k \geq 1$ we have

$$b_k(x,\xi,\lambda) = \sum_{l \geq \frac{k+m}{m}} \frac{b_{kl}(x,\xi)}{[a_0(x,\xi)-\lambda]^l} , \tag{4.13}$$

where the sum is finite and $b_{kl}(x,\xi)$ are polynomials of the functions $a_0, \ldots, a_m$ and their derivatives. Consequently, they are polynomials in ξ with coefficients from $B^\infty(\mathbb{R}^n)$ independent of x for large $|x|$. They are homogeneous in ξ of degree $m(l-1)-k$, which explains the inequality $l \geq (k+m)/m$. The derivatives $\partial_x^\alpha \partial_\xi^\beta b_k$ have the similar structure.

We can now introduce the operators

$$B_k(\lambda) f(x) = \int \mathrm{e}^{\mathrm{i}x\cdot\xi} b_k(x,\xi,\lambda)(Ff)(\xi) đ\xi \quad (0 \neq \lambda \in \mathcal{L}) \tag{4.14}$$

and define the parametrix of order N for $A - \lambda I$ by

$$B^{(N)}(\lambda) = B_0(\lambda) + \ldots + B_N(\lambda) . \tag{4.15}$$

However, by this definition the parametrix is not defined for all λ with sufficiently small $|\lambda|$, which is not convenient when considering the powers of A^z (see Sect. 5.3 below). Therefore we shall proceed in the following way. Suppose that $\theta(\tau)$ is a function from $C^\infty(\mathbb{R}_+)$ equal to 0 for $\tau \leq 1/2$ and 1 for $\tau \geq 1$. We put

$$b_{\theta k}(x,\xi,\lambda) = \theta(|\xi|^2 + |\lambda|^{2/m}) b_k(x,\xi,\lambda) \tag{4.16}$$

and define a set $\mathcal{L}_{(\rho)}$, $\rho > 0$, by

$$\mathcal{L}_{(\rho)} = \mathcal{L} \cup \{\lambda : |\lambda| \leq \rho\} . \tag{4.17}$$

The functions (4.16) are well-defined for $\lambda \in \mathcal{L}_{(\rho)}$, where ρ is so small that $a_0(x,\xi) - \lambda \neq 0$ for $|\lambda| \leq \rho$, $|\xi|^2 + |\lambda|^{2/m} \geq 1/2$, and they coincide with $b_k(x,\xi,\lambda)$ for $|\lambda| \geq 1$. For $\lambda \in \mathcal{L}_{(\rho)}$ we now put

$$B_{\theta k}(\lambda) f(x) = \int e^{ix\cdot\xi} b_{\theta k}(x,\xi,\lambda)(Ff)(\xi) đ\xi , \tag{4.18}$$

and instead of (4.15) we put

$$B^{(N)}(\lambda) = B_\theta^{(N)}(\lambda) = B_{\theta 0}(\lambda) + \ldots + B_{\theta N}(\lambda) . \tag{4.19}$$

Below we shall give an estimate of the norm of this operator (see (4.32)).

b. We turn to the case when A is a pseudodifferential operator with symbol

$$a(x,\xi) \sim a_0(x,\xi) + a_1(x,\xi) + \ldots . \tag{4.8'}$$

If we try to define $b_0, b_1, \ldots$ in the same way as in the case of differential operators, then we encounter the following obstacle. In general, for $k \geq 1$ $b_{kl}(x,\xi)$ is different from 0 for $l \geq 2$, and if $m(l-1) - k < 0$, then these functions have singularities at $\xi = 0$, which excludes the possibility of defining $B_k(\lambda)$ or $B_{\theta k}(\lambda)$ by the above formulas even for $\lambda \in \mathcal{L}$ large in modulus. Therefore we have to perform a deeper 'repair' of the construction than in (4.16), (4.18-19), beginning with the functions $a_j(x,\xi)$. For simplicity we shall assume that $\mathbb{R}_-$ *is the bisectrix of the angle* $\mathcal{L}$. Using the same function $\theta(\tau)$ as above, we put

$$a_{\theta j}(x,\xi) = \theta(|\xi|) a_j(x,\xi) \quad (j = 1, 2, \ldots) \tag{4.20}$$

and

$$a_{\theta 0}(x,\xi) = \theta(|\xi|) a_0(x,\xi) + 1 - \theta(|\xi|) . \tag{4.21}$$

Completing the definition of these functions for $\xi = 0$ by $a_{\theta 0}(x,0) = 1$, $a_{\theta j}(x,0) = 0$ $(j = 1, 2, \ldots)$, we have

$$a_{\theta j}(x,\xi) \in S^{m-j}(\mathbb{R}^n \times \mathbb{R}_n) \quad (j = 0, 1, \ldots) , \tag{4.22}$$

$$a(x,\xi) \sim a_{\theta 0}(x,\xi) + a_{\theta 1}(x,\xi) + \ldots , \tag{4.23}$$

$$a_{\theta j}(x,\tau\xi) = \tau^{m-j} a_{\theta j}(x,\xi) \qquad (|\xi| \geq 1, \quad \tau \geq 1) , \tag{4.24}$$

$$a_{\theta 0}(x,\xi) - \lambda \neq 0 \quad \text{for all} \quad (x,\xi) \in \mathbb{R}^n \times \mathbb{R}_n, \quad \lambda \in \mathcal{L}_{(\rho)} , \tag{4.25}$$

where ρ is a sufficiently small positive number. We shall now construct the functions $b_{\theta k}(x,\xi,\lambda)$ rational in λ with the homogeneity property

$$b_{\theta k}(x,\tau\xi,\tau^m\lambda) = \tau^{-m-k} b_{\theta k}(x,\xi,\lambda) \qquad (|\xi| \geq 1, \quad \tau \geq 1) , \tag{4.26}$$

starting from the relation

$$\sum_{j=0}^{\infty} [a_{\theta j}(x,\xi) - \lambda] \circ \sum_{k=1}^{\infty} b_{\theta k}(x,\xi,\lambda) \sim 1 . \tag{4.27}$$

For $b_{\theta k}$ we obtain the formulas analogous to (4.12_0), (4.12_1), ..., with $a_0, a_1, \ldots$ replaced by $a_{\theta 0}, a_{\theta 1}, \ldots$. (So we should keep in mind that $b_{\theta k}(x,\xi,\lambda)$ is defined in *different* ways depending on whether A is a differential operator or not.) The analogue of the formula (4.13) has the form

$$b_{\theta k}(x,\xi,\lambda) = \sum_{l \geq 2} \frac{b_{\theta k l}(x,\xi)}{[a_{\theta 0}(x,\xi) - \lambda]^l} , \tag{4.28}$$

where $b_{\theta kl}$ are some polynomials of the functions $a_{\theta 0}, a_{\theta 1}, \ldots$ and their derivatives. In addition,

$$b_{\theta kl} \in S^{(l-1)m-k}(\mathbb{R}^n \times \mathbb{R}_n) , \tag{4.29}$$

$$b_{\theta kl}(x,\tau\xi) = \tau^{(l-1)m-k} b_{\theta kl}(x,\xi) \qquad (|\xi| \geq 1, \quad \tau \geq 1) , \tag{4.30}$$

and even the function $b_{\theta k2} = -a_{\theta k}$ is in general different from 0 for arbitrarily large k, therefore in (4.28) we have $l \geq 2$ unlike (4.13). We now define the operators $B_{\theta k}(\lambda)$ and $B^{(N)}(\lambda) = B_\theta^{(N)}(\lambda)$ by formulas of the form (4.18), (4.19) with new $b_{\theta k}$. As above, $\lambda \in \mathcal{L}_{(\rho)}$.

The symbol $b_\theta^{(N)} = b_{\theta 0} + \ldots + b_{\theta N}$ of the operator $B^{(N)}(\lambda)$ satisfies the inequalities

$$|\partial_x^\alpha \partial_\xi^\beta b_\theta^{(N)}(x,\xi,\lambda)| \leq C_{\alpha\beta N}(1+|\xi|)^{-|\beta|}(1+|\xi|^m+|\lambda|)^{-1} \tag{4.31}$$

for all α, β, where $C_{\alpha\beta N}$ does not depend on (x,ξ,λ), $\lambda \in \mathcal{L}_{(\rho)}$. (In the case of differential operators these inequalities can be strengthened. Cf. Sect. 4.4 below.) They allow us to prove that for all s and $f \in H_s(\mathbb{R}^n)$, $\lambda \in \mathcal{L}_{(\rho)}$, we have

$$\|B^{(N)}(\lambda)f\|_{s+m} + |\lambda| \, \|B^{(N)}(\lambda)f\|_s \leq C_s \|f\|_s , \tag{4.32}$$

where C_s does not depend on f, λ (compare with the formulation of Theorem 1.1.2 on boundedness of pseudodifferential operators).

c. Passing now to a pseudodifferential operator A on a manifold M, elliptic with parameter in $\mathcal{L}$, we define a parametrix $B^{(N)}(\lambda)$ of order N for $A - \lambda I$ by

$$B^{(N)}(\lambda) = B_\theta^{(N)}(\lambda) = \sum_{k=1}^{K} \psi_k B_{\theta k}^{(N)}(\lambda)(\varphi_k \cdot)\ , \tag{4.33}$$

where $\{\varphi_k\}$, $\{\psi_k\}$ are the same systems of functions as in (2.3) and $B_{\theta k}^{(N)}$ is a parametrix of order N for $A_k - \lambda I$ in $\mathbb{R}^n$; A_k have the same meaning as in (2.3). The estimate of the form (4.32) remains true on M. For the operator $T_N(\lambda)$ (see (4.6)) we can prove the estimate

$$\|T_N(\lambda)f\|_{s+N+1,M} + |\lambda|^{\frac{N+1}{m}} \|T_N(\lambda)f\|_{s,M} \leq C_s' \|f\|_{s,M} \tag{4.34}$$

if A is a differential operator, and the estimate

$$\|T_N(\lambda)f\|_{s+N+1,M} + |\lambda|\, \|T_N(\lambda)f\|_{s+N+1-m,M} \leq C_s' \|f\|_{s,M} \tag{4.35}$$

in the general case. This allows us to prove the invertibility of the operator $A - \lambda I$ for $\lambda \in \mathcal{L}$ sufficiently large in modulus and to establish (using additionally the ellipticity of A) the estimate

$$\|R(\lambda)f\|_{s+m,M} + |\lambda|\, \|R(\lambda)f\|_{s,M} \leq C_s \|f\|_{s,M}\ , \tag{4.36}$$

which concludes the proof of Theorem 4.1.2 in the case $A(\lambda) = A - \lambda I$.

The estimate (4.34) for differential operators is stronger than the estimate (4.35) in the general case. As corollaries we obtain different estimates for $R_A(\lambda) - B_\theta^{(N)}(\lambda)$ (where $R_A(\lambda)$ is the resolvent $(A - \lambda I)^{-1}$). In the case of a differential operator we have

$$\|[R_A(\lambda) - B_\theta^{(N)}(\lambda)]f\|_{s+m+N+1,M} + |\lambda|^{\frac{m+N+1}{m}} \|[R_A(\lambda) - B_\theta^{(N)}(\lambda)]f\|_{s,M} \leq C_s' \|f\|_{s,M}\ ; \tag{4.37}$$

and in the general case

$$\|[R_A(\lambda) - B_\theta^{(N)}(\lambda)]f\|_{s+m+N+1,M} + |\lambda|^2 \|[R_A(\lambda) - B_\theta^{(N)}(\lambda)]f\|_{s-m+N+1,M} \leq C_s' \|f\|_{s,M}\ . \tag{4.38}$$

In particular, we can see that in the former case the order of the approximation in λ of the resolvent by the parametrix improves unboundedly as N grows, but we do not have that in the latter case. However, the approximation 'in smoothness' in the latter case improves with the growth of N with no limit, as in the former case.

4.3. Generalizations

a. We can assume that a pseudodifferential operator A on M is a matrix operator with dimensions of the matrix, say, $p \times p$. The condition of ellipticity with parameter in the angle $\mathcal{L}$ for operators A and $A - \lambda I$ means that the eigenvalues of the principal symbol $a_0(x, \xi)$ do not belong to $\mathcal{L}$. Under this assumption we can construct a parametrix, and as a corollary we obtain an analogue of Theorem 4.1.2. It is clear that in the construction we have to employ the matrix $[a_0(x,\xi) - \lambda E]^{-1}$ instead of dividing by $a_0(x,\xi) - \lambda$ and take into account the fact that matrices do not commute in general. However, the structure of the formulas remains in principle the same as in the scalar case.

b. We can consider more general matrix pseudodifferential operators $A = (A_{jk})_{j,k=1}^{p}$ on M. We assume that $m > 0$ and the real numbers $m_1, \ldots, m_p$ are given. Define the numbers l_j by $l_j = m - m_j$ (so that $l_j + m_j = m$ for all j) and assume that A is a pseudodifferential operator of type $\{l_j, m_k\}$ (see 3.2b). Considering $A - \lambda I$, we can assign to the parameter the 'weight' m with respect to differentiation. The condition of ellipticity with parameter in $\mathcal{L}$ is that the eigenvalues of the principal symbol $a_0(x,\xi) = (a_{jk0}(x,\xi))$ do not belong to $\mathcal{L}$. We call this condition *ellipticity with parameter in the sense of Douglis–Nirenberg.* (From it the ellipticity in the sense of Douglis–Nirenberg follows for any fixed λ, see 3.2b.)

This case can be reduced to the previous one by the substitution $u_k = \Lambda^{-m_k} v_k$, where Λ is an invertible elliptic pseudodifferential operator of order 1, and by the action of Λ^{m_j} on the j-th equation of the system $Au - \lambda u = f$ ($j, k = 1, \ldots, p$; cf. Remark 3.2.2). The ellipticity with parameter in $\mathcal{L}$ is preserved since the new principal symbol is similar to the principal symbol of the original operator. Therefore, we obtain the results analogous to those stated above. In particular, the analogue of Theorem 4.1.2 is valid with a correction concerning the choice of the spaces. The a priori estimate has the following form:

$$\sum_{k=1}^{p} \|u_k\|_{s+m_k+m,M} + |\lambda| \sum_{k=1}^{p} \|u_k\|_{s+m_k,M} \le C_s \sum_{j=1}^{p} \|f_j\|_{s+m_j,M} \,. \tag{4.39}$$

c. We consider a scalar operator $A(\lambda)$ of the form (4.1). Suppose that it is elliptic with parameter in the angle $\mathcal{L}$. We assume, without loss of generality, that $A_0 = -I$. The substitution

$$U_k = \lambda^{k-1} u \quad (k = 1, \ldots, p), \quad U = (U_1, \ldots, U_p)' \tag{4.40}$$

reduces the equation $A(\lambda)u = f$ to the system

$$(A - \lambda I)U = F \,, \tag{4.41}$$

where $F = (0, \ldots, 0, f)'$ and

$$A = \begin{pmatrix} 0 & 1 & 0 & \ldots & 0 & 0 \\ 0 & 0 & 1 & \ldots & 0 & 0 \\ \vdots & \vdots & \vdots & \ddots & \vdots & \vdots \\ 0 & 0 & 0 & \ldots & 0 & 1 \\ A_p & A_{p-1} & A_{p-2} & \ldots & A_2 & A_1 \end{pmatrix} . \tag{4.42}$$

This pseudodifferential operator is elliptic with parameter in the sense of Douglis–Nirenberg with

$$m_1 = m(p-1), \dots, m_p = 0 \,, \quad l_j = m - m_j \,, \tag{4.43}$$

which leads to Theorem 4.1.2 in full generality.

d. The next stage of generalization is that we consider the operator (4.1) on M with matrix coefficients. All A_k are $q \times q$ matrices of scalar pseudodifferential operators of order mk, and A_0 is the operator of multiplication by a matrix. The condition of ellipticity with parameter in $\mathcal{L}$ takes the form

$$\det \sum_{l=0}^{p} \lambda^{p-l} a_{l0}(x,\xi) \neq 0 \quad \text{for} \quad (x,\xi) \in T^*M \,,$$

$$\lambda \in \mathcal{L}, \quad (\xi, \lambda) \neq 0 \,. \tag{4.44}$$

Without loss of generality we can assume that $A_0 = -I$. The analogue of Theorem 4.1.2 is proved by the reduction to the case of the operator $A - \lambda I$, where A is the matrix of the form (4.42) but all its elements are $q \times q$ blocks. It is easy to verify that A is now elliptic with parameter in $\mathcal{L}$ in the sense of Douglis–Nirenberg.

e. Finally, we can consider a matrix pseudodifferential operator on M of the most general structure polynomially depending on λ and elliptic with parameter in $\mathcal{L}$ in the sense of Douglis–Nirenberg. Let $m > 0$, $m_1, \dots, m_q$ be fixed numbers and suppose that $A(\lambda)$ has the form (4.1), where

$$A_l = (A_{ljk})_{j,k=1}^{q} \,, \quad A_{ljk} \in \Psi_{\mathrm{ph}}^{ml+m_k-m_j}(M) \,, \quad A_0 = -I \,.$$

Ellipticity condition with parameter in $\mathcal{L}$ takes the form

$$\det \sum_{l=0}^{p} \lambda^{p-l} a_{l0}(x,\xi) \neq 0 \quad \text{for} \quad (x,\xi) \in T^*M \backslash 0, \, 0 \neq \lambda \in \mathcal{L} \,, \tag{4.45}$$

where a_{l0} is the principal symbol of A_l, i.e. it is the matrix of principal symbols of A_{ljk}. This case can be reduced to the previous one by the mode used above: we put $u_k = \Lambda^{-m_k} v_k$ and act by the operator Λ^{m_j} on the j-th equation. Ellipticity with parameter is preserved, and we obtain the analogue of Theorem 4.1.2 with the estimate

$$\sum_{k=1}^{q} \|u_k\|_{s+mp+m_k,M} + |\lambda|^p \sum_{k=1}^{q} \|u_k\|_{s+m_k,M} \le C_s \sum_{j=1}^{q} \|f_j\|_{s+m_j,M} \,. \tag{4.46}$$

4.4. Calculus of Pseudodifferential Operators with a Parameter. Some details of the construction of a parametrix described in Sect. 4.2 are essential when we consider functions of pseudodifferential operators, and the reductions considered in Sect. 4.3 are useful in the spectral theory of nonselfadjoint elliptic pseudodifferential operators. Now, abstracting from these special problems, we outline the calculus of pseudodifferential operators with a parameter which allows us to construct a parametrix and prove theorems of the type of Theorem 4.1.2 under quite general assumptions. The definitions and results are mainly taken from the book of Grubb (1986), but we shall make some simplifying assumptions and diverge from (Grubb 1986) in notation.

We denote the parameter by μ, and for simplicity we assume that it has the weight 1 with respect to the differentiation and varies along $\overline{\mathbb{R}}_+$. (In the notation of the previous sections, $\lambda = \mu^m$ and $\mathcal{L} = \overline{\mathbb{R}}_+$.) Also for simplicity we shall only consider scalar pseudodifferential operators in $\mathbb{R}^n$ assuming that their symbols do not depend on x for large $|x|$ (this assumption is implicit in what follows).

We first define the required classes of symbols and pseudodifferential operators. We put

$$< \xi, \mu >= (1 + |\xi|^2 + \mu^2)^{1/2} . \tag{4.47}$$

Let t, ν be real numbers and $a(x, \xi, \mu)$ a function from $C^\infty(\mathbb{R}^n \times \mathbb{R}_n \times \overline{\mathbb{R}}_+)$. We shall write $a \in S^{t,\nu} = S^{t,\nu}(\mathbb{R}^n \times \mathbb{R}_n \times \overline{\mathbb{R}}_+)$ if for any $\alpha, \beta \in \mathbb{Z}^n_+$, $\gamma \in \mathbb{Z}_+$ we have the estimate

$$\begin{aligned} &|\partial_x^\alpha \partial_\xi^\beta \partial_\mu^\gamma a(x, \xi, \mu)| \\ &\leq C_{\alpha\beta\gamma}(< \xi >^{\nu-|\beta|} + < \xi, \mu >^{\nu-|\beta|}) < \xi, \mu >^{t-\nu-\gamma} , \end{aligned} \tag{4.48}$$

where $C_{\alpha\beta\gamma}$ does not depend on (x, ξ, μ). Obviously, for a fixed $\mu \in \overline{\mathbb{R}}_+$ such a symbol belongs to the class $S^t(\mathbb{R}^n \times \mathbb{R}_n)$. The numbers t and ν are respectively called *order* and *regularity* of the symbol $a(x, \xi, \mu)$. To explain the role of regularity we note that the estimate (4.48) is equivalent to the two following estimates:

$$|\partial_x^\alpha \partial_\xi^\beta \partial_\mu^\gamma a(x, \xi, \mu)| \leq C'_{\alpha\beta\gamma} < \xi, \mu >^{t-|\beta|-\gamma} \quad \text{for} \quad |\beta| \leq \nu , \tag{4.49}$$

$$|\partial_x^\alpha \partial_\xi^\beta \partial_\mu^\gamma a(x, \xi, \mu)| \leq C'_{\alpha\beta\gamma} < \xi >^{\nu-|\beta|} < \xi, \mu >^{t-\nu-\gamma} \quad \text{for} \quad |\beta| > \nu . \tag{4.50}$$

Comparing these estimates, we can see that the first one is optimal (the parameter μ can be considered as the complementary variable ξ_{n+1}) and the second is worse; the deterioration occurs for $|\beta| > \nu$.

For the corresponding pseudodifferential operator $A(\mu) = a(x, D, \mu)$ we shall write $A(\mu) \in \Psi^{t,\nu} = \Psi^{t,\nu}(\mathbb{R}^n, \mathbb{R}_+)$.

As regards the presence of the derivatives of the symbol with respect to μ, we note that these derivatives are not used in the formulas of symbolic calculus (i.e. in the formulas for the symbol of the composition of two pseudodifferential operators, etc.). But their presence in (4.48-50) selects a class of symbols with a 'good' behavior in μ, which turns out to be invariant with respect to the operations of the symbolic calculus. We note that the case $\nu = \infty$ was in fact considered in the book of Shubin (1978) without any assumptions on smoothness of the symbols with respect to the parameter. The variant with $\nu = \infty$ is simpler (see (Shubin 1978)) and is convenient, as we shall soon explain, when differential operators are considered.

We note that if $a(x,\xi,\mu) \in S^{t,\nu}$, then $\partial_x^\alpha \partial_\xi^\beta \partial_\mu^\gamma a(x,\xi,\mu) \in S^{t-|\beta|-\gamma,\nu-|\beta|}$.

We put

$$S^{-\infty,\nu-\infty} = \bigcap_{N\in\mathbb{N}} S^{-N,\nu-N} . \tag{4.51}$$

The symbols $c(x,\xi,\mu)$ from this class satisfy the inequalities

$$|\partial_x^\alpha \partial_\xi^\beta \partial_\mu^\gamma c(x,\xi,\mu)| \leq C_{\alpha\beta\gamma N} < \xi >^{-N} (1+\mu)^{-\nu-\gamma} \tag{4.52}$$

for any α, β, γ, and N. The pseudodifferential operators from the corresponding class $\Psi^{-\infty,\nu-\infty}$ are integral operators

$$\int K(x,y,\mu)u(y)dy \tag{4.53}$$

with infinitely smooth kernels satisfying the inequalities of the form

$$|\partial_x^\alpha \partial_\xi^\beta \partial_\mu^\gamma K(x,y,\mu)| \leq C'_{\alpha\beta\gamma N}(1+|x-y|)^{-N}(1+\mu)^{-\nu-\gamma} . \tag{4.54}$$

Suppose that the symbol $a(x,\xi,\mu)$ belongs to $S^{t,\nu}$, and let $\{a_j(x,\xi,\mu)\}_0^\infty$ be a sequence of symbols from $S^{t-j,\nu-j}$. We shall write

$$a \sim a_0 + a_1 + \ldots \tag{4.55}$$

if $a - (a_0 + \ldots + a_{N-1}) \in S^{t-N,\nu-N}$ for any $N \in \mathbb{N}$. We note that the relation $b \sim a_0 + a_1 + \ldots$ in $S^{t,\nu}$ is valid simultaneously with (4.55) if and only if $a - b \in S^{-\infty,\nu-t-\infty}$.

Theorem 4.4.1. *For any sequence of symbols $a_j \in S^{t-j,\nu-j}$ $(j = 0,1,\ldots)$ there is a symbol $a \in S^{t,\nu}$ such that the relation* (4.55) *is satisfied.*

We now define the classes of polyhomogeneous symbols and polyhomogeneous pseudodifferential operators with a parameter. Suppose that $a(x,\xi,\mu) \in S^{t,\nu}$ and there exists a sequence of functions $\{a_j(x,\xi,\mu)\}_0^\infty$ belonging to C^∞ for $(\xi,\mu) \neq 0$ and positively homogeneous in (ξ,μ) of degree $t-j$ such that

$$a(x,\xi,\mu) \sim \theta((|\xi|^2+\mu^2)^{1/2})[a_0(x,\xi,\mu) + \ldots] . \tag{4.56}$$

Then we shall write $a \in S_{\text{ph}}^{t,\nu}$ and $A(\mu) = a(x,D,\mu) \in \Psi_{\text{ph}}^{t,\nu}$; and in addition, we shall take the liberty to write $a \sim a_0 + \ldots$. The function $a_0(x,\xi,\mu)$ will be called the *principal symbol* of the pseudodifferential operator $A(\mu)$.

As illustration, we shall now go through the symbols from Sect. 4.2 assuming that $\mathcal{L} = \overline{\mathbb{R}}_+$. If A is a differential operator of order m with coefficients from $B^\infty(\mathbb{R}^n)$ and symbol $a(x,\xi)$, then $a(x,\xi) - \mu^m \in S_{\text{ph}}^{m,+\infty}$. Regularity is infinite here for the simple reason that $\partial_\xi^\beta[a(x,\xi) - \mu^m] = 0$ for $|\beta| > m$. Next, the symbols (4.16), as is easy to see, belong to $S_{\text{ph}}^{-m-k,\infty}$. But if A is a pseudodifferential operator from Ψ_{ph}^n, then in the general case for any N

$$a_{\theta 0}(x,\xi)+\ldots+a_{\theta N}(x,\xi)-\mu^m\in S_{\mathrm{ph}}^{m,m}$$

and

$$b_{\theta 0}(x,\xi,\mu^m)+\ldots+b_{\theta N}(x,\xi,\mu^m)\in S_{\mathrm{ph}}^{-m,m}\,.$$

The theorems of the calculus of pseudodifferential operators with a parameter will be given for polyhomogeneous symbols and operators. For more general symbols from $S^{t,\nu}$ and operators from $\Psi^{t,\nu}$ we have analogous assertions.

Besides symbols, we can also consider amplitude functions $p(x,y,\xi,\mu)$ depending on a parameter. The definition of the class $S_{\mathrm{ph}}^{t,\nu}=S_{\mathrm{ph}}^{t,\nu}(\mathbb{R}^{2n}\times\mathbb{R}^n\times\overline{\mathbb{R}}_+)$ for amplitude functions is analogous to the definition of the class $S_{\mathrm{ph}}^{t,\nu}$ for symbols; we only have to replace x by (x,y) everywhere.

Theorem 4.4.2. *Suppose that $p(x,y,\xi,\mu)$ is an amplitude function from $S_{\mathrm{ph}}^{t,\nu}$. Then the corresponding operator*

$$A(\mu)u(x)=\int\!\!\int e^{i(x-y)\cdot\xi}p(x,y,\xi,\mu)u(y)dy đ\xi$$

($u\in S(\mathbb{R}^n)$) is a pseudodifferential operator from $\Psi_{\mathrm{ph}}^{t,\nu}$ with symbol

$$a(x,\xi,\mu)\sim\sum\frac{1}{\alpha!}\partial_\xi^\alpha D_y^\alpha p(x,y,\xi,\mu)|_{y=x}\,. \tag{4.57}$$

Theorem 4.4.3. *Suppose that $A(\mu)$ is a pseudodifferential operator from the space $S_{\mathrm{ph}}^{t,\nu}$ with symbol $a(x,\xi,\mu)$. Then the operator $[A(\mu)]^*$ formally adjoint to $A(\mu)$ is a pseudodifferential operator from $S_{\mathrm{ph}}^{t,\nu}$ with symbol*

$$a^{(*)}(x,\xi,\mu)\sim\sum\frac{1}{\alpha!}\partial_\xi^\alpha D_x^\alpha\overline{a(x,\xi,\mu)}\,. \tag{4.58}$$

Theorem 4.4.4. *Suppose that $A(\mu)$, $B(\mu)$ are pseudodifferential operators from $S_{\mathrm{ph}}^{t_1,\nu_1}$, $S_{\mathrm{ph}}^{t_2,\nu_2}$ with symbols $a(x,\xi,\mu)$, $b(x,\xi,\mu)$. Then $C(\mu)=A(\mu)B(\mu)$ is a pseudodifferential operator from $S_{\mathrm{ph}}^{t_1+t_2,\nu(\nu_1,\nu_2)}$ with symbol*

$$c(x,\xi,\mu)\sim\sum\frac{1}{\alpha!}\partial_\xi^\alpha a(x,\xi,\mu)\cdot D_x^\alpha b(x,\xi,\mu)\,, \tag{4.59}$$

and here $\nu(\nu_1,\nu_2)=\min(\nu_1,\nu_2,\nu_1+\nu_2)$.

Ellipticity with parameter in $\mathcal{L}=\overline{\mathbb{R}}_+$ of a pseudodifferential operator $A(\mu)\in\Psi_{\mathrm{ph}}^{t,\nu}$ with principal symbol $a_0(x,\xi,\mu)$ is defined by the condition $a_0(x,\xi,\mu)\neq 0$ for $(\xi,\mu)\neq 0$, $\mu\in\overline{\mathbb{R}}_+$. Here $t\in\mathbb{R}$, so t may also be negative.

Theorem 4.4.5. *Suppose that $A(\mu)$ is a pseudodifferential operator from the space $\Psi_{\mathrm{ph}}^{t,\nu}$ elliptic with parameter in $\mathcal{L}=\overline{\mathbb{R}}_+$. Then there is an elliptic with parameter in $\mathcal{L}$ pseudodifferential operator $B(\mu)\in\Psi_{\mathrm{ph}}^{-t,\nu}$ such that*

$$A(\mu)B(\mu) = I + T_1(\mu), \quad B(\mu)A(\mu) = I + T_2(\mu) , \tag{4.60}$$

where $T_j(\mu) \in \Psi^{-\infty,\nu-\infty}$ *($j = 1, 2$).*

The asymptotic expansion of the symbol of $B(\mu)$ is first constructed here by the asymptotic expansion (4.56) of the symbol of $A(\mu)$.

We note that for $t > 0$, $\nu \geq 0$, using the inequalities of the form (4.49-50) for b instead of a and with $-t$ instead of t, we obtain the inequalities

$$|\partial_x^\alpha \partial_\xi^\beta b(x, \xi, \mu)| \leq C_{\alpha\beta} < \xi >^{-t-|\beta|} , \tag{4.61}$$

$$\mu^t |\partial_x^\alpha \partial_\xi^\beta b(x, \xi, \mu)| \leq C_{\alpha\beta} < \xi >^{-|\beta|} , \tag{4.62}$$

where $C_{\alpha\beta}$ does not depend on (x, ξ, μ). These inequalities allow us to prove the estimate

$$\|B(\mu)f\|_{s+t} + \mu^t \|B(\mu)f\|_s \leq C_s \|f\|_s \tag{4.63}$$

for any s, where C_s does not depend on f, μ (cf. Theorem 1.1.2).

Next, we obtain the following analogue of Theorem 4.1.2:

Theorem 4.4.6. *Suppose that the operator* $A(\mu)$ *belongs to* $\Psi_{\text{ph}}^{t,\nu}$ *with* $t > 0$, $\nu > 0$, *and is elliptic with parameter in* $\mathcal{L} = \overline{\mathbb{R}}_+$. *Then for any* s *this operator has bounded inverse* $A^{-1}(\mu)$ *for sufficiently large* μ. *Moreover, we have the estimate*

$$\|u\|_{s+t} + \mu^t \|u\|_s \leq C_s' \|A(\mu)u\|_s , \tag{4.64}$$

where C_s' *does not depend on* u, μ.

We conclude with a remark that for the pseudodifferential operators with a parameter considered in this section the analogue of Theorem 1.4.1 on the change of variables is also true, which allows us to pass to the study of pseudodifferential operators with a parameter on a compact manifold M. The principal symbol $a_0(x, \xi, \mu)$ turns out here to be a function on the cotangent bundle for any fixed μ. Theorem 4.4.6 can be transferred to the case of pseudodifferential operators $A(\mu)$ on M elliptic with parameter.

§5. Functions of Elliptic Operators

5.1. Functions of Pseudodifferential Operators of Zero Order

a. Let first A be a bounded operator in a Hilbert space H, so that its spectrum $\sigma(A)$ is a bounded closed set. We denote by $\mathcal{F}(A)$ the set of all functions $\Phi(\lambda)$ holomorphic in a neighborhood U_Φ of the spectrum $\sigma(A)$ of A. Let Γ be a positively oriented contour lying in U_Φ and going around the $\sigma(A)$. (More precisely, Γ consists of no more than a finite number of Jordan closed rectifiable curves $\Gamma_1, \ldots, \Gamma_m$, each Γ_j is the boundary of a subdomain $U^{(j)} \subset U_\Phi$ and is positively oriented with respect to $U^{(j)}$, $U^{(j)} \cap U^{(k)} = \varnothing$ ($j \neq k$), and $\sigma(A) \subset \bigcup U^{(j)}$.) Then the operator $\Phi(A)$ is given by

$$\Phi(A) = -\frac{1}{2\pi \mathrm{i}} \int_\Gamma \Phi(\lambda) R_A(\lambda) d\lambda \,. \tag{5.1}$$

The operator $\Phi(A)$ is bounded, and it can be easily verified that for Φ_1, $\Phi_2 \in \mathcal{F}(A)$, $\alpha, \beta \in \mathbb{C}$ we have

$$(\alpha\Phi_1 + \beta\Phi_2)(A) = \alpha\Phi_1(A) + \beta\Phi_2(A) \,, \tag{5.2}$$

$$(\Phi_1\Phi_2)(A) = \Phi_1(A)\Phi_2(A) \,. \tag{5.3}$$

Next, we have the *spectral mapping theorem*: if $\Phi \in \mathcal{F}(A)$, then

$$\sigma(\Phi(A)) = \Phi(\sigma(A)) \,. \tag{5.4}$$

In addition, if $\Phi_1 \in \mathcal{F}(A)$ and $\Phi_2 \in \mathcal{F}(\Phi_1(A))$, then

$$\Phi_3(\lambda) = \Phi_2(\Phi_1(\lambda)) \in \mathcal{F}(A) \quad \text{and} \quad \Phi_3(A) = \Phi_2(\Phi_1(A)) \,. \tag{5.5}$$

All that can be found, for example, in the book (Dunford and Schwartz 1958, Chap. VII).

If A is a selfadjoint operator, $A = A^*$, with the spectral representation

$$A = \int \lambda dE_\lambda \,, \tag{5.6}$$

where $\{E_\lambda\}$ is the corresponding partition of unity on $\mathbb{R}$, then

$$\Phi(A) = \int \Phi(\lambda) dE_\lambda \,. \tag{5.7}$$

This formula permits us to define operators $\Phi(A)$ for $\Phi(\lambda)$ more general than above, for example, for functions continuous on the real axis in a neighborhood of the spectrum.

b. Suppose that A is a pseudodifferential operator of zero order on M. Then it is a bounded operator, for example, in $H_0(M)$, and we face the question of under what conditions $\Phi(A)$ is also pseudodifferential. In (Seeley 1967a) the following theorem is proved.

Theorem 5.1.1. *Suppose that $A \in \Psi^0_{\mathrm{ph}}(M)$ and $\Phi(\lambda)$ is a function holomorphic in a neighborhood of the spectrum $\sigma(A)$. Then $\Phi(A) \in \Psi^0_{\mathrm{ph}}(M)$ and the principal symbol of this pseudodifferential operator is equal to $\Phi(a_0(x,\xi))$, where $a_0(x,\xi)$ is the principal symbol of A.*

The analogous result is true for $\mathbb{R}^n$ in place of M, and in this case we can indicate the full asymptotic expansion of the symbol of $\Phi(A)$:

$$\varphi(x,\xi) \sim \sum_{k=0}^{\infty} \int_\Gamma \Phi(\lambda) b_k(x,\xi,\lambda) d\lambda \,, \tag{5.8}$$

where the functions b_k are defined by relation (4.9). This formula explains the approach of Seeley to the proof of the theorem. We note that there is no need to assume that A is elliptic since for $\lambda \notin \sigma(A)$, $A - \lambda I$ is automatically elliptic.

Theorem 5.1.2. *Suppose that A is a pseudodifferential operator from the space $\Psi^0_{\text{ph}}(M)$ selfadjoint in $H_0(M)$ and $\Phi(\lambda)$ is a C^∞ function in a neighborhood of the spectrum $\sigma(A)$ on the axis $\mathbb{R}$. Then $\Phi(A)$ is a pseudodifferential operator from $\Psi^0_{\text{ph}}(M)$ with the principal symbol $\Phi(a_0(x,\xi))$, where $a_0(x,\xi)$ is the principal symbol of A.*

This theorem (and also its generalization to the case of the function of some pairwise commuting pseudodifferential operators of zero order) was proved in various ways in (Dunau 1977) and (Guillemin and Sternberg 1979) (in the latter paper it is indicated that the result is due to Melrose, Seeley, and Uhlmann).

5.2. Positive Operators in a Hilbert Space and Some Functions of Them. In this section we give some preparations which will enable us to consider functions of elliptic operators of positive order.

a. An operator A in a Hilbert space H is called *positive*, or *of positive type* (see Krasnosel'skij and Zabrejko 1966), if the resolvent set $\rho(A)$ contains the semi-axis $\overline{\mathbb{R}}_-$ and

$$\|R_A(\lambda)\| \le C(1+|\lambda|)^{-1} \tag{5.9}$$

for $\lambda \in \overline{\mathbb{R}}_-$. If $A + cI$ is positive only for sufficiently large $c > 0$, then we call A *positive in a weakened form.*

A positive (in general, in a weakened form) operator is an abstract model of a pseudodifferential operator of positive order elliptic with parameter in $\mathcal{L} = \overline{\mathbb{R}_-}$ (see Sects. 4.1–4.2). This selection of a direction is convenient to consider some functions of A, for example A^z.

If A is invertible and the estimate (5.9) also holds in the angle $\mathcal{L} = \mathcal{L}_\varphi$ of measure 2φ with the bisector $\mathbb{R}_-$, then we call A a *φ-positive* operator. If this is true only for $\lambda \in \mathcal{L}$ sufficiently large in modulus, then we say that A is *φ-positive in a weakened form.* In the former case the resolvent exists and the estimate (5.9) is preserved in $\mathcal{L}_{(\varepsilon)}$ (see (4.17)) for sufficiently small $\varepsilon > 0$. In the latter case the estimate is satisfied for sufficiently large $\varepsilon > 0$ in

$$\mathcal{L}_{(-\varepsilon)} = \mathcal{L} \backslash \{\lambda : |\lambda| < \varepsilon\}\,. \tag{5.10}$$

Remark 5.2.1. The resolvent $R_A(\lambda)$ cannot decay in norm at infinity faster than $|\lambda|^{-1}$. More precisely, if $\varepsilon > 0$, then $\|R_A(\lambda)\| \ge (1-\varepsilon)|\lambda|^{-1}$ for sufficiently large $|\lambda|$. In this connection in the papers of Agmon (see for example (Agmon 1962)) the rays coming out of the origin, along which the inequality of the form (5.9) holds for λ sufficiently large in modulus, are called the rays of minimal growth of the norm of the resolvent. In our opinion, they should be called the rays of maximal decay of the norm of the resolvent.

b. To define $\Phi(A)$ when A is a nonselfadjoint unbounded operator, we need an analogue of the formula (5.1). But now the spectrum $\sigma(A)$ is in general an unbounded set, and therefore we have to consider unbounded contours imposing on $\Phi(\lambda)$ a condition at infinity which ensures the convergence of the integral.

Obviously, the contours can be taken in various forms. For example, to define the power A^z for $\mathrm{Re}\, z < 0$ one usually takes the contour consisting of two rays symmetric with respect to $\mathbb{R}_-$ with the ends connected by a circle of small radius with the center at the origin. We introduce a contour $\Gamma(\delta, \psi)$ similar in form, consisting of two rays $\{\lambda : \arg(\lambda - \delta) = \pm\psi\}$ oriented upwards. Here $0 < \psi < \pi$ and $\delta \in \mathbb{R}$. Denote the open angles bounded by this contour, the left and the right, by $G^-_{\delta,\psi}$ and $G^+_{\delta,\psi}$, respectively.

Let $\Phi(\lambda)$ be a function holomorphic in $G^+_{\delta',\psi}$ ($\delta' \in \mathbb{R}$, $0 < \psi < \pi$) and satisfying in this domain the inequality

$$|\Phi(\lambda)| \le C(1 + |\lambda|)^{-h}, \tag{5.11}$$

where positive numbers C and h do not depend on λ. On the other hand, suppose that $\rho(A)$ contains the domain $G^-_{\delta'',\psi}$ for some $\delta'' > \delta'$ and in this domain the inequality of the form (5.9) is satisfied (so that A is ψ-positive, at least in a weakened form). Assuming that $\delta' < \delta < \delta''$, we put

$$\Phi(A) = \frac{1}{2\pi i} \int_{\Gamma(\delta,\psi)} \Phi(\lambda) R_A(\lambda)\, d\lambda\, . \tag{5.12}$$

Obviously, the integral absolutely converges, i.e.

$$\int_{\Gamma(\delta,\psi)} |\Phi(\lambda)| \cdot \|R_A(\lambda)\|\, |d\lambda| < \infty\, ,$$

and so (5.12) is a bounded operator. If A has discrete spectrum, then $\Phi(A)$ is compact (this can be easily verified by approximating the integral (5.12) by an integral sum corresponding to the integral over a finite part of the contour). The spectrum of $\Phi(A)$ consists of zero and the eigenvalues $\Phi(\lambda_j)$, where λ_j are the eigenvalues of A. Formulas of the form (5.2), (5.3) are valid for two functions Φ_1, Φ_2 with the properties we have just mentioned if they can be defined by means of the same contour $\Gamma(\delta, \psi)$. Below we give the analogues of (5.5) for some concrete Φ_1, Φ_2 only.

c. Suppose now that A is a normal operator with discrete spectrum consisting of eigenvalues λ_j ($j = 1, 2, \ldots$), and let $\{e_j\}_1^\infty$ be the corresponding orthonormal basis of eigenvectors of A. If $\Phi(\lambda)$ is any numerical function defined at all points λ_j, then the operator $\Phi(A)$ can be defined by

$$\Phi(A)f = \sum_{j=1}^{\infty} \Phi(\lambda_j)(f, e_j)e_j\, . \tag{5.13}$$

This operator is bounded if $\{\Phi(\lambda_j)\}$ is a bounded sequence and is compact if $\Phi(\lambda_j) \to 0$ as $j \to \infty$. If the sequence $\{\Phi(\lambda_j)\}$ is unbounded, then the domain $D_{\Phi(A)}$ of $\Phi(A)$ is determined by the condition

$$\sum |\Phi(\lambda_j)|^2 |(f, e_j)|^2 < \infty .$$

The operator $\Phi(A)$ is normal, and it is selfadjoint if all $\Phi(\lambda_j)$ are real.

In particular, we introduce the operator $|A|$ by putting

$$|A|f = \sum |\lambda_j|(f, e_j)e_j . \tag{5.14}$$

This is a selfadjoint operator with domain $D_{|A|} = D_A$.

Definitions (5.12) and (5.13) are consistent: if the conditions for both of them are fulfilled, then both formulas yield $\Phi(A)e_j = \lambda_j e_j$.

d. Suppose that the operator A is φ-positive. For Re$z < 0$ the operator A^z is defined by the formula of the form (5.12), i.e. by the formula

$$A^z = \frac{1}{2\pi \mathrm{i}} \int_{\Gamma(\delta,\psi)} \lambda^z R_A(\lambda) d\lambda , \tag{5.15}$$

where $\psi - (\pi - \varphi)$ and δ are small positive numbers, and

$$\lambda^z = \exp\bigl[z(\log|\lambda| + \mathrm{i}\arg\lambda)\bigr] , \quad -\pi < \arg\lambda \leq \pi . \tag{5.16}$$

For $z = -k$, where k is a natural number, we obtain the operator A^{-k} inverse to A^k. For Re$z_1 < 0$, Re$z_2 < 0$ we have

$$A^{z_1} \cdot A^{z_2} = A^{z_1+z_2} . \tag{5.17}$$

Next, for Re$z > 0$ we put

$$A^z f = g \quad \text{if} \quad A^{-z} g = f , \tag{5.18}$$

and we finally define A^z for Re$z = 0 <$ Rez_1 as the closure of the operator $A^{z-z_1}A^{z_1}$ with domain $D_{A^{z_1}}$. It can be verified that A^z does not depend on the choice of z_1 with Re$z_1 > 0$ and that $A^0 = I$. (We have shown one of the possible approaches to the definition of A^z for Re$z \geq 0$.)

We give a list of some properties of A^z.

1) $0 \leq$ Re$z_1 <$ Re$z_2 \Longrightarrow D_{A^{z_1}} \supset D_{A^{z_2}}$; in addition $D_{A^{z_2}}$ is dense in $D_{A^{z_1}}$ and A^{z_1} is equal to the closure of its restriction to $D_{A^{z_2}}$.
2) If Re$z_3 > \max($Rez_1, Rez_2, Re$(z_1 + z_2))$, then the relation (5.17) is valid on $D_{A^{z_3}}$.
3) If A is a normal operator with discrete spectrum, then with the same notation as in (5.13) we have

$$A^z f = \sum \lambda_j^z (f, e_j)e_j . \tag{5.19}$$

For $\mathrm{Re}z \leq 0$ this is a bounded operator. For $\mathrm{Re}z > 0$ its domain is determined by the condition

$$\sum |\lambda_j|^{2\mathrm{Re}z} |(f, e_j)|^2 < \infty \tag{5.20}$$

and does not depend on $\mathrm{Im}z$.

4) For $\mathrm{Re}z < 0$ the operator A^z is a holomorphic function of z with values in the space of bounded operators in H. For $\mathrm{Re}z < h$, $h > 0$, and $f \in D_{A^{z_1}}$, $\mathrm{Re}z_1 = h$, $A^z f$ is a holomorphic function of z with values in H.
5) Suppose that A is φ-positive and $0 < s(\pi - \varphi) < \pi$. Then A^s is ψ-positive, where $\psi = \pi - s(\pi - \varphi)$, and $(A^s)^z = A^{sz}$.
6) Suppose that A is φ-positive and $0 < s < 1$. Then in $\mathcal{L}_\varphi$ we have

$$\|A^s R_A(\mu)\| \leq C(1 + |\mu|)^{s-1} , \tag{5.21}$$

where C does not depend on μ.

e. Suppose that A is $\pi/2$-positive at least in a weakened form. We put

$$\mathrm{e}^{-tA} = \frac{1}{2\pi \mathrm{i}} \int_{\Gamma(\delta,\psi)} \mathrm{e}^{-t\lambda} R_A(\lambda) d\lambda \quad (t > 0) . \tag{5.22}$$

Here ψ is slightly smaller than $\pi/2$, and δ is a point in $\mathbb{R}$ such that the spectrum $\sigma(A)$ lies to the right of the contour $\Gamma(\delta, \psi)$.

The operator (5.22) is bounded, and if A has discrete spectrum, then e^{-tA} is compact. Comparing definitions (5.15) and (5.22), we note that what we need now is that A be $(\pi/2)$-positive in a weakened form: the function $\mathrm{e}^{-t\lambda}$ of λ is holomorphic everywhere, but the angle ψ in (5.22) has to be acute to ensure the absolute convergence.

For any $u_0 \in H$ the function $u(t) = \mathrm{e}^{-tA} u_0$ is the solution to the *Cauchy problem*

$$u'(t) + Au(t) = 0 \quad (t > 0) , \tag{5.23}$$

$$u(0) = u_0 . \tag{5.24}$$

In particular, $\mathrm{e}^{-tA} u_0 \longrightarrow u_0$ in H as $t \to 0$. Equation (5.23) is an abstract analogue of the heat equation $\partial_t u(t,x) - \Delta_x u(t,x) = 0$, and under our assumptions on A we can call (5.23) a *parabolic* equation.

Formula (5.22) still makes sense for complex t with sufficiently small $|\arg t|$, namely for $|\arg t| < \varphi_0 - (\pi/2)$, if A is φ-positive in a weakened form for $\varphi < \varphi_0$, where $\varphi_0 > \pi/2$. Moreover, e^{-tA} is a holomorphic function of t with values in the space of bounded operators in H, and $\mathrm{e}^{-t_1 A} \cdot \mathrm{e}^{-t_2 A} = \mathrm{e}^{-(t_1+t_2)A}$, so that here we have an analytic semigroup.

If A is a normal operator with discrete spectrum, then with the same notation as in (5.13) we have

$$e^{-tA}f = \sum e^{-t\lambda_j}(f, e_j)e_j \ . \tag{5.25}$$

If A is φ-positive only for $\varphi < \varphi_0 \leq \pi/2$, then we can define e^{-tA^s} for sufficiently small s, say, for $s = 1/2$. Namely,

$$e^{-tA^s} = \frac{1}{2\pi i} \int_{\Gamma(\delta,\psi)} e^{-t\lambda^s} R_A(\lambda) d\lambda \tag{5.26}$$

with suitable δ and ψ. We note that for such s and any $f \in H$ we have

$$e^{-tA^s} f \longrightarrow f \quad \text{in} \quad H \quad (t \to 0) \ . \tag{5.27}$$

f. In (5.15) and (5.22) the operators A^z ($\text{Re}z < 0$) and e^{-tA} ($t > 0$) are expressed by means of $R_A(\lambda)$. We shall indicate other relations between these operators.

We assume that A is φ-positive for $\varphi < \varphi_0$. Then for $|\arg\lambda - \pi| < \varphi_0$, $0 < c < 1$, we have

$$R_A(\lambda) = \frac{1}{2i} \int_{\text{Re}z=c} \frac{(-\lambda)^{z-1}}{\sin \pi z} A^{-z} dz \ , \tag{5.28}$$

where $(-\lambda)^{z-1}$ is a holomorphic function of z and λ coinciding with $e^{(z-1)\log|\lambda|}$ for $\lambda \in \mathbb{R}_-$.

Next, if $\varphi_0 > \pi/2$, then for $|\arg t| < \varphi_0 - (\pi/2)$, $c > 0$ we have

$$e^{-tA} = \frac{1}{2\pi i} \int_{\text{Re}z=c} t^{-z} \Gamma(z) A^{-z} dz \ . \tag{5.29}$$

For $\varphi_0 > \pi/2$, $\text{Re}\lambda \leq 0$ (we restrict ourselves to these λ) we have

$$R_A(\lambda) = \int_0^\infty e^{-\tau(A-\lambda I)} d\tau \ . \tag{5.30}$$

In all these formulas the integrals converge absolutely.

We also give an analogue of the formula (5.28) for $[R_A(\lambda)]^q$ with $q \in \mathbb{N}$:

$$[R_A(\lambda)]^q = \frac{1}{2i} \frac{(-1)^{q-1}}{(q-1)!} \int_{\text{Re}z=c} \frac{(z-1)\dots(z-q+1)(-\lambda)^{z-q}}{\sin \pi z} A^{-z} dz \tag{5.31}$$

for $0 < c < q$, $|\arg\lambda - \pi| < \varphi_0$.

5.3. Powers of an Elliptic Operator. Suppose that A is a pseudodifferential operator from $\Psi^m_{\text{ph}}(M)$, scalar at first, where $m > 0$ and M is a closed manifold. Hence, as we already know, A can be considered as a closed operator in $H_0(M)$ with dense domain $H_m(M)$. Suppose that this operator is positive. (Hence it follows that it is elliptic with parameter in an angular neighborhood of the ray $\mathbb{R}_-$.) Then, as we saw in 5.2d, the powers A^z, $z \in \mathbb{C}$, are well-defined.

Theorem 5.3.1 (Seeley 1967a). *Under above conditions A^z is an elliptic pseudodifferential operator from $\Psi^m_{\mathrm{ph}}(M)$ with the principal symbol $[a_0(x,\xi)]^z$, where $a_0(x,\xi)$ is the principal symbol of A.*

The principal symbol a_0^z, as well as further terms of the expansion of the local complete symbol of A^z, in general have complex degrees of homogeneity mz, $mz-1,\ldots$. The possibility of considering such polyhomogeneous pseudodifferential operators was mentioned in 3.1d.

A proof of the theorem is first carried out for $\mathrm{Re}z<0$ employing the approximation of the resolvent in (5.15) by the parametrix constructed in Sect. 4.2. For the local complete symbol of A^z we obtain a full asymptotic expansion in functions

$$\begin{aligned} c_{\theta jz}(x,\xi) &= \int_{\Gamma(\delta,\psi)} \lambda^z b_{\theta j}(x,\xi,\lambda)d\lambda \\ &= -\operatorname{Res}_{\lambda=a_{\theta 0}(x,\xi)}\{\lambda^z b_{\theta j}(x,\xi,\lambda)\} \end{aligned} \tag{5.32}$$

($j=0,1,\ldots$), where δ, ψ are sufficiently small numbers. The latter equality is obtained by replacing $\Gamma(\delta,\psi)$ with a closed contour going around the $a_{\theta 0}(x,\xi)$. In particular, it is clear that the principal symbol is equal to a_0^z. For the subprincipal symbol we obtain the following formula:

$$\mathrm{sub}A^z = za_0^{z-1}\mathrm{sub}A\,. \tag{5.33}$$

The list of properties of the power A^z given in 5.2d in the case of an elliptic pseudodifferential operator A can be extended:

7) For $\mathrm{Re}z\geq 0$, the domain of A^z as an operator in $H_s(M)$ coincides with $H_{s+m\mathrm{Re}z}(M)$, and consequently does not depend on $\mathrm{Im}z$. In particular, for $\mathrm{Re}z=0$, A^z is a bounded operator in any $H_s(M)$.
8) For $m\mathrm{Re}z<h$, A^z is a holomorphic function of z with values in the space of bounded operators from $H_{s+h}(M)$ to $H_s(M)$.

From Theorem 5.3.1 we also have

Corollary 5.3.2. *Suppose that A_0 is an elliptic pseudodifferential operator of order m on M, and suppose that A_0 is positive in $H_0(M)$. Let τ be an arbitrary real number. Then $A_0^{\tau/m}$ defines a continuous isomorphism from $H_\tau(M)$ to $H_0(M)$, and hence the norm in $H_\tau(M)$ is equivalent to the norm*

$$\|A_0^{\tau/m}u\|_{0,M}\,. \tag{5.34}$$

If, in particular, A_0 is a selfadjoint operator of order 1 with positive eigenvalues λ_j and with orthonormal basis $\{e_j(x)\}$ in $H_0(M)$ of eigenfunctions, $A_0e_j=\lambda_je_j$ ($j=1,2,\ldots$), then the norm (5.34) can be written in the form

$$\left\{\sum_{j=1}^{\infty}\lambda_j^{2\tau}|(u,e_j)_{0,M}|^2\right\}^{1/2}, \tag{5.35}$$

and the corresponding scalar product can be easily shown. For example, if on M we have a Riemannian metric and Δ is the corresponding Laplace–Beltrami operator, then as A_0 we can take the operator

$$\Lambda = (I - \Delta)^{1/2} . \tag{5.36}$$

For $\tau < -n$, as we have mentioned in Sect. 1.1, A_0^τ is an integral operator with continuous kernel $K_\tau(x, y)$ which satisfies

$$K_\tau(x, y) = \sum_{j=1}^{\infty} \lambda_j^\tau e_j(x)\overline{e_j(y)} , \tag{5.37}$$

where the convergence is uniform by the well-known Mercer theorem. Putting $x = y$ and integrating over M, we obtain

$$\sum_{j=1}^{\infty} \lambda_j^\tau = \int_M K_\tau(x, x)dx < \infty \quad (\tau < -n) . \tag{5.38}$$

Hence, assuming that $\{\lambda_j\}$ is a nondecreasing sequence, we obtain an important estimate of eigenvalues:

$$\lambda_j^{-1} = o\left(j^{-\frac{1}{n}+\varepsilon}\right) \quad (j \to \infty) \tag{5.39}$$

for any $\varepsilon > 0$. Actually, the right-hand side can be replaced by $O(j^{-1/n})$ and, moreover, $\lambda_j \sim cj^{1/n}$ with some $c > 0$ (see 6.1b below).

As is explained in the next section, the estimate (5.39) has important applications to general linear operators acting in Sobolev spaces over M.

5.4. Estimates of s-numbers, Integral Representation, and Traces of Operators of Negative Order

a. First we recall some definitions and facts from the general theory of linear operators (see for example (Gohberg and Krejn 1965), (Dunford and Schwartz 1963), and (Reed and Simon 1978)). Let A be a compact operator in a Hilbert space H. Then the nonnegative compact operators $(A^*A)^{1/2}$ and $(AA^*)^{1/2}$ are well-defined. They have the same system of eigenvalues (taking into account the multiplicities) which are denoted by $s_j(A)$ $(j = 1, 2, \ldots)$. These eigenvalues are called *singular numbers* or *s-numbers* of the operator A. They are numbered in the nonincreasing order: $s_1(A) \geq s_2(A) \geq \ldots$.

From what has been said above it is clear that $s_j(A) = s_j(A^*)$. We also note that if H_1 and H_2 are two Hilbert spaces, A is a compact operator in H_1, and B_1, B_2 are bounded operators from H_1 to H_2 and from H_2 to H_1 respectively, so that B_1AB_2 is a compact operator in H_2, then

$$s_j(B_1AB_2) \leq \|B_1\| \, \|B_2\| s_j(A) . \tag{5.40}$$

Let the eigenvalues of A be numbered (taking into account multiplicities) in the order with nonincreasing moduli: $|\lambda_1(A)| \geq |\lambda_2(A)| \geq \ldots$. It is well known that

$$\sum_{j\leq k} |\lambda_j(A)|^p \leq \sum_{j\leq k} [s_j(A)]^p \quad (k = 1, 2, \ldots;\ p > 0) \tag{5.41}$$

(the Weyl inequalities) and for any $q > 0$

$$s_j(A) = O(j^{-q}) \Longrightarrow \lambda_j(A) = O(j^{-q}) \, . \tag{5.42}$$

By $\mathfrak{S}_p$ $(p > 0)$ we denote the class of all compact operators A such that

$$\sum [s_j(A)]^p < \infty \, . \tag{5.43}$$

This is a two-sided ideal in the space of all bounded operators in H. It grows when p increases. For $p \geq 1$ it is a Banach space with the norm

$$|A|_p = \left(\sum [s_j(A)]^p\right)^{1/p} \, . \tag{5.44}$$

The operators A from $\mathfrak{S}_1$ are called *trace class* (or *nuclear*) *operators*, and the quantity $|A|_1$ is called the *trace class* (or *nuclear*) *norm*. The number

$$\mathrm{tr} A = \sum_j \lambda_j(A) \tag{5.45}$$

is called the *trace* of the trace class operator A; it is a finite sum or an absolutely convergent series (see (5.41)). If $\{e_j\}_1^\infty$ is an orthonormal basis of H, then, by Lidskij's theorem, we have

$$\mathrm{tr} A = \sum (Ae_j, e_j) \, , \tag{5.45'}$$

hence it is clear that tr is a linear functional on $\mathfrak{S}_1$. It is continuous since $|\mathrm{tr} A| \leq |A|_1$.

The operators A from $\mathfrak{S}_2$ are called *Hilbert–Schmidt operators*. The product of two Hilbert–Schmidt operators is a trace class operator. If A is a Hilbert–Schmidt operator, then for any orthonormal basis $\{e_j\}_1^\infty$ of H

$$|A|_2^2 = \mathrm{tr}(AA^*) = \sum_j \|Ae_j\|^2 = \sum_{j,k} |(Ae_j, e_k)|^2 \, . \tag{5.46}$$

And conversely, if the right-hand side is finite for some orthonormal basis $\{e_j\}$, then A is a Hilbert–Schmidt operator. The space $\mathfrak{S}_2$ is a Hilbert space with the scalar product

$$< A, B > = \mathrm{tr}(AB^*) \, . \tag{5.47}$$

If $H = L_2(\Omega)$, where Ω is a metric space with a positive measure dx, then the operator A in H is Hilbert–Schmidt if and only if it admits the representation

$$Au(x) = \int_{\Omega} K(x,y)u(y)dy \ , \tag{5.48}$$

where $K \in L_2(\Omega \times \Omega)$. In addition we have $|A|_2 = \|K\|_{L_2}$.

b. We pass on to operators that are defined in Sobolev spaces of functions on an n-dimensional closed manifold M and improve the smoothness of functions.

Proposition 5.4.1. *Suppose that B is a bounded operator from $H_s(M)$ to $H_{s+\tau}(M)$ for some $s \in \mathbb{R}$ and $\tau > 0$, so that it is a compact operator in $H_s(M)$. Then its s-numbers satisfy the estimate*

$$s_j(B) = o\left(j^{-\frac{\tau}{n}+\varepsilon}\right) \tag{5.49}$$

for any $\varepsilon > 0$.

This follows from the formula (5.38) for eigenvalues of the operator Λ (see (5.36)) and the inequalities of the form (5.40). For example, for $s = 0$ it is sufficient to write B in the form $\Lambda^{-\tau}(\Lambda^{\tau}B)$. In (5.49), as well as in (5.39), we can in fact replace the right-hand side by $O(j^{-\tau/h})$ (cf. 6.1b). But from (5.49) we already obtain the following quite precise result.

Theorem 5.4.2. *Under assumptions of Proposition 5.4.1 the operator B belongs to $\mathfrak{S}_p$ if $\tau > n/p$. In particular, if $\tau > n/2$, then B is a Hilbert–Schmidt operator, and if $\tau > n$, it is a trace class operator.*

c. It is now a good moment to give theorems on integral representation of operators of sufficiently low order. We give analogues of theorems from the paper of Agmon (1965) for operators on n-dimensional closed manifold M (Agmon considered operators in $\mathbb{R}^n$ and in a domain in $\mathbb{R}^n$). We assume that the local coordinates that we use on M are consistent with a positive density given on M. By $\|A\|_{\tau_1,\tau_2}$ we denote the norm of A as an operator from $H_{\tau_1}(M)$ to $H_{\tau_2}(M)$.

Theorem 5.4.3. *Suppose that A is a bounded operator from $H_0(M)$ to $H_\tau(M)$, where $\tau > n/2$. Then A is an integral operator,*

$$Au(x) = \int_M K(x,y)u(y)dy \ , \tag{5.50}$$

where $K(x,y)$ is a function of y belonging to $H_0(M)$ for each x. In addition

$$\left\{\int |K(x,y)|^2 dy\right\}^{1/2} \le C_1 \|A\|_{0,\tau}^{n/2\tau} \|A\|_{0,0}^{1-(n/2\tau)} \ , \tag{5.51}$$

so that

$$\left\{\int\int |K(x,y)|^2 dxdy\right\}^{1/2} \le C_2 \|A\|_{0,\tau}^{n/2\tau} \|A\|_{0,0}^{1-(n/2\tau)} \ ; \tag{5.52}$$

moreover, if $0 < \mu < \min(\tau - (n/2), 1)$, then

$$\left\{\int |K(x,y) - K(\widetilde{x},y)|^2 dy\right\}^{1/2} \leq C_3|x - \widetilde{x}|^\mu \|A\|_{0,\tau} . \tag{5.53}$$

The constants in the above and in the following inequalities depend only on M, τ, and μ. The modulus $|x - \widetilde{x}|$ is computed in local coordinates for sufficiently close x and $\widetilde{x}$. The proof is based on the analysis of the embedding of $H_\tau(\mathbb{R}^n)$ into the space of bounded continuous functions for $\tau > n/2$. From (5.52) we again find that A is a Hilbert–Schmidt operator.

Theorem 5.4.4. *Suppose that A is a bounded operator from $H_{-\tau}(M)$ to $H_\tau(M)$, where $\tau > n/2$. Then A is an integral operator in $H_0(M)$ of the form* (5.50) *with continuous kernel $K(x,y)$ which satisfies the following inequalities:*

$$|K(x,y)| \leq C_4 \|A\|_{-\tau,\tau} , \tag{5.54}$$

$$|K(x,y) - K(\widetilde{x},y)| \leq C_5|x - \widetilde{x}|^\mu \|A\|_{-\tau,\tau} , \tag{5.55}$$

$$|K(x,y) - K(x,\widetilde{y})| \leq C_6|y - \widetilde{y}|^\mu \|A\|_{-\tau,\tau} \tag{5.56}$$

for $0 < \mu < \min(\tau - (n/2), 1)$.

In fact, we have here some smoothness of the kernel. Obviously, this smoothness grows with the growth of τ.

Theorem 5.4.5. *Suppose that the operators A and A^*, the adjoint to A in $H_0(M)$, are bounded from $H_0(M)$ to $H_\tau(M)$ for some $\tau > n$. Then A is an integral operator of the form* (5.50) *with continuous kernel satisfying*

$$|K(x,y)| \leq C_7(\|A\|_{0,\tau} + \|A^*\|_{0,\tau})^{n/\tau} \|A\|_{0,0}^{1-(n/\tau)} . \tag{5.57}$$

In concrete situations, such as for pseudodifferential operators of order $-\tau$, continuity is known or can be deduced from explicit formulas. But this cannot be always obtained sufficiently easily, and besides, it is sometimes useful to know the forms of the right-hand sides of the above inequalities.

d. Under the assumptions of Theorem 5.4.4 or 5.4.5 the operator A is of trace class in $H_0(M)$.

Theorem 5.4.6 (see e.g. (Duistermaat 1981)). *If the integral operator* (5.50) *is of trace class in $H_0(M)$ and has continuous kernel, then*

$$\operatorname{tr} A = \int_M K(x,x) dx . \tag{5.58}$$

Continuity of the kernel $K(x,y)$ is not sufficient itself for the operator (5.50) to be of trace class. Consider, for example, an integral operator on the circle S

$$Au(x) = \int_{-\pi}^{\pi} f(x-y) u(y) dy ,$$

where $f(x)$ is a real continuous 2π-periodic function whose Fourier series with respect to the system $\{e^{inx}\}_{-\infty}^{\infty}$ does not converge everywhere. It is a self-adjoint operator, and its eigenvalues are the Fourier coefficients of f in this system. Obviously, this operator is not of the trace class.

5.5. Meromorphic Continuation of the Kernel and the Trace of a Power of a Positive Elliptic Pseudodifferential Operator. Let A be an elliptic pseudodifferential operator of order $m > 0$, positive in $H_0(M)$. Then for any complex z, A^{-z} is a pseudodifferential operator from $\Psi_{\text{ph}}^{-mz}(M)$ of order $-m\text{Re}z$ in the scale of Sobolev spaces on M (see Sect. 5.3). For $m\text{Re}z > n$ it is an integral operator with kernel continuous in (x, y); we denote it by $K_{-z}(x, y)$. Using (5.15) and the approximation of the resolvent by the parametrix constructed in Sect. 4.2, the following theorem can be proved (Seeley 1967a).

Theorem 5.5.1. *1) The kernel $K_{-z}(x, y)$ is continuous in (x, y, z) and holomorphic in z for* $\text{Re}z > n/m$.

2) The restriction of $K_{-z}(x, y)$ to the complement of the diagonal in $M \times M$ can be continued to an entire analytic function of z, continuous in (x, y, z).

3) For $x = y$ this kernel can be continued to a meromorphic function $K_{-z}(x, x)$ of z on the complex plane, which can have poles only at the points $z_j = (n - j)/m$ ($j \in \mathbb{Z}_+$), and all these poles are simple. The residue at z_j is equal to $-\rho_j(x)$, where

$$\rho_j(x) = \frac{1}{(2\pi)^n m} \sum_{q=1}^{K} \varphi_q(x) \int_{|\omega|=1} \text{Res}_{\lambda=a_0^{(q)}(x,\omega)} \left\{ \lambda^{\frac{j-n}{m}} b_j^{(q)}(x, \omega, \lambda) \right\} dS . \tag{5.59}$$

In particular,

$$\rho_0(x) = \frac{-1}{(2\pi)^n m} \sum_{q=1}^{K} \varphi_q(x) \int_{|\omega|=1} [a_0^{(q)}(x, \omega)]^{-\frac{n}{m}} dS . \tag{5.60}$$

4) For $z = 0$ the function $K_{-z}(x, x)$ does not have a pole, and

$$K_0(x, x) = \frac{1}{(2\pi)^n m} \sum_{q=1}^{K} \varphi_q(x) \int_{|\omega|=1} dS \int_0^{\infty} b_0^{(q)}(x, \omega, -\sigma) d\sigma . \tag{5.61}$$

We give some explanations. In these formulas we use a partition of unity $\{\varphi_q\}_1^K$ subordinated to a covering of the manifold M by coordinate neighborhoods O_q. The symbols $a_0^{(q)}$, $b_j^{(q)}$ are computed in local coordinates in O_q which are assumed to be consistent with the density given on M. By dS we denote the 'surface' element (or 'volume' element) on the unit sphere in $\mathbb{R}_n$:

$$d\xi = |\xi|^{n-1} d|\xi| dS . \tag{5.62}$$

For brevity of the formulation of the theorem, we have not included in it the smoothness properties with respect to the system of variables. We add them now:

1) outside the diagonal in $M \times M$ the continued function $K_{-z}(x,y)$ is C^∞ with respect to (x,y,z);
2) $K_{-z}(x,x)$ is C^∞ with respect to (x,z) for z different from the poles;
3) if z_j is a pole, then $K_{-z}(x,x)(z-z_j)$ is C^∞ with respect to (x,z) for z close to z_j.

If A is a differential operator, then K additionally has the following properties:

5) $K_{-z}(x,x)$ does not have poles at $z=-l$ ($l=1,2,\ldots$), and

$$K_l(x,x) = \frac{(-1)^l}{(2\pi)^n m} \sum_{q=1}^{K} \varphi_q(x) \int_{|\omega|=1} dS \int_0^\infty \sigma^l b^{(q)}_{ml+n}(x,\omega,-\sigma)d\sigma\,. \tag{5.63}$$

6) If $n \geq 2$, then $K_{-z}(x,x)$ does not have poles at $z_j = (n-j)/m$ with odd j.

7) If $n \geq 3$ and n is odd, then $K_l(x,x) = 0$ ($l \in \mathbb{Z}_+$).

The function

$$\zeta(z;A) = \mathrm{tr}A^{-z} = \int_M K_{-z}(x,x)dx = \sum_{j=1}^\infty \lambda_j^{-z} \qquad \left(\mathrm{Re}z > \frac{n}{m}\right) \tag{5.64}$$

is called the *zeta-function of the operator A* (sometimes $\zeta(-z;A)$ bears this name). Here and in what follows λ_j are eigenvalues of A numbered so that the multiplicities are taken into account. If M is the circle S and A is of the form (2.31) with $a(x,n)=n$ for $n>0$ and $a(x,n)=0$ for $n \leq 0$ (cf. 3.1i), then

$$\zeta(z;A) = \sum_{j=1}^\infty j^{-z}\,, \tag{5.65}$$

which is the classical Riemann zeta-function. From Theorem 5.5.1 it follows that in the general case $\zeta(z;A)$ is holomorphic in z for $\mathrm{Re}z > n/m$ and has a meromorphic continuation on the complex plane with possible poles at z_j ($j \in \mathbb{Z}_+$), where all poles are simple. The residue at z_j is equal to $-\rho_j$, where ρ_j is the integral of $\rho_j(x)$ over M. At $z=0$ we do not have a pole, and the value $\zeta(0;A)$ is equal to the integral of $K_0(x,x)$ over M. If A is a differential operator, then we do not have poles at $z=-l$, $l \in \mathbb{N}$, and $\zeta(-l;A)$ is equal to the integral of $K_l(x,x)$ over M. Besides, if A is a differential operator and $n \geq 2$, then $\zeta(z;A)$ does not have poles at z_j with odd j, and if in addition $n \geq 3$ is odd, then $\zeta(l;A)=0$.

We now give an analytic formula for the index $\kappa(A)$ of an arbitrary elliptic operator of order $m>0$. We put

$$L_1 = I + A^*A\,, \quad L_2 = I + AA^*\,. \tag{5.66}$$

These are selfadjoint elliptic operators with discrete spectrum, and it can be easily verified that they have the eigenvalue 1 with multiplicities $\dim \operatorname{Ker} A$ and $\dim \operatorname{Ker} A^*$, respectively, while the remaining eigenvalues of L_1 and L_2 have equal multiplicities. The situation is similar for (normal) operators L_1^{-z} and L_2^{-z}, hence $\kappa(A) = \zeta(z; L_1) - \zeta(z; L_2)$. By a holomorphic continuation at $z = 0$ we obtain the desired formula

$$\kappa(A) = \zeta(0; L_1) - \zeta(0; L_2)\,. \tag{5.67}$$

From this formula we cannot directly see that the index of an elliptic operator depends only on its principal symbol. But this formula, or an analogous formula (5.82) given below, is used in derivation of the final formula for the index (see the survey (Fedosov 1990)).

We shall also give an estimate for $K_{-z}(x, x)$ in vertical bands of the z-plane. We denote by U_δ the union of disks with centers at poles of $K_{-z}(x, x)$ and small radius δ.

Theorem 5.5.2. *Suppose that a pseudodifferential operator $A \in \Psi_{\mathrm{ph}}^m(M)$ is φ-positive for $\varphi < \varphi_0$ and $\pi - \varphi_0 < \psi < \pi$, $a < b$. Then for $a \le \operatorname{Re} z \le b$, $z \notin U_\delta$, we have the estimate*

$$|K_{-z}(x, x)| \le C e^{\psi |\operatorname{Im} z|}\,, \tag{5.68}$$

where C does not depend on (x, z).

In the case of a selfadjoint pseudodifferential operator with positive spectrum the estimate is essentially stronger: the right-hand side of (5.68) can be replaced by $C(1 + |z|)^h$ with a certain h (see 5.8d).

5.6. The Exponent e^{-tA} of an Elliptic Pseudodifferential Operator A; Asymptotics of Its Kernel and Trace. Suppose that A is an elliptic pseudodifferential operator from $\Psi_{\mathrm{ph}}^m(M)$, $m > 0$, $\pi/2$-positive in $H_0(M)$. Using (5.22), it can be easily verified that for $t > 0$ and any $k \in \mathbb{Z}_+$ the operator $A^k e^{-tA}$ is bounded in any $H_s(M)$. Hence e^{-tA} ($t > 0$) is an operator of order $-\infty$ in the scale of Sobolev spaces on M and, consequently, it is an integral operator:

$$e^{-tA} u(x) = \int_M \Theta(t, x, y) u(y) dy\,. \tag{5.69}$$

The kernel Θ is infinitely smooth with respect to (t, x, y) in $\mathbb{R}_+ \times M \times M$. It turns out that for $x = y$ and $t \to +0$ it has a full asymptotic expansion in simple functions of t:

Theorem 5.6.1. *If $t \to +0$, then we have, uniformly in $x \in M$,*

$$\Theta(t, x, x) \sim \sum_{j \in \mathbb{Z}_+ \cup J} \eta_j(x) t^{\frac{j-n}{m}} + \sum_{l=1}^{\infty} \widetilde{\eta}_l(x) t^l \log t\,, \tag{5.70}$$

where

$$J = \{j = n + lm : l \in \mathbb{Z}_+\}. \tag{5.71}$$

The formulas for the coefficients $\eta_j(x)$, $\widetilde{\eta}_l(x)$ are given below. If A is a differential operator, then $\widetilde{\eta}_l(x) = 0$ for all l.

In general, when A is a pseudodifferential operator, then to obtain the expansion (5.70) we can employ formula (5.29) and combine it with the results presented in the previous section. From (5.29) we obtain the formula

$$\Theta(t, x, x) = \frac{1}{2\pi \mathrm{i}} \int_{\mathrm{Re} z = c} t^{-z} \Gamma(z) K_{-z}(x, x) dz \tag{5.72}$$

with $c > n/m$. Here for $K_{-z}(x, x)$ we have the estimate (5.68), where ψ can be taken slightly smaller than $\pi/2$, and for the Γ-function in any vertical band of the z-plane outside some neighborhoods of its poles we have

$$|\Gamma(z)| \le C(\omega) \mathrm{e}^{-\omega |\mathrm{Im} z|} \tag{5.73}$$

for $0 < \omega < \pi/2$, where $C(\omega)$ does not depend on z. Therefore the modulus of the expression under the integral sign in (5.72) decays exponentially as $|\mathrm{Im} z| \to \infty$, and we can shift the line of integration to the left, 'jumping' over the poles of the expression under the integral. For any $c' < c$ different from the poles we obtain

$$\Theta(t, x, y) = \sum \mathrm{Res}\{t^{-z} \Gamma(z) K_{-z}(x, x)\} + \int_{\mathrm{Re} z = c'} t^{-z} \Gamma(z) K_{-z}(x, x) dz ,$$

where the sum is taken over the poles of the expression in curly brackets between c' and c. The terms of this sum will just enter into the right-hand side of the asymptotics (5.70). We pass on to details.

1) Suppose that $j \in \mathbb{Z}_+$ and $j \notin J$. To these j there correspond the terms of the first sum on the right in (5.70) with

$$\eta_j(x) = -\Gamma\left(\frac{n-j}{m}\right) \rho_j(x) , \tag{5.74}$$

where $\rho_j(x)$ is defined by (5.59). In particular,

$$\eta_0(x) = \frac{\Gamma(n/m)}{(2\pi)^n m} \sum_{q=1}^{K} \varphi_q(x) \int_{|\omega|=1} [a_0^{(q)}(x, \omega)]^{-\frac{n}{m}} dS . \tag{5.75}$$

2) Suppose that $j = n + lm$, where $l \in \mathbb{Z}_+$ and either $j \notin \mathbb{Z}_+$ (so that $m \notin \mathbb{N}$) or $j \in \mathbb{Z}_+$ but $K_{-z}(x, x)$ does not have a pole at $z = -l$. To these j there correspond the terms of the first sum on the right in (5.70) with

$$\eta_j(x) = \frac{(-1)^l}{l!} K_l(x, x) ; \tag{5.76}$$

in the second sum we have

$$\widetilde{\eta}_l(x) = 0 \quad \text{for the same } l\,. \tag{5.77}$$

In these two cases we obtain the terms of the right-hand side of (5.70) corresponding to simple poles of the expression under the integral in (5.72); they are the poles of either only K_{-z} or only the Γ-function. The remaining terms correspond to the points z where both K_{-z} and $\Gamma(z)$ have a simple pole, so that the expression under the integral in (5.72) has poles of the second order.

3) Suppose that simultaneously $j \in \mathbb{Z}_+$ and $j = n + lm$ with $l \in \mathbb{Z}_+$, and suppose that $K_{-z}(x,x)$ has a pole at $z = -l$. (Consequently, $l \geq 1$ and m is rational.) To those j there correspond the terms of the second sum on the right hand-side of (5.70) with

$$\widetilde{\eta}_l(x) = \frac{(-1)^l}{l!}\rho_j(x) \tag{5.78}$$

and the terms of the first sum with

$$\eta_j(x) = \frac{(-1)^l}{l!}\widetilde{K}_l(x) - \rho_j(x)\widetilde{\Gamma}(-l)\,. \tag{5.79}$$

Here $\widetilde{K}_l(x)$ and $\widetilde{\Gamma}(-l)$ are the values at the point $z = -l$ of the holomorphic parts of the functions $K_{-z}(x,x)$ and $\Gamma(z)$:

$$\begin{aligned} \widetilde{K}_l(x) &= \left[K_{-z}(x,x) + \frac{\rho_j(x)}{z+l}\right]_{z=-l}, \\ \widetilde{\Gamma}(-l) &= \left[\Gamma(z) - \frac{(-1)^l}{l!(z+l)}\right]_{z=-l}. \end{aligned} \tag{5.80}$$

Suppose now that A is a differential operator. Then the case 3) is impossible, so the second sum in (5.70) is absent. In (5.76) $K_l(x,x)$ can be computed by (5.63). Besides, $\eta_j(x) = 0$ if n and j are odd, $n \geq 3$.

We now obtain the asymptotic expansion of the trace $\theta(t) = \operatorname{tr} e^{-tA}$ by integrating the expansion (5.70) in x:

$$\theta(t) = \operatorname{tr} e^{-tA} = \sum_{j=1}^{\infty} e^{-t\lambda_j} \sim \sum_{j\in\mathbb{Z}_+\cup J} \eta_j t^{\frac{j-n}{m}} + \sum_{l=1}^{\infty} \widetilde{\eta}_l t^l \log t\,, \tag{5.81}$$

where η_j, $\widetilde{\eta}_l$ are integrals of $\eta_j(x)$, $\widetilde{\eta}_l(x)$ over M.

We also turn our attention to the following fact. If A is not a differential operator, then all coefficients in (5.70) and (5.81), except $\eta_j(x)$ and η_j with $j = n + lm$, $l \in \mathbb{N}$, are computed by means of a finite number of terms of the asymptotic expansion for the complete local symbol of A. If $j = n + lm$, $l \in \mathbb{N}$, and A is not a differential operator, then such computation is impossible in principle. For let P be the Riesz projection (see (6.62) below) on the root subspace of A corresponding to any eigenvalue λ_0. Then P is an operator of order $-\infty$, hence A and $A+P$ have the same complete symbols. On the other hand, when we pass from A to $A+P$, the trace $\operatorname{tr} e^{-tA}$ changes by the term $\left[e^{-t(\lambda_0+1)} - e^{-t\lambda_0}\right]\nu$, where ν is the multiplicity of the eigenvalue λ_0, and it can be easily verified that for that reason the coefficients at t^l on the right of (5.81) are changed ($l \in \mathbb{N}$).

We also note that similarly to (5.67) we can obtain the following formula for the index of the operator A:

$$\kappa(A) = \eta_n(L_1) - \eta_n(L_2) ; \tag{5.82}$$

here $\eta_n(L_j)$ are the coefficients at t^0 in the formula of the form (5.81) for $\operatorname{tr} e^{-tL_j}$ and the operators L_j ($j = 1, 2$) are defined by (5.66).

5.7. Asymptotics of the Kernel and the Trace of the Resolvent. Suppose that A is a pseudodifferential operator from $\Psi^m_{\mathrm{ph}}(M)$, elliptic with parameter in the angle $\mathcal{L}_\varphi$ of measure 2φ with bisector $\mathbb{R}_-$, and let $m > n = \dim M$. We know that in this case the resolvent $R_A(\lambda) = (A - \lambda I)^{-1}$ exists for $\lambda \in \mathcal{L}_\varphi$ sufficiently large in modulus and it is an integral operator with kernel continuous in (x, y). Denote it by $K(x, y, \lambda)$:

$$R_A(\lambda)f(x) = \int_M K(x, y, \lambda)f(y)dy . \tag{5.83}$$

The kernel $K(x, y, \lambda)$ is in fact continuous in (x, y, λ) and holomorphic in λ. The parametrix constructed in Sect. 4.2 also is an integral operator with continuous kernel $K_N(x, y, \lambda)$, for which explicit formulas can be easily indicated. Approximating K by K_N, we can establish that $K(x, y, \lambda)$ has proper asymptotics for $x = y$, $\lambda \to \infty$ in $\mathcal{L}_\varphi$, and compute the coefficients of this asymptotics.

A complete asymptotic expansion of $K(x, x, \lambda)$ can be obtained in this way in the case of a differential operator. Using inequalities (4.37) in combination with inequalities of the form (5.54) for the difference $K - K_N$, we arrive at the following result.

Theorem 5.7.1. *Suppose that A is a differential operator of order $m > n$, elliptic with parameter in $\mathcal{L}_\varphi$. Then*

$$K(x, x, \lambda) \sim \sum_{j=0}^{\infty} c_j(x)(-\lambda)^{\frac{n-j}{m}-1} \quad \text{as} \quad \lambda \to \infty \quad \text{in} \quad \mathcal{L}_\varphi . \tag{5.84}$$

Here the function $(-\lambda)^{\frac{n-j}{m}-1}$ is holomorphic for $|\arg\lambda - \pi| \le \varphi$ and coincides with $|\lambda|^{\frac{n-j}{m}-1}$ on $\mathbb{R}_-$. In the notation of Sect. 4.2 we have

$$c_j(x) = \sum_{q=1}^{K} \varphi_q(x) \int_{\mathbb{R}_n} b_j^{(q)}(x, \xi, -1)đ\xi . \tag{5.85}$$

In particular,

$$c_0(x) = \int_{T_x^*M} [a_0(x,\xi)+1]^{-1} đ\xi\ , \tag{5.86}$$

$$c_1(x) = -\int_{T_x^*M} (\mathrm{sub}A)(x,\xi)[a_0(x,\xi)+1]^{-2} đ\xi\ . \tag{5.87}$$

We also note that for $n \geq 2$ all $c_j(x)$ with odd j are equal to 0, since in this case all integrands in (5.85) are odd in ξ.

If A is a pseudodifferential operator (not a differential operator), then in this way we can only obtain the beginning of the asymptotic expansion of the kernel $K(x,x,\lambda)$ with accuracy $O(|\lambda|^{-2})$, which is limited by the estimate (4.38). Namely, we obtain the formula

$$K(x,x,\lambda) = \sum_{j=0}^{n+[m]} c_j(x)(-\lambda)^{\frac{n-j}{m}-1} + O(|\lambda|^{-2})$$

for noninteger m (where $[m]$ is the integer part of m) and the formula

$$K(x,x,\lambda) = \sum_{j=0}^{n+m-1} c_j(x)(-\lambda)^{\frac{n-j}{m}-1} + \tilde{c}(x)(-\lambda)^{-2}\log(-\lambda) + O(|\lambda|^{-2})$$

for integer m. The function $\log(-\lambda)$ is holomorphic for $|\arg\lambda - \pi| < \pi$ and coincides with $\log|\lambda|$ on $\mathbb{R}_-$, and

$$\tilde{c}(x) = \frac{-1}{(2\pi)^n m}\sum_{q=1}^{K} \varphi_q(x)\int_{|\omega|=1} a_{n+m}^{(q)}(x,\omega)dS$$

(in the case of a differential operator we have $a_j = 0$ for $j > m$). The coefficients $c_j(x)$ can be computed by the same formulas (5.85).

A full asymptotic expansion for $K(x,x,\lambda)$ when A is a pseudodifferential operator can be obtained by a different method, similar to that used to get (5.70). Namely, assuming without loss of generality that A is φ-positive, we employ the following formula which can be deduced from (5.28):

$$K(x,x,\lambda) = \frac{1}{2\mathrm{i}}\int_{\mathrm{Re}z=c} \frac{(-\lambda)^{z-1}}{\sin\pi z} K_{-z}(x,x)dz \tag{5.88}$$

with $n/m < c < 1$. The estimate (5.68) (where we take ψ slightly smaller than $\pi - \varphi$) and the estimate

$$|\sin\pi z|^{-1} \leq C\mathrm{e}^{-\pi|\mathrm{Im}z|} \tag{5.89}$$

allow us to shift the line of integration to the left jumping over the poles of the integrand. When doing this, we add the residues at these poles, and these residues give the terms of the asymptotics.

Theorem 5.7.2. *Suppose that A is a pseudodifferential operator of order $m > n$ elliptic with parameter in $\mathcal{L}_\varphi$. Then*

$$K(x,x,\lambda) \sim \sum_{j\in\mathbb{Z}_+\cup J} c_j(x)(-\lambda)^{\frac{n-j}{m}-1} + \sum_{l=1}^{\infty} \widetilde{c_l}(x)(-\lambda)^{-l-1}\log(-\lambda) \quad (5.90)$$

as $\lambda \to \infty$, $\lambda \in \mathcal{L}_\varphi$, where J is defined in (5.71) *and the formulas for the coefficients $c_j(x)$, $\widetilde{c_l}(x)$ are given below.*

1) Suppose that $j \in \mathbb{Z}_+$ and $j \notin J$. To these j there correspond the terms of the first sum on the right-hand side of (5.90) with

$$c_j(x) = \pi \left(\sin \pi \frac{j-n}{m}\right)^{-1} \rho_j(x)\,, \quad (5.91)$$

where $\rho_j(x)$ is defined in (5.59).

2) Suppose that $j = n + lm$, where $l \in \mathbb{Z}_+$ and either $j \notin \mathbb{Z}_+$ (so that $m \notin \mathbb{N}$) or $j \in \mathbb{Z}_+$ but $K_{-z}(x,x)$ does not have a pole at $z = -l$. To these j there correspond the terms of the first sum on the right in (5.90) with

$$c_j(x) = (-1)^l K_l(x,x)\,; \quad (5.92)$$

in the second sum we have

$$\widetilde{c_l}(x) = 0 \quad \text{for the same } l\,. \quad (5.93)$$

3) Suppose that simultaneously $j \in \mathbb{Z}_+$ and $j = n + lm$ with $l \in \mathbb{Z}_+$, and suppose that $K_{-z}(x,x)$ has a pole at $z = -l$. (Consequently, $l \geq 1$ and m is rational.) To these j there correspond the terms of the first sum on the right-hand side of (5.90) with

$$c_j(x) = (-1)^l \widetilde{K}_l(x) \quad (5.94)$$

and the terms of the second sum with

$$\widetilde{c_l}(x) = (-1)^{l+1}\rho_j(x)\,. \quad (5.95)$$

The quantity $\widetilde{K}_l(x)$ is defined in (5.80).

If A is a differential operator, then the case 3) is impossible. Clearly, in this case two methods of computation of the coefficients $c_j(x)$ yield the same result.

Integrating (5.84) or (5.90) in x, we obtain the asymptotics of the trace of the resolvent. In the general case we have

$$\mathrm{tr} R_A(\lambda) = \sum (\lambda_j - \lambda)^{-1}$$

$$\sim \sum_{j \in Z_+ \cup J} c_j(-\lambda)^{\frac{n-j}{m}-1} + \sum_{l=1}^{\infty} \widetilde{c}_l(-\lambda)^{l-1}\log(-\lambda)\,, \tag{5.96}$$

where

$$c_j = \int_M c_j(x)dx\,, \quad \widetilde{c}_l = \int_M \widetilde{c}_l(x)dx\,. \tag{5.97}$$

If A is a differential operator, all coefficients of the asymptotics, except $c_j(x)$ and c_j for $j = n + lm$, $l \in \mathbb{N}$, can be expressed in terms of the local complete symbol of A. This is impossible for the excluded coefficients if A is not a differential operator.

If A is a differential operator, then for $x \neq y$ the kernel $K(x, y, \lambda)$ rapidly decays as $\lambda \to \infty$ in $\mathcal{L}$. In the general case this does not hold: $K(x, y, \lambda)$ has power asymptotics (as a result of the presence of the poles of $(\sin \pi z)^{-1}$). But this is not very interesting.

If $A = A^* > 0$, then the asymptotics for the kernel and the trace of the resolvent can be obtained outside a 'parabolic' neighborhood of the ray $\mathbb{R}_+$ of the form

$$\left\{\lambda : \mathrm{Re}\lambda > 0\,, |\mathrm{Im}\lambda| \le C(\mathrm{Re}\lambda)^{1-\frac{\theta}{m}}\right\}\,, \tag{5.98}$$

where θ is an arbitrary number between 0 and 1. An indirect method of proving this result will be indicated below in 5.8d.

If the condition $m > n$ is not fulfilled, then instead of $K(x, x, \lambda)$ we can consider the kernel of sufficiently high natural power of the resolvent. (If we raise to a natural power the operator A itself instead of the resolvent, then the angle of ellipticity with parameter may disappear.) This can be done by means of (5.31), more precisely, by means of its analogue for the corresponding kernels. We restrict ourselves to showing the principal term of the asymptotics. Suppose that $q > n/m$, so that $[R_A(\lambda)]^q$ is an integral operator; we denote its kernel by $K^{(q)}(x, y, \lambda)$.

Theorem 5.7.3. *If $\lambda \to \infty$ in $\mathcal{L}_\varphi$, then*

$$K^{(q)}(x, x, \lambda) = (-\lambda)^{\frac{n}{m}-q} c_0^{(q)}(x) + O\left(|\lambda|^{\frac{n-1}{m}-q}\right) \tag{5.99}$$

uniformly in x and $\arg\lambda$, *where*

$$c_0^{(q)}(x) = \int_{T_x^* M} [a_0(x, \xi) + 1]^{-q} d\!\!{}^-\xi\,. \tag{5.100}$$

5.8. The Exponent e^{-itA} for an Elliptic Pseudodifferential Operator A of the First Order. The function e^{-itA}, which we shall consider now, is not a pseudodifferential operator. It belongs to a broader class of *Fourier integral operators*.

a. Suppose that A is an elliptic pseudodifferential operator of order 1 on M, for the time being scalar, with the *real* principal symbol $a_0(x, \xi)$. This is equivalent to the fact that $A = A_0 + T$, where $A_0 = \mathrm{Re}A = (A + A^*)/2$ is selfadjoint (and has the same principal symbol) and $\mathrm{ord}\, T \leq 0$. If $n \geq 2$, then $a_0(x, \xi)$ keeps the sign, since T^*M is connected, and we shall assume that $a_0(x, \xi) > 0$. We put

$$\Omega = \mathbb{R} \times M \ , \quad \Omega_T = [-T, T] \times M \quad (T > 0) \ , \tag{5.101}$$

and we consider the *Cauchy problem*

$$D_t u(t, x) + Au(t, x) = 0 \quad ((t, x) \in \Omega) \ , \tag{5.102}$$

$$u(0, x) = u_0(x) \quad (x \in M) \ . \tag{5.103}$$

Equation (5.102) can be called *strictly hyperbolic* (cf. (Taylor 1981)), because the principal symbol a_0 is real. The Cauchy problem (5.102)–(5.103) is well-posed. Moreover, the following assertions are valid.

1) The solution $u \in C^\infty(\Omega)$ to this problem satisfies the estimate

$$\|u(t, x)\|_{s,M} \leq C_s \mathrm{e}^{c_s |t|} \|u_0(x)\|_{s,M} \tag{5.104}$$

for any s, where C_s, c_s do not depend on u, t.

Let $s \in \mathbb{R}$ and $u_0(x) \in H_s(M)$. A function $u(t, x)$ continuous in t with values in $H_s(M)$ is called a *strong solution* to the problem (5.102)–(5.103) in $H_s(M)$ if $u(0, x) = u_0(x)$ and if there exists a sequence $u_n(t, x) \in C^\infty(\Omega)$ $(n = 1, 2, \ldots)$ satisfying (5.102) such that

$$\|u_n(t, x) - u(t, x)\|_{s,M} \to 0 \quad (n \to \infty)$$

uniformly on any segment $[-T, T]$.

2) For any $s \in \mathbb{R}$, $u_0 \in H_s(M)$ there is a unique strong solution to the problem (5.102)–(5.103) in $H_s(M)$.
3) (*Theorem on improved smoothness*). Suppose that $u(t, x)$ is a strong solution to (5.102)–(5.103) in $H_s(M)$ and $u_0(x) \in H_{s'}(M)$, where $s' > s$. Then $u(t, x)$ is a strong solution to this problem in $H_{s'}(M)$.
4) For simplicity we take $s \in \mathbb{Z}_+$. Assume that $u(t, x)$ is a strong solution to the problem (5.102)–(5.103) in $H_s(M)$. Then $u(t, x) \in H_s(\Omega_T)$ for any $T > 0$ and

$$\|u(t, x)\|_{s,\Omega_T} \leq C_{s,T} \|u_0(x)\|_{s,M} \ , \tag{5.105}$$

where $C_{s,T}$ does not depend on $u(t, x)$. Here

$$\|u\|_{s,\Omega_T} = \left(\sum_{\alpha + |\beta| \leq s} \int_{\Omega_T} |\partial_t^\alpha \partial_x^\beta u(t, x)|^2 dt dx \right)^{1/2}$$

From what has been said it follows that for $u_0(x) \in C^\infty(M)$ the problem has a unique solution $u(t, x)$ and it belongs to $C^\infty(\Omega)$.

We now write the solution to the problem (5.102)–(5.103) in the form

$$u(t,x) = \mathrm{e}^{-\mathrm{i}tA}u_0(x) = U(t)u_0(x) \ . \tag{5.106}$$

Thus we have defined an operator $U(t) = \mathrm{e}^{-\mathrm{i}tA}$; some properties of it have just been described.

We shall say a few words about the proofs. Assertion 1) for $s = 0$ can be proved by differentiating the scalar product $(u,u)_{0,M}$ in t, employing the fact that the operator T is bounded in $H_0(M)$, and using the well-known Gronwall lemma (see, for example, (Taylor 1981)). For $s \neq 0$ we additionally use the elliptic pseudodifferential operator Λ^s (see Sect. 5.3), and we apply the result for $s = 0$ to $\Lambda^s A \Lambda^{-s}$ in place of A and $v = \Lambda^s u$ in place of u. A traditional proof of assertion 2) for $s = 0$ is based on proving the fact that weak and strong solution coincide and showing the uniqueness of the adjoint problem, which follows from 1). Alternatively, in the case when A is an elliptic pseudodifferential operator, one can use the completeness of the system of root vectors of this operator (see Sect. 6.2 below) and combine it with the estimate (5.104). Assertion 3) follows from 2). Assertion 4) can be easily proved using permutability of D_t and A .

If A is selfadjoint, then $U(t) = \mathrm{e}^{-\mathrm{i}tA}$ is a unitary operator in $H_0(M)$ acting by the formula

$$U(t)u_0(x) = \sum_{j=1}^{\infty} \mathrm{e}^{-\mathrm{i}\lambda_j t}(u_0, e_j)e_j(x) \ , \tag{5.107}$$

where $\{e_j(x)\}_1^\infty$ is the orthonormal basis of eigenfunctions of A and λ_j are the corresponding eigenvalues. The operator $U(t)$ remains unitary in $H_s(M)$ if we choose there a suitable scalar product. For example, if all λ_j are nonzero, then we may put

$$(u,v)_{s,M} = (|A|^s u, |A|^s v)_{0,M} \ , \tag{5.108}$$

where $|A|$ is defined by a formula of the form (5.14).

We note that in principle we can define an analogue of $U(t)$ for an abstract operator A in a Hilbert space, selfadjoint or close to it. We shall not dwell on that.

b. The operator $U(t)$ can be considered as a solution to the *Cauchy problem*

$$D_t U(t) + AU(t) = 0 \quad (t \in \mathbb{R}) \ , \quad U(0) = I \ . \tag{5.109}$$

We do not have an explicit formula for $U(t)$ is general case. But we can construct a quite precise approximation $\Phi(t)$ of this operator, i.e. a parametrix for the Cauchy problem (5.109), for sufficiently small $|t|$, in the form of a Fourier integral operator depending on parameter t. The operator $U(t)$ is itself a Fourier integral operator depending on t. An introduction to the theory of such operators and references to the literature can be found in the papers (Hörmander 1968, 1971), (Egorov 1988), (Egorov and Shubin 1988b) and in the books (Egorov 1984), (Shubin 1978), (Hörmander 1985b), (Kumano-go 1981), (Taylor 1981), and (Trèves 1980, vol.2). The space does not permit us to discuss this theory in detail, and we outline a construction of $\Phi(t)$ in elementary terms (cf. (Hörmander 1968) and (Hörmander 1985b); a local procedure of constructing $\Phi(t)$ originates in the paper (Lax 1957)).

We construct $\Phi(t)$ as a solution to the *Cauchy problem*

$$D_t\Phi(t) + A\Phi(t) = T(t) \quad (|t| < \delta) , \quad \Phi(0) = I , \tag{5.109'}$$

where $T(t)$ should be an integral operator with infinitely smooth kernel defined in $(-\delta, \delta) \times M \times M$.

We assume that $n \geq 2$ and $a_0(x, \xi) > 0$ on $T^*M\backslash 0$.

We fix a covering $\{O_k\}_1^K$ of the manifold M by coordinate neighborhoods and subordinated partition of unity $\{\varphi_k(x)\}_1^K$, and we set

$$\Phi(t)u = \sum_{k=1}^{K} \Phi_k(t)[\varphi_k u] \quad (u \in C^\infty(M)) , \tag{5.110}$$

where $\Phi_k(t)$ is a Fourier integral operator in $\mathbb{R}^n$ depending on parameter t. The support of $\Phi_k(t)[\varphi_k u]$ for sufficiently small $|t|$ does not go beyond the chart of the neighborhood O_k, which allows us to consider each term in (5.110) as a function on $(-\delta, \delta) \times M$. The Fourier integral operator $\Phi_k(t)$ is constructed in the form

$$\Phi_k(t)v(x) = \int e^{i\psi(t,x,\xi)} q_k(t, x, \xi)(Fv)(\xi) đ\xi . \tag{5.111}$$

Here ψ is a real *phase function* (more precisely, 'incomplete' phase function; a complete phase function is equal to $\psi(t, x, \xi) - (y \cdot \xi)$); q_k is the *amplitude function*, which should be infinitely smooth in (t, x, ξ), belong to $S^0_{\mathrm{ph}}(\mathbb{R}^n \times \mathbb{R}_n)$, and be equal to 0 for $x \notin U_k'$, where U_k' is a subdomain of U_k with the closure $\overline{U_k'} \subset U_k$. The functions ψ_k and q_k are assumed to satisfy the initial conditions

$$\psi_k(0, x, \xi) = x \cdot \xi , \tag{5.112}$$

$$q_k(0, x, \xi) = \chi_k(x) , \tag{5.113}$$

where $\chi_k(x)$ is a function from $C_0^\infty(\mathbb{R}^n)$ equal to 1 in a neighborhood of the support of φ_k and 0 outside a larger neighborhood with closure in U_k. Thus $\Phi_k(0)$ is a pseudodifferential operator, and, as can be easily verified, $\Phi(0) = I$. For $t \neq 0$ the operators $\Phi_k(t)$ and $\Phi(t)$ are no longer pseudodifferential, and the main difference is that now the phase function ψ_k is not equal to $x \cdot \xi$. Nevertheless, $\Phi_k(t)$ and $\Phi(t)$ are zero-order operators in the scale of Sobolev spaces in $\mathbb{R}^n$ and M respectively (see (5.104) and Proposition 5.8.1 below).

To compute ψ and the terms of the asymptotic expansion of the amplitude q_k, we apply the operator $D_t + A$ (where A is written in local coordinates after multiplying on the left by a suitable cutting off factor) to (5.111) under the integral sign. This is possible for $v \in C_0^\infty(\mathbb{R}^n)$, $\mathrm{supp} v \subset U_k$. We obtain the integral (the index k is omitted)

$$\int e^{i\psi}[\partial_t\psi \cdot q + D_t q + r](Fv)(\xi)d\xi \;, \tag{5.114}$$

where

$$r(t,x,\xi) \sim \sum_\nu \frac{1}{\nu!}(\partial_\xi^\nu a)(x,\partial_x\psi(t,x,\xi))D_y^\nu[q(t,y,\xi)e^{i\widetilde{\psi}(t,x,y,\xi)}]|_{y=x} \;,$$

and

$$\widetilde{\psi}(t,x,y,\xi) = \psi(t,y,\xi) - \psi(t,x,\xi) - \partial_x\psi(t,x,\xi)\cdot(y-x) \;.$$

Here we have used the asymptotic 'formula for action of a pseudodifferential operator on the exponential function' (more precisely, on the function of the form $e^{i\psi}q$) which can be derived by means of the stationary phase method. We do not give this formula separately; see, for example, (Shubin 1978) or (Trèves 1980, Vol. 2).

We equate to 0 the terms of degree $1, 0, \ldots$ of homogeneity in the amplitude function of the operator (5.114) and successively obtain, first, the *Hamilton–Jacobi equation* for the phase function ψ

$$\partial_t\psi(t,x,\xi) + a_0(x,\partial_x\psi(t,x,\xi)) = 0 \;, \tag{5.115}$$

and next, linear equations to determine the terms q_j of the asymptotic expansion $q \sim q_0 + q_1 + \ldots$ of the amplitude function. They all have the form

$$Lq_j(t,x,\xi) = f_j(t,x,\xi) \quad (j = 0,1,\ldots) \;, \tag{5.116}$$

where

$$L = \partial_t + \sum_{\nu=1}^{n}(\partial_{\xi_\nu}a_0)(x,\partial_x\psi(t,x,\xi))\partial_{x^\nu} + b(t,x,\xi) \;, \tag{5.117}$$

$$b(t,x,\xi) = ia_1(x,\partial_x\psi(t,x,\xi)) + \sum_{|\delta|=2}\frac{1}{\delta!}(\partial_\xi^\delta a_0)(x,\partial_x\psi(t,x,\xi))\partial_x^\delta\psi(t,x,\xi) \;.$$

The function f_0 is equal to 0. The further functions f_j are positively homogeneous in ξ of degrees $-j$ and can be assumed known if ψ, $q_0, \ldots, q_{j-1}$ are already defined. Equation (5.115) is accompanied by the initial condition (5.112) and equations (5.116) by the initial conditions following from (5.113): $q_0|_{t=0} = \chi$, $q_j|_{t=0} = 0$ $(j = 1,2,\ldots)$. Briefly speaking, ψ, q_0, $q_1, \ldots$ are successively uniquely determined for $|t| < \delta$ with the same number δ, which is limited by the condition of unique solvability of the Cauchy problem (5.115), (5.112). Having the asymptotic expansion for q, we construct q itself.

Thus, $\Phi(t)$ is constructed for $|t| < \delta$ with sufficiently small $\delta > 0$. One can prove the following

Proposition 5.8.1. *The operator $U(t) - \Phi(t)$ has infinitely smooth kernel for $|t| < \delta$.*

c. Suppose that $\rho(\lambda)$ is a function from $S(\mathbb{R})$ whose Fourier transform $(F\rho)(t)$ has a compact support. We put $đt = dt/2\pi$, and we introduce the operator

$$\rho(\lambda I - A) = \int e^{i\lambda t}(F\rho)(t)e^{-itA}đt\,. \tag{5.118}$$

Proposition 5.8.2. *The operator $\rho(\lambda I - A)$ has order $-\infty$. It is an integral operator with kernel $K_\rho(x, y, \lambda)$ which is an infinitely smooth function of (x, y, λ) and an entire analytic function of λ.*

Using the above formulas for the operator $\Phi(t)$ and the stationary phase method, one can obtain full asymptotic expansion of $K_\rho(x, y, \lambda)$ for $x = y$, $\lambda \to +\infty$ if $\operatorname{supp} F\rho \subset (-\delta, \delta)$. We recall that A is assumed to be a scalar elliptic pseudodifferential operator of the first order with positive principal symbol a_0 on a closed manifold M of dimension $n \geq 2$.

Theorem 5.8.3. *Under the above conditions we have*

$$K_\rho(x, x, \lambda) \sim \sum_{l=0}^{\infty} \omega_l(x)\lambda^{n-l-1} \quad (\lambda \to +\infty) \tag{5.119}$$

uniformly in $x \in M$. If $(F\rho)(t) = 1$ in a neighborhood of the origin, then the coefficients $\omega_l(x)$ do not depend on $\rho(\lambda)$. In particular, we have

$$\omega_0(x) = n \int_{a_0(x,\xi)\leq 1} đ\xi\,, \tag{5.120}$$

$$\omega_1(x) = -n(n-1) \int_{a_0(x,\xi)\leq 1} (\operatorname{sub} A)(x, \xi)đ\xi\,. \tag{5.121}$$

Unfortunately, we do not have here enough space to explain the algorithm of computation of further coefficients. We only mention that in the computations we use the values of the functions ψ, $q_0, \ldots$ and their derivatives at $t = 0$; to find these values, we do not need to know the solutions ψ, $q_0, \ldots$ to (5.115), (5.116) for $t \neq 0$. Of course, the formulas for $\omega_l(x)$ get more complicated as l grows. For details we refer to the paper of Duistermaat and Guillemin (1975), where, however, the (full) phase function of the operator $\Phi_k(t)$ is constructed in a slightly different way; it is linear in t there. However, we shall show below a simple connection between $\omega_l(x)$ and the functions (5.59) for $A = A^* > 0$.

If we do not assume that $(F\rho)(t) = 1$ in a neighborhood of the origin, then in the right-hand side of (5.120) we have to add the factor $(F\rho)(0)$, while the expression for $\omega_1(x)$ will depend on $\partial_t^j(F\rho)(0)$, $j = 0, 1, 2$.

The kernel $K_\rho(x, x, \lambda)$ rapidly decays in modulus as $\lambda \to -\infty$.

Integrating the expansion (5.119) with respect to x, we obtain the full asymptotic expansion of the trace of the operator $\rho(\lambda I - A)$.

Remark 5.8.4. Theorem 5.8.3 admits a generalization: one can replace e^{-itA} by the composition Be^{-itA}, where B is a polyhomogeneous pseudodifferential operator of zero order. In this case the coefficient in the leading term of the asymptotics has the form

$$n \int_{a_0(x,\xi)\le 1} b_0(x,\xi) đ\xi \,, \tag{5.122}$$

where b_0 is the principal symbol of B.

d. Theorem 5.8.3 allows us to obtain again and to complete some of the results of Sect. 5.5 in the case of a scalar $A = A^* > 0$. We first assume for simplicity that $m = \text{ord}A = 1$. Suppose that $\rho(\lambda)$ is a function whose Fourier transform $(F\rho)(t)$ belongs to $C_0^\infty(\mathbb{R})$ and is equal to 1 in a neighborhood of the origin and 0 outside a larger neighborhood lying inside $(-\delta, \delta)$. Next, let $\chi(\lambda)$ be a function from $C^\infty(\mathbb{R})$ equal to 0 for $\lambda < \lambda_1/2$ and 1 for $\lambda \ge \lambda_1$, where λ_1 is the first eigenvalue of A, and we put $\chi_z(\lambda) = \chi(\lambda)\lambda^z$. One can show that for $\text{Re}z > n$ the kernels of the operators A^{-z} and $\rho(\lambda I - A)$ are connected in the following way:

$$K_{-z}(x,x) = \int K_{-z}(x,x,\lambda)\chi_{-z}(\lambda)d\lambda + \dots \,,$$

where the dots denote a kernel which can be continued to an entire analytic function of z such that in any band $\{z : c_1 \le \text{Re}z \le c_2\}$ it can be majorized in modulus by a polynomial of z of order not exceeding $c_1 + \text{const}$ for large $|c_1|$.

Hence we can again deduce that $K_{-z}(x,x)$ continues to a meromorphic function on the complex plane with simple poles at known points; in addition, $\rho_k(x) = -\omega_k(x)$, while if A has order m, then

$$m\rho_k(x) = -\omega_k(x) \,. \tag{5.123}$$

Moreover, it can be seen that if $A = A^* > 0$, then the kernel $K_{-z}(x,x)$ does not exceed in modulus a polynomial of z of order not greater than $c_1 + \text{const}$ in any vertical band $\{z : c_1 \le \text{Re}z \le c_2\}$ outside some neighborhoods of the poles.

The last statement allows us to improve Theorems 5.7.1 and 5.7.2: the asymptotic expansion of the resolvent turns out to be valid outside the domain of the form (5.98) for $0 < \theta < 1$.

5.9. Generalizations and Variants

a. We first of all mention that after some natural amendments the results of Sects. 5.5–5.7 extend to matrix pseudodifferential operators with the same highest order m in all entries in the matrix. (Of course, in the formulas for the traces of operators the matrix entries are replaced by the traces of these matrices.) This is possible since, as we mentioned in 4.3a, a parametrix for the resolvent can be constructed for an elliptic with parameter pseudodifferential matrix operator of the indicated structure. (We also recall that elliptic with parameter pseudodifferential operators of more general structure reduce to them, see 4.3b.) We emphasize that here we do not have to impose any restrictions on multiplicities of eigenvalues of the principal symbol $a_0(x, \xi)$. The situation with the results of Sect. 5.8 is different. Here we have to require that the principal symbol be a Hermitian matrix with constant multiplicities of eigenvalues (this is necessary to construct the operator $\Phi(t)$). However Ivrii (1982, 1984) proposed a method which allows us to manage without Fourier integral operators and without those restrictions. For description of this method see (Rozenblyum et al. 1989); we shall say a couple of words about that after Theorem 6.1.7.

b. We shall now mention some results related to more general functions of operators.

In the paper of Pattisier (1977) there are considered functions of elliptic pseudodifferential operators $A \in \Psi^m(M)$ defined by a formula somewhat more general than (5.12):

$$\Phi(A) = \frac{1}{2\pi i} A^q \int_\Gamma \lambda^{-q} \Phi(\lambda) R_A(\lambda) d\lambda \,. \tag{5.124}$$

Here Γ is a contour close in shape to $\Gamma(\delta, \psi)$ and the insertion of $\lambda^{-q} A^q$ allows us to assume that $\Phi(\lambda)$ satisfies the inequality of the form $|\Phi(\lambda)| \leq C(1+|\lambda|)^h$ in $G^- \cup \Gamma$, where, in general, $h > 0$. In addition to the corresponding generalization of (5.3) we can establish that $\Phi(A)$ is a pseudodifferential operator of order not greater than mh with the principal symbol $\Phi(a_0(x, \xi))$. This is a direct generalization of Seeley's results for $\Phi(\lambda) = \lambda^z$.

Strichartz (1972) and Dunau (1977) consider in various ways functions $\Phi(A)$ of selfadjoint pseudodifferential operators $A \in \Psi^m(M)$, $m > 0$, elliptic in the sense of definition of 3.1b, under the assumption that $\Phi \in S^t(\mathbb{R})$. It is shown that then $\Phi(A)$ is a pseudodifferential operator from $\Psi^{mt}(M)$. Strichartz considers a scalar pseudodifferential operator A and obtains full asymptotic expansion of the local complete symbol of $\Phi(A)$. Dunau deals with A acting in sections of a finite-dimensional vector bundle over M, and this pseudodifferential operator has, in general, infinitely many eigenvalues of both signs. The variant $\Phi(\lambda) \in S^t_{\mathrm{ph}}(\mathbb{R})$ and the case of $m < 0$ are also considered. In these papers functions of a finite number of pairwise commuting formally selfadjoint pseudodifferential operators are also studied.

c. If A is a selfadjoint elliptic pseudodifferential operator of the first order on M, then for a broad class of functions $f(A)$ the following representation analogous to (5.118) is possible:

$$f(A) = \int (Ff)(t) e^{-itA} dt \,. \tag{5.125}$$

In the book of Taylor (1981, Chap.XII, §1) one can find some generalizations of the Strichartz theorem based on this formula and on an investigation of functions obtainable by (5.125).

§6. Spectral Properties of Elliptic Operators of Nonzero Order on M

Here we discuss the questions of localization of the spectrum, asymptotics of eigenvalues, completeness of root functions, and summability of the Fourier series with respect to them. As regards the questions that are presented in detail in (Rozenblyum et al. 1989), we restrict ourselves to brief formulations.

6.1. Selfadjoint Elliptic Pseudodifferential Operators

a. We consider a pseudodifferential operator A on M of order $m \neq 0$ formally selfadjoint relative to the scalar product in $H_0(M)$, and we first assume that A is scalar. Then A is a selfadjoint operator in $H_0(M)$ with domain $H_m(M)$ for $m > 0$. Its eigenfunctions belong to $C^\infty(M)$, and there exists an orthonormal basis $\{e_j(x)\}_1^\infty$ in $H_0(M)$ consisting of eigenfunctions. Any function $u(x) \in H_0(M)$ can be expanded into the Fourier series with respect to this basis with the convergence in $H_0(M)$:

$$u(x) = \sum_{j=1}^{\infty} c_j(u) e_j(x) \,, \tag{6.1}$$

where $c_j(u) = (u, e_j)_{0,M}$. The kernel $\operatorname{Ker} A$ is finite-dimensional. We may assume that $\operatorname{Ker} A = \{0\}$, without loss of generality in questions we are concerned with. We shall also assume that $m > 0$. We define the scalar product in $H_s(M)$ by the formula of the form (5.108) with s/m instead of s. Then A remains selfadjoint in $H_s(M)$, and if $Ae_j = \lambda_j e_j$, then $\{|\lambda_j|^{-s/m} e_j\}$ is an orthonormal basis of eigenfunctions in $H_s(M)$. If $u \in H_s(M)$, then the series (6.1) remains to be the Fourier series for $u(x)$ with respect to this basis in $H_s(M)$, and this series converges in $H_s(M)$ to $u(x)$. In addition, if $s < 0$, then the formula for the Fourier coefficients in (6.1) can be understood in the following way: $c_j(u) = (|A|^{s/m} u, |A|^{-s/m} e_j)_{0,M}$.

If $u \in H_t(M)$, where $t > s$, then the rate of convergence of the Fourier series in $H_s(M)$ can be estimated in the following way: for $l \leq l'$ we have

$$\left\| \sum_{j=l}^{l'} c_j(u) e_j \right\|_{s,M} \leq \max_{l \leq j \leq l'} |\lambda_j|^{\frac{s-t}{M}} \|u\|_{t,M} \,. \tag{6.2}$$

The question of convergence or summability of the series (6.1) in other metrics is very rich in content. We restrict ourselves to references: see the surveys (Alimov et al. 1976), (Bitsadze et al. 1987) and the paper (Hörmander 1966). We only add that some simple results can be obtained from what has been said above by means of the embedding theorems for Sobolev spaces.

b. We pass on to spectral asymptotics. The principal symbol $a_0(x, \xi)$ preserves the sign on $T^*M \backslash 0$ if $n \geq 2$. We assume that $a_0(x, \xi) > 0$ on $T^*M \backslash 0$.

Then A is semibounded from below, and all its eigenvalues, except possibly finitely many, are positive. Suppose that they are numbered, taking into account the multiplicities, in the nondecreasing order $\lambda_1 \le \lambda_2 \le \dots$. We denote by $N(\lambda)$ and $e(x,y,\lambda)$ the *distribution function of eigenvalues* and the *spectral function* of A:

$$N(\lambda) = N(\lambda; A) = \mathrm{card}\{j : \lambda_j \le \lambda\} = \max_{\lambda_j \le \lambda} j \; ; \tag{6.3}$$

$$e(x,y,\lambda) = \sum_{\lambda_j \le \lambda} e_j(x)\overline{e_j(y)} \, . \tag{6.4}$$

Clearly, we have

$$N(\lambda) = \int_M e(x,x,\lambda)dx \, . \tag{6.5}$$

We first give a rough result on the asymptotics of $e(x,x,\lambda)$ and $N(\lambda)$.

Theorem 6.1.1. *Suppose that $A = A^*$ is an elliptic pseudodifferential operator of order $m > 0$ with positive principal symbol $a_0(x,\xi)$ on an n-dimensional closed manifold M. Then*

$$e(x,x,\lambda) = \alpha_0(x)\lambda^{\frac{n}{m}} + o\left(\lambda^{\frac{n}{m}}\right) \quad (\lambda \to +\infty) \, ,$$

where

$$\alpha_0(x) = \int_{a_0(x,\xi)\le 1} đ\xi \, , \tag{6.6}$$

and

$$N(\lambda) = \alpha_0\lambda^{\frac{n}{m}} + o\left(\lambda^{\frac{n}{m}}\right) \, , \quad \textit{where} \quad \alpha_0 = \int_M \alpha_0(x)dx \, . \tag{6.7}$$

These asymptotic formulas can be obtained, using (to some extent) the results presented in Sects. 5.5–5.7, by three various methods connected with three classical Tauberian theorems. For simplicity of notation we shall explain this on the example of formula (6.7). Without loss of generality we may assume that $A > 0$, i.e. $\lambda_1 > 0$. The function $N(\lambda; A)$ is defined on $\overline{\mathbb{R}}_+$, is right continuous at the points of discontinuity, and (which is an essential condition in all Tauberian theorems) does not decrease.

1) The zeta-function $\zeta(z; A)$ is the *Mellin–Stieltjes transform* of $N(\lambda)$:

$$\zeta(z; A) = \int_0^\infty \lambda^{-z} dN(\lambda) \, . \tag{6.8}$$

This integral converges, $\zeta(z; A)$ is holomorphic for $\mathrm{Re}z > n/m$, and $\zeta(z; A) + \rho_0(z - n/m)^{-1}$ can be continuously extended on the half-plane $\{z : \mathrm{Re}z \ge n/m\}$ (see Sect. 5.5). Hence, by the Ikehara Tauberian theorem (see, for example, (Shubin 1978)) we obtain

$$\lambda^{-\frac{n}{m}}N(\lambda) \to -\frac{m}{n}\rho_0 \quad (\lambda \to +\infty) , \tag{6.9}$$

which, as can be easily seen, is equivalent to (6.7).

2) The function $\theta(t)$ is the *Laplace–Stieltjes transform* of $N(\lambda)$:

$$\theta(t) = \int_0^\infty e^{-t\lambda} dN(\lambda) . \tag{6.10}$$

This integral converges for $t > 0$, and $t^{n/m}\theta(t) \to \eta_0$ as $t \to +0$ (see Sect. 5.6). Hence, by the Karamata Tauberian theorem (see, for example, (Taylor 1981)) we obtain

$$\lambda^{-\frac{n}{m}}N(\lambda) \to \frac{\eta_0}{\Gamma(1+nm^{-1})} \quad (\lambda \to +\infty) , \tag{6.11}$$

which coincides with (6.9). This approach to finding spectral asymptotics is called the *method of parabolic equation.*

3) Suppose that q is a natural number such that $mq > n$. The trace $\mathrm{tr}[R_A(\lambda)]^q$ coincides with the *Stieltjes transform* (it would be more correct to say Stieltjes–Stieltjes transform) of order q of $N(\lambda)$:

$$S_q(\zeta) = \int_0^\infty (\zeta - \lambda)^{-q} dN(\lambda) . \tag{6.12}$$

This integral converges, in particular, for $\zeta \in \mathbb{R}_-$ large in modulus, and, as $\zeta \to \infty$ along $\mathbb{R}_-$, we have

$$(-\zeta)^{q-\frac{n}{m}} S_q(\zeta) \to c_0^{(q)} , \tag{6.13}$$

where $c_0^{(q)}$ is the integral of (5.100) over M. Hence, by the Hardy–Littlewood Tauberian theorem (see, for example, (Subkhankulov 1976) or (Widder 1941)) it follows that

$$\lambda^{-\frac{n}{m}}N(\lambda) \to \left[\frac{n}{m}B\left(\frac{n}{m}, q - \frac{n}{m}\right)\right]^{-1} c_0^{(q)} , \tag{6.14}$$

where $B(\alpha, \beta)$ is the Euler beta-function, and here the right-hand side coincides with the right-hand side of (6.9), which can be easily directly verified. This approach to finding the asymptotics of the spectrum is called the *resolvent method.*

Formula (6.7) is equivalent to the following asymptotic formula for the eigenvalues λ_j:

$$\lambda_j = \beta_0 j^{\frac{m}{n}} + o\left(j^{\frac{m}{n}}\right) , \quad \text{where} \quad \beta_0 = \alpha_0^{-\frac{n}{m}} . \tag{6.15}$$

In particular, we find that in (5.39) we can replace $o\left(j^{-\frac{1}{n}+\varepsilon}\right)$ by $O\left(j^{-\frac{1}{n}}\right)$. As a corollary, we can replace the right-hand side of (5.49) by $O(j^{-\frac{\tau}{n}})$.

Formulas (6.6) and (6.7) and their analogues for matrix elliptic pseudodifferential operators and elliptic pseudodifferential operators acting in sections of vector bundles were obtained in full generality by Seeley (1967a) by means of the Ikehara Tauberian theorem. A great number of authors dealt with refinement of the estimate of the remainder in these formulas (and in analogous formulas for elliptic boundary problems) using strengthened versions of the Tauberian theorem mentioned above, so called 'Tauberian theorems with remainder' (they are systematically presented in the book of Subkhankulov (1976)). The most precise result for scalar pseudodifferential operators was obtained by Hörmander (1968):

Theorem 6.1.2. *Under assumptions of Theorem 6.1.1, for* $n \geq 2$ *we have*

$$e(x,x,\lambda) = \alpha_0(x)\lambda^{\frac{n}{m}} + O\left(\lambda^{\frac{n-1}{m}}\right) \quad (\lambda \to +\infty) \tag{6.16}$$

uniformly in x, *and*

$$N(\lambda) = \alpha_0\lambda^{\frac{n}{m}} + O\left(\lambda^{\frac{n-1}{m}}\right) \quad (\lambda \to +\infty)\,. \tag{6.17}$$

The latter relation is equivalent to the following asymptotic formula for λ_j:

$$\lambda_j = \beta_0 j^{\frac{n}{m}} + O\left(j^{\frac{n-1}{m}}\right)\,, \tag{6.18}$$

where β_0 is the same as in (6.15).

Hörmander used the *method of hyperbolic equation*: he investigated the beginning of the expansion (5.119) and employed an argument which is tantamount to applying the following Tauberian theorem (cf. Ivrii (1982, 1984)). For a function $\rho(\lambda)$ from $S(\mathbb{R})$ whose Fourier transform has compact support, we adopt the notation

$$\Psi_\rho(\lambda) = \int_0^\infty \rho(\lambda-\mu)dN(\mu)\,, \tag{6.19}$$

assuming that $N(\lambda)$ is a nondecreasing piecewise constant function and that this integral absolutely converges.

Theorem 6.1.3. *Suppose that there is a* $\delta > 0$ *such that for any function* $\rho(\lambda)$ *from* $S(\mathbb{R})$ *with* $\operatorname{supp}(F\rho) \subset [-\delta,\delta]$ *we have*

$$\Psi_\rho(\lambda) = O(\lambda^{n-1}) \quad (\lambda \to +\infty) \tag{6.20}$$

and for a certain function $\rho_1(\lambda)$ *from* $S(\mathbb{R})$ *with* $\operatorname{supp}(F\rho_1) \subset [-\delta,\delta]$ *and* $(F\rho_1)(0) = 1$ *we have*

$$\Psi_{\rho_1}(\lambda) = n\alpha\lambda^{n-1} + O(\lambda^{n-2}) \quad (\lambda \to +\infty)\,, \tag{6.21}$$

where $\alpha > 0$, $n > 1$ *(*n *does not have to be an integer). Then*

$$N(\lambda) = \alpha\lambda^n + O(\lambda^{n-1}) \quad (\lambda \to +\infty)\,. \tag{6.22}$$

This theorem combined with Theorem 5.8.3 gives the required results (6.16) and (6.17) if $m = 1$, and to this case we can easily reduce the case of arbitrary $m > 0$.

The fact that the estimate (6.17) is the best possible can be verified on the example of the operator $-\Delta$ on the sphere S^n (see page 40). The multiplicities of its eigenvalues $k(k+n-1)$ are so large that in this case the estimate of the remainder term in the asymptotics of $N(\lambda)$ cannot be improved. It is also impossible to single out the second term of the asymptotics from the remainder term.

The second term in the asymptotics of $N(\lambda)$ can be singled out under an additional assumption about the principal symbol $a_0(x, \xi)$ of an elliptic pseudodifferential operator $A = A^* > 0$, more precisely, about the Hamiltonian flow defined by the principal symbol. To formulate this assumption, we consider the *bicharacteristics* of $a_0(x, \xi)$, i.e. the solutions $x = x(t)$, $\xi = \xi(t)$ of the system

$$\dot{x} = \partial_\xi a_0(x, \xi) \, , \quad \dot{\xi} = -\partial_x a_0(x, \xi) \, . \tag{6.23}$$

It can be easily verified that $a_0(x, \xi)$ is the first integral of this system:

$$\partial_t a_0[x(t), \xi(t)] \equiv 0 \, . \tag{6.24}$$

Therefore, the bicharacteristics belong to the bundle

$$S^*M = \{(x, \xi) \in T^*M : a_0(x, \xi) = 1\} \, , \tag{6.25}$$

if they belong to S^*M for some t. We will consider the bicharacteristics lying on S^*M.

We call the point $(y, \eta) \in S^*M$ *periodic* if the bicharacteristic that leaves it at instant $t = 0$ returns to this point at some $t = T > 0$; more precisely, we then call (y, η) a T-periodic point. By uniqueness of solutions to the Cauchy problem for system (6.23), we then have $x(t+T) \equiv x(t)$, $\xi(t+T) \equiv \xi(t)$.

On T^*M we have an invariantly defined density $dx d\xi$, and it defines the measure and the integral on S^*M. Namely, if $f(x, \xi)$ is a function given on S^*M, then extending it to T^*M as a function positive homogeneous of degree 0, we set

$$\int\!\!\int_{S^*M} f(x, \xi) dS dx \cdot \int_0^1 r^{n-1} dr = \int\!\!\int_{a_0(x,\xi) \le 1} f(x, \xi) dx d\xi \, ,$$

i.e.

$$\int\!\!\int_{S^*M} f(x, \xi) dS dx = n \int\!\!\int_{a_0(x,\xi) \le 1} f(x, \xi) dx d\xi \, . \tag{6.26}$$

In particular, if V is a subset of S^*M, then its measure $\mu(V)$ is defined as the integral over S^*M of its characteristic function (and V is measurable if this integral exists).

We put

$$\alpha_{-1} = \frac{1}{(2\pi)^n} \int\int_{S^*M} (\operatorname{sub} A)(x,\xi) dS dx \; . \tag{6.27}$$

The following theorem is due to Duistermaat and Guillemin (1975).

Theorem 6.1.4. *Suppose that the measure of the set of all periodic points of S^*M is equal to* 0. *Then*

$$N(\lambda) = \alpha_0 \lambda^{\frac{n}{m}} - \alpha_{-1} \lambda^{\frac{n-1}{m}} + o\left(\lambda^{\frac{n-1}{m}}\right) \tag{6.28}$$

as $\lambda \to +\infty$.

A proof is also given in Hörmander's book (1985b); it employs the beginning of the expansion (5.119) as well as Remark 5.8.4. We note that the argument can be given in the form of the application of the following Tauberian theorem (cf. Ivrii (1982, 1984); the initial assumptions about $N(\lambda)$ and the notation are the same as in Theorem 6.1.3).

Theorem 6.1.5. *Suppose that there is a* $\delta > 0$ *such that for any function* $\rho(\lambda)$ *from* $S(\mathbb{R})$ *with* $\operatorname{supp}(F\rho) \subset [-\delta, \delta]$ *we have*

$$\Psi_\rho(\lambda) = n\alpha(F_\rho)(0)\lambda^{n-1} + o(\lambda^{n-1}), \tag{6.29}$$

and suppose that for a function $\rho_1(\lambda)$ *from* $S(\mathbb{R})$ *with* $\operatorname{supp}(F\rho_1) \subset [-\delta, \delta]$ *and* $(F\rho_1)(t) = 1$ *in a neighborhood of the origin we have*

$$\Psi_\rho(\lambda) = n\alpha\lambda^{n-1} + (n-1)\gamma\lambda^{n-2} + o(\lambda^{n-2}) \qquad (\lambda \to +\infty) \; , \tag{6.30}$$

where $\alpha > 0$, $n > 1$, *and* $\gamma \in \mathbb{R}$. *Moreover, suppose that for any function* $\rho(\lambda)$ *from* $S(\mathbb{R})$ *whose Fourier transform has compact support not containing* 0 *we have*

$$\Psi_\rho(\lambda) = o(\lambda^{n-1}) \; . \tag{6.31}$$

Then

$$N(\lambda) = \alpha\lambda^n + \gamma\lambda^{n-1} + o(\lambda^{n-1}) \qquad (\lambda \to +\infty) \; . \tag{6.32}$$

Besides this theorem, some essential properties of Fourier integral operators find application in the proof of Theorem 6.1.4; we have no space to dwell on that here.

We note that if A is a scalar elliptic differential operator (of even order), then the coefficient α_{-1} is equal to 0, since it is the integral of a function odd in ξ.

In some cases the estimate of the remainder term in (6.28) can be improved. In this direction there are, in particular, the results of Bérard (1977) and Volovoj (1987). We only mention that in the case of the operator $(-\Delta)^{1/2}$ on the n-dimensional torus $\mathbb{T}^n$ (where Δ is the Laplace–Beltrami operator) we have the following result:

$$N(\lambda) = \alpha_0 \lambda^n + O(\lambda^{n-1-\alpha}) \; , \tag{6.33}$$

where $\alpha = \alpha(n) \geq 0$ and $\alpha(n) \to 1$ as $n \to \infty$ (see, for example, (Beals et al. 1983)).

c. The *Riesz means of order* α ($\alpha > 0$) of the function $N(\lambda)$ are defined by the following equality:

$$N^{(\alpha)}(\lambda) = \int_0^\lambda \left(1 - \frac{\mu}{\lambda}\right)^\alpha dN(\mu) \quad (\lambda > 0) . \tag{6.34}$$

This function is the result of some 'smoothing out' of $N(\lambda)$, the better the larger α is, and this suggests that $N^{(\alpha)}(\lambda)$ could have asymptotics with number of terms growing as α grows. This in fact is true, but we want to emphasize that the key point is not the smoothing out itself but the fact that, say, the Stieltjes transform of $N(\lambda)$, i.e. the trace of the resolvent $R_A(\lambda)$, has full asymptotic expansion in the powers of λ. This is typical just for elliptic operators.

We restrict ourselves to the case $\alpha \in \mathbb{N}$, $m > n$. Put

$$J_1(\alpha) = \{j \in \mathbb{Z}_+ : j \leq \alpha\} \cup \{j = n + lm : l \in \mathbb{Z}_+, j < \alpha + 1\} , \tag{6.35}$$

$$J_2(\alpha) = \{l \in \mathbb{N} : n + lm < \alpha + 1\} . \tag{6.36}$$

Let $\theta \in (0, 1)$. We choose this number sufficiently close to 1 so that $j \in J_1(\alpha) \Rightarrow j < \theta(\alpha + 1)$.

Theorem 6.1.6 (Agranovich 1987). *Let $\alpha \in \mathbb{N}$ and suppose that $A = A^*$ is an elliptic pseudodifferential operator of order $m > n$ with positive principal symbol. Then*

$$N^{(\alpha)}(\lambda) = \sum_{j \in J_1(\alpha)} h_{\alpha_j} \lambda^{\frac{n-j}{m}} + \sum_{l \in J_2(\alpha)} \widetilde{h}_{\alpha l} \lambda^{-l} \log \lambda + O\left(\lambda^{\frac{n - \theta(\alpha+1)}{m}}\right) \tag{6.37}$$

as $\lambda \to +\infty$.

A method of the computation of the coefficients is shown below.

To obtain this asymptotic formula, one can use the following inequality (Sadov 1987). Let $N(\lambda)$ be a piecewise constant nondecreasing function on $\overline{\mathbb{R}}_+$, and let $S(\zeta) = S_1(\zeta)$ be its Stieltjes transform of order 1 (see (6.12) for $q = 1$); we assume that it is defined for $\zeta \notin \overline{\mathbb{R}}_+$. Then if $\Gamma(\zeta)$ is a rectifiable oriented curve linking the points $\overline{\zeta} = \lambda - i\eta$ and $\zeta = \lambda + i\eta$ ($\lambda > 0$, $\eta > 0$) and having no common points with $\overline{\mathbb{R}}_+$, then

$$\left| N^{(\alpha)}(\lambda) - \frac{1}{2\pi i} \int_{\Gamma(\zeta)} S(z) \left(1 - \frac{z}{\lambda}\right)^\alpha dz \right| \leq \frac{1}{\alpha\pi} \frac{\eta^{\alpha+1}}{\lambda^\alpha} |S(\zeta)| . \tag{6.38}$$

This inequality can be regarded as a generalization of the *Pleijel inequality* (Pleijel 1963) which corresponds to the case $\alpha = 0$. In this case $N^{(\alpha)}(\lambda)$

is replaced by $N(\lambda)$, and $1/\alpha\pi$ on the right-hand side can be replaced by $(\pi^{-2}+2^{-2})^{1/2}$ (Sadov 1987)[5].

To obtain (6.37), we have to substitute the terms of the asymptotic expansion of the trace of the resolvent $R_A(\zeta)$ in place of $S(\zeta)$ in (6.38), assuming that $\Gamma(\zeta)$ is the arc of the circle with the center at the origin. We recall that this expansion in the case in question is valid outside the domain (5.98). We obtain (6.37) with the coefficients which at first have the form of integrals along a nonclosed contour. We can close the contour without violating the estimate of the remainder term in (6.37). The final formulas for the coefficients have the following form (see (5.71); we take the clockwise orientation in the integrals):

$$h_{\alpha j} = \frac{c_j}{2\pi \mathrm{i}} \int_{|z|=1} (-z)^{\frac{n-j}{m}-1}(1-z)^\alpha dz \quad (j \notin J)\,, \tag{6.39}$$

$$h_{\alpha j} = \frac{c_j}{2\pi \mathrm{i}} \int_{|z|=1} (-z)^{-l-1}(1-z)^\alpha dz + \frac{\widetilde{c}_l}{2\pi \mathrm{i}} \int_{|z|=1} (-z)^{-l-1}(1-z)^\alpha \log(-z) dz \tag{6.40}$$

$$(j = n + lm\,, \quad l \in \mathbb{Z}_+)\,,$$

$$\widetilde{h}_{\alpha l} = \frac{\widetilde{c}_l}{2\pi \mathrm{i}} \int_{|z|=1} (-z)^{-l-1}(1-z)^\alpha dz \quad (l \in \mathbb{N})\,. \tag{6.41}$$

Analogous results can be obtained for the Riesz means $\mathrm{e}^{(\alpha)}(x,x,\lambda)$ of the spectral function $e(x,x,\lambda)$.

d. We turn to matrix elliptic selfadjoint pseudodifferential operators of order $m > 0$. All what has been said at the beginning of this section about the basis of eigenfunctions remains valid; of course, they will be now vector-valued functions. The principal symbol $a_0(x,\xi)$ can have positive and negative eigenvalues, and then A has infinite sets of positive and negative eigenvalues. We denote by $N_+(\lambda)$ and $N_-(\lambda)$ $(\lambda > 0)$ the number of eigenvalues of A on $[0,\lambda]$ and $[-\lambda,0)$, respectively, taking into account the multiplicities, and by $e_+(x,y,\lambda)$ and $e_-(x,y,\lambda)$ the kernels of the orthogonal projections in $H_0(M)$ on the invariant subspaces corresponding to these eigenvalues. Next, we denote by $\lambda_j^+(x,\xi)$ and $\lambda_j^-(x,\xi)$, respectively, all positive and negative eigenvalues of the principal symbol $a_0(x,\xi)$, and we put

$$\alpha_0^\pm(x) = \sum \int_{\pm\lambda_j^\pm(x,\xi)\le 1} đ\xi\,, \quad \alpha_0^\pm = \int_M \alpha_0^\pm(x) dx\,. \tag{6.42}$$

Theorem 6.1.2 can be generalized in the following way (Ivrii (1982, 1984)).

Theorem 6.1.7. *If* $\lambda \to +\infty$, *then*

[5] In (Sadov 1987) another generalization of the Pleijel inequality was also obtained: for $N(\lambda)$ and $S_q(\xi)$ instead of $S(\xi)$. See also (Agranovich 1987).

$$e_\pm(x,x,\lambda) = \alpha_0^\pm(x)\lambda^{\frac{n}{m}} + O\left(\lambda^{\frac{n-1}{m}}\right) \tag{6.43}$$

uniformly in X and

$$N_\pm(\lambda) = \alpha_0^\pm(x)\lambda^{\frac{n}{m}} + O\left(\lambda^{\frac{n-1}{m}}\right) . \tag{6.44}$$

For a brief description of a proof see (Rozenblyum et al. 1989). We only mention that Ivrii, using the tools of the theory of hyperbolic equations, first shows that the singularity of the distribution $U(t,x,x)$ for $t=0$ is 'normal', where $U(t,x,y)$ is the Schwartz kernel of the operator e^{-itA}. This means that for any j and α, $(tD_t)^j(tD_x)^\alpha U(t,x,x) \in H_s((-t_0,t_0)\times M)$ for some $t_0>0$ and s. Next, Ivrii computes this singularity (finds an asymptotic expansion) by the method of successive approximations. The changes of multiplicities of the eigenvalues of the principal symbol occur on a set of zero measure and do not prevent these calculations. Additionally, it is useful to write the orthogonal projections on the subspaces invariant with respect to A, that correspond to all positive and all negative eigenvalues, in the form of pseudodifferential operators. Also two-term asymptotic formulas for $N_\pm(\lambda)$ are derived under supplementary assumptions about the principal symbol. The results remain valid for elliptic pseudodifferential operators acting in sections of vector bundles.

6.2. Weak Perturbations of Selfadjoint Elliptic Pseudodifferential Operators. Here we consider an elliptic pseudodifferential operator A of order $m>0$ on M, $n=\dim M \geq 2$, assuming that $\operatorname{ord}(A-A^*)<m$. Consequently, A has a real principal symbol $a_0(x,\xi)$ and

$$A = A_0 + A_1 , \tag{6.45}$$

where $A_0 = \operatorname{Re} A = (A+A^*)/2$ is selfadjoint and has the same principal symbol $a_0(x,\xi)$, while A_1 has order $m_1<m$, $m-m_1\in\mathbb{N}$. We first assume that A is scalar and $a_0(x,\xi)>0$ on $T^*M\backslash 0$. Without loss of generality, we may assume that A is invertible and $A_0>0$. We recall that we enumerate $\lambda_j=\lambda_j(A_0)$ by natural numbers in the nondecreasing order and that according to (6.18) we have

$$\lambda_j = \beta_0 j^p + O(j^r) , \tag{6.46}$$

where

$$\beta_0 > 0 , \quad p = \frac{m}{n} > r = \frac{m-1}{n} . \tag{6.47}$$

The operator A_1 is subordinated to the q-th power of A_0 in the sense that

$$\|A_1 A_0^{-q}\| = b < \infty , \tag{6.48}$$

where

$$q = \frac{m_1}{m} . \tag{6.49}$$

A natural abstract model of this situation is an operator A of the form (6.45) in a Hilbert space H, where A_0 is a selfadjoint positive operator with discrete spectrum and the asymptotics (6.46) of eigenvalues, while A_1 is an operator subordinated to the q-th power of A_0 for some $q < 1$. It is convenient to formulate almost all assertions of this section in abstract terms first. We note that the case $-\infty \leq q < 0$ is quite meaningful (in Example 2.2.3, if we replace A by $-A$, we have $q = -\infty$). Condition (6.46) will be used only after Theorem 6.2.3.

a. We introduce the following notation:

$$O_r(\zeta) = O(\zeta; r) \text{ is an open disk of radius } r \text{ with center at } \zeta\,;$$

$$O_r = O_r(0)\,; \quad \sigma = \operatorname{Re}\lambda\,, \quad \tau = \operatorname{Im}\lambda\,;$$

$$d(\lambda) \text{ is the distance of } \lambda \text{ from the spectrum } \sigma(A_0)\,.$$

Proposition 6.2.1. *Under condition* (6.48), *the spectrum of an operator A in a Hilbert space H is discrete and is contained in the union of the disks $O(\lambda_j; b\lambda_j^q)$. For sufficiently small $\delta > 0$, outside the union Ω_δ of the disks $O(\lambda_j; b(1+\delta)\lambda_j^q)$ we have*

$$\|R_A(\lambda)\| \leq \frac{1+\delta}{\delta}\|R_{A_0}(\lambda)\| = \frac{1+\delta}{\delta}\frac{1}{d(\lambda)}\,. \tag{6.50}$$

The proof reduces to the use of the following simple argument. First, at the points $\lambda \in \varrho(A_0)$ such that $\|A_1 R_{A_0}(\lambda)\| \leq \theta < 1$ there exists

$$R_A(\lambda) = R_{A_0}(\lambda)[I + A_1 R_{A_0}(\lambda)]^{-1}, \quad \text{and} \quad \|R_A(\lambda)\| \leq (1-\theta)^{-1}\|R_{A_0}(\lambda)\|\,. \tag{6.51}$$

Second, for any κ and any $\lambda \in \varrho(A_0)$ we have

$$\|A_0^\kappa R_{A_0}(\lambda)\| = \sup_j \lambda_j^\kappa |\lambda_j - \lambda|^{-1}\,, \tag{6.52}$$

since $A_0^\kappa R_{A_0}(\lambda)$ is a normal operator with eigenvalues $\lambda_j^\kappa(\lambda_j - \lambda)^{-1}$.

From this proposition we can deduce

Theorem 6.2.2. *Suppose that* (6.48) *is satisfied with some $q < 1$. Additionally, if $q < 0$, we assume that $\lambda_{j+1}\lambda_j^{-1} \to 1$ $(j \to \infty)$. Then there is a $C(q) > 0$ such that for $0 < \theta < 1$ and $b' > b\theta^{-1}C(q)$ the spectrum $\sigma(A)$ is contained inside a 'parabolic neighborhood' of the ray $\mathbb{R}_+$:*

$$\Omega(\mathbb{R}_+) = O_{r_0} \cup \left\{\lambda : \sigma > \frac{1}{2}r_0,\ |\tau| < b'C(q)\sigma^q\right\}\,, \tag{6.53}$$

where $r_0 > 0$ is sufficiently large, and moreover, outside $\Omega(\mathbb{R}_+)$ we have the relations $\|A_1 R_{A_0}(\lambda)\| \leq \theta$ and (6.51).

Theorem 6.2.3. *Suppose that condition* (6.48) *is satisfied for some* $q < 1$ *and* $A_0 \in \mathfrak{S}_\gamma$ *for some* $\gamma > 0$. *Then the system of root vectors of* A *is complete in* H.

This theorem is a corollary to a more general completeness theorem that we give in 6.4a.

We shall now discuss some improvements of this result. To this end we have to decompose the domain $\Omega(\mathbb{R}_+)$ by means of vertical segments Γ_l ($l = 1, 2, \ldots$) into bounded parts $\Omega^{(0)}$, $\Omega^{(1)}, \ldots$ in such a way that on Γ_l not only there are no points of $\sigma(A)$ and $\sigma(A_0)$, but also some optimal estimates for the resolvents $R_A(\lambda)$ and $R_{A_0}(\lambda)$ and their difference are satisfied. We now assume that conditions (6.48) and (6.46) are satisfied; for simplicity we impose on r the condition $p - 1 \leq r$, which is not restrictive for us. The abscissas α_l of the segments Γ_l can be chosen in intervals

$$\Delta_l = (l^{\varrho p},\ (l+1)^{\varrho p}) \tag{6.54}$$

assuming that

$$\varrho \geq (p - r)^{-1} . \tag{6.55}$$

This can be done quite easily if

$$p(1 - q) > 1 , \tag{6.56}$$

since in this case there exist points $\alpha_l \in \Delta_l$ for sufficiently large l that do not belong to any of the disks $O(\lambda_j; b(1+\delta)\lambda_j^q)$. In addition to (6.51) we obtain on Γ_l the estimate

$$\|R_{A_0}(\lambda)\| = d^{-1}(\lambda) \leq C l^{-\varrho(p-1)} . \tag{6.57}$$

The case

$$p(1 - q) \leq 1 \tag{6.58}$$

is considerably more difficult, since in this case in general we do not have 'gaps' between the disks. In this case we have to use quite deep results from the theory of functions of complex variable and some facts concerning the determinants $\det[I + T]$, where T is a trace class operator. Imposing on ϱ a supplementary condition

$$\varrho > [p(1 - q)]^{-1} \tag{6.59}$$

and putting

$$\gamma_0 = p^{-1} + q - 1 , \tag{6.60}$$

we can show that there exist points $\alpha_l \in \Delta_l$ such that on Γ_l

$$\|R_A(\lambda)\| \leq C_1 C_2^{\alpha_l^\gamma} \alpha_l^{-q} ,$$

$$\|R_A(\lambda)\| \leq C_1 \alpha_l^{\gamma - q} , \quad \|A_1 R_A(\lambda)\| \leq C_1 C_2^{\alpha_l^\gamma} , \tag{6.61}$$

where C_1, C_2 are some positive constants and $\gamma = \gamma_0$.

Let $\mathcal{T}_l$ be the boundary of $\Omega^{(l)}$ with positive orientation. We put

$$P_l = -\frac{1}{2\pi \mathrm{i}} \int_{\mathcal{T}_l} R_A(\lambda) d\lambda \,, \quad Q_l = -\frac{1}{2\pi \mathrm{i}} \int_{\mathcal{T}_l} R_{A_0}(\lambda) d\lambda \,. \tag{6.62}$$

These are, so called, *Riesz projections* corresponding to the contours $\mathcal{T}_l$ and the operators A and A_0, respectively. Both systems $\{P_l\}$ and $\{Q_l\}$ are disjunctive ($P_l P_k = \delta_{lk} P_l$, $Q_l Q_k = \delta_{lk} Q_l$). The subspaces $P_l H = \mathcal{L}_l$ and $Q_l H = \mathfrak{M}_l$ are finite-dimensional and invariant with respect to A and A_0, respectively. The operators Q_l are orthogonal projections, and $\{\mathfrak{M}_l\}_0^\infty$ is an *orthogonal basis of subspaces* of H, i.e. for any $f \in H$ we have

$$f = \sum_{l=0}^{\infty} Q_l f \tag{6.63}$$

and here the terms are pairwise orthogonal. The series

$$\sum_{l=0}^{\infty} P_l f \,, \tag{6.64}$$

so far formal, is obtained from the Fourier series of the vector f with respect to the system of root vectors of A by an arrangement of parentheses such that the l-th pair of parentheses selects terms corresponding to all eigenvalues of A that lie in $\Omega^{(l)}$. We also put

$$P_{l\gamma}(t) = -\frac{1}{2\pi \mathrm{i}} \int_{\mathcal{T}_l} \mathrm{e}^{-\lambda^\gamma t} R_A(\lambda) d\lambda \,, \tag{6.65}$$

$$Q_{l\gamma}(t) = -\frac{1}{2\pi \mathrm{i}} \int_{\mathcal{T}_l} \mathrm{e}^{-\lambda^\gamma t} R_{A_0}(\lambda) d\lambda \,, \tag{6.66}$$

where $\lambda^\gamma = \mathrm{e}^{\mathrm{i}\gamma \arg\lambda} |\lambda|^\gamma$ for $|\arg\lambda| < \pi$, $l \geq 1$. For $l = 0$ the factor $\mathrm{e}^{-\lambda^\gamma t}$ under the integral sign may be replaced by 1.

b. We can now formulate the main theorems on summability of the Fourier series with respect to the root vectors of A. In these theorems we assume that the spaces $\mathcal{L}_l$ are constructed as above with the corresponding restrictions on ϱ. The larger $p(1-q)$ is, the stronger the results are.

Theorem 6.2.4. *Suppose that conditions* (6.46) *and* (6.48) *are satisfied with some* $q < 1$, *and suppose that* $p(1-q) < 1$. *If* $\gamma > \gamma_0 = p^{-1} - (1-q)$, *then for any* $f \in H$ *the series*

$$\sum_{l=0}^{\infty} P_{l\gamma}(t) f \tag{6.67}$$

converges in H *and its sum* $f(t)$ *converges to* f *as* $t \to +0$.

By definition, this means that $\{\mathcal{L}_l\}_0^\infty$ is an *Abel basis of order* γ *of subspaces* in H, and we shall call the system of root vectors of A an *Abel basis of order* γ *with parentheses* in H.

In the proof of the theorem we use the estimates of the differences $P_{l\gamma}(t) - Q_{l\gamma}(t)$ following from (6.61) and the Hilbert formula

$$R_{A_0}(\lambda) - R_A(\lambda) = R_{A_0}(\lambda) A_1 R_A(\lambda) .$$

Theorem 6.2.5. *Suppose that conditions* (6.46) *and* (6.48) *are satisfied with some* $q < 1$, *and suppose that* $p(1-q) = 1$. *For* $q < 0$ *we additionally assume that*

$$\|A_0^{-q/2} A_1 A_0^{q/2}\| < \infty \tag{6.68}$$

and $\varrho p > 1$. *Then the series* (6.64) *converges to* f *for any* $f \in H$ *and the convergence is preserved after any rearrangement of the terms.*

By definition, this means that $\{\mathcal{L}_l\}_0^\infty$ is an *unconditional basis of subspaces* in H. The system of root vectors in such a situation is called an *unconditional basis with parentheses* in H.

We also mention that the system $\{H^{(l)}\}_0^\infty$ of subspaces in H is called a *Riesz basis of subspaces* in H if there is a linear bounded invertible operator B such that $\{BH^{(l)}\}_0^\infty$ is an orthogonal basis of subspaces in H. The system $\{H^{(l)}\}$ is a Riesz basis if and only if it is an unconditional basis (Gel'fand's theorem, see the references in (Gohberg and Krejn 1965)). Thus, Theorem 6.2.5 gives sufficient conditions for the system $\{\mathcal{L}_l\}_0^\infty$ to be a Riesz basis.

Theorem 6.2.6. *Suppose that conditions* (6.46) *and* (6.48) *are satisfied with some* $q < 1$, *and suppose that* $p(1-q) > 1$. *Assume that* ϱ *satisfies conditions* (6.55),

$$\begin{aligned} 2\varrho[p(1-q) - 1] > 1 \quad &\textit{for} \quad -p^{-1} < q < 1 , \\ 2\varrho p > 1 \quad &\textit{for} \quad q \le -p^{-1} . \end{aligned} \tag{6.69}$$

Then

$$\sum \|P_l - Q_l\|^2 < \infty . \tag{6.70}$$

By definition, this means that $\{\mathcal{L}_l\}_0^\infty$ is a *Bary basis of subspaces* in H or the basis of subspaces *quadratically close to orthogonal.* In such a situation the system of the root vectors of A is called a *Bary basis with parentheses* in H. If a system is a Bary basis with parentheses, then it is a Riesz basis with parentheses.

c. From these theorems we directly obtain the corresponding results for a pseudodifferential operator A (see the beginning of this section) in $H = H_s(M)$ for any s (with the scalar product of the form (5.108) with $A_0^{1/m}$ instead of $|A|$). We only need to take into account that in this case

$$p(1-q) = \frac{m - m_1}{n} . \tag{6.71}$$

The corresponding spaces $\mathcal{L}_l$ lie in $H_\infty(M)$; in the cases $m - m_1 < n$, $m - m_1 = n$, and $m - m_1 > n$ they form, respectively, an Abel basis of order $\gamma > [n - (m - m_1)]/m$, a Riesz basis, and a Bary basis of subspaces in any $H_s(M)$.

We also give estimates of the rate of convergence of the series $\sum P_l f$ (cf. (6.2)) and $\sum(P_l - Q_l)f$ taking into account the smoothness of f. These estimates can be also formulated in abstract situation, but we present them for elliptic operators. We put

$$\eta = \varrho\left[\frac{m - m_1}{n} - 1\right], \quad \omega = \frac{s_1}{n}. \tag{6.72}$$

Theorem 6.2.7. *For $f \in H_{s+s_1}(M)$, $s_1 \geq 0$, and $k \leq k'$ we have*

$$k^{\eta+\omega\varrho}\left\|\sum_k^{k'}(P_l - Q_l)f\right\|_s \leq C_1\|f\|_{s+s_1} \tag{6.73}$$

if $m - m_1 > n$, and

$$k^{\omega\varrho}\left\|\sum_k^{k'} P_l f\right\|_s \leq C_2\|f\|_{s+s_1} \tag{6.74}$$

if $m - m_1 \geq n$, where C_1, C_2 do not depend on k, k', f.

Here we assume that ϱ satisfies the conditions of Theorems 6.2.6 and 6.2.5, respectively.

Thus the rate of convergence of the series $\sum P_l f$ and $\sum(P_l - Q_l)f$ is the higher, the higher is the additional smoothness of f. Moreover, the rate of convergence of the latter series is the higher, the lower the order of the operator $A - A_0$ is for fixed orders of A and A_0.

d. Besides the above theorems, in the theory of weak perturbations of self-adjoint operators there is a deep theorem on preservation of the asymptotics of the spectrum when we pass from A_0 to A. The theorem is due to Markus and Matsaev (1982), see also (Markus 1986). When it is applied to elliptic operators, it yields the following result. We denote by $N_1(\lambda; A)$ the distribution function of the moduli of eigenvalues of a pseudodifferential operator A:

$$N_1(\lambda; A) = \operatorname{card}\{j : |\lambda_j| \leq \lambda\}. \tag{6.75}$$

Theorem 6.2.8. *Suppose that A is an elliptic pseudodifferential operator of order $m > 0$ on an n-dimensional closed manifold with positive principal symbol $a_0(x, \xi)$. Then $N_1(\lambda; A)$ has asymptotics coinciding with* (6.17)*:*

$$N_1(\lambda; A) = \alpha_0\lambda^{\frac{n}{m}} + O\left(\lambda^{\frac{n-1}{m}}\right). \tag{6.76}$$

If A has positive principal symbol and real subprincipal symbol and if for $N(\lambda; A_0)$, where $A_0 = \operatorname{Re} A$, we have the asymptotic formula (6.28)*, then the same formula holds for $N_1(\lambda; A)$:*

$$N_1(\lambda; A) = \alpha_0 \lambda^{\frac{n}{m}} - \alpha_{-1} \lambda^{\frac{n-1}{m}} + o\left(\lambda^{\frac{n-1}{m}}\right) . \tag{6.77}$$

If instead of (6.28) we have for $N_1(\lambda; A)$ a formula with improved estimate of the remainder, then the same formula can be obtained for $N_1(\lambda; A)$ in case A is sufficiently close to A_0; for details see (Markus and Matsaev 1982). We also mention that in (6.76) and (6.77) we can replace $N_1(\lambda; A)$ by

$$N_2(\lambda; A) = \text{card}\{j : \operatorname{Re} \lambda_j \geq \lambda\} . \tag{6.78}$$

e. We shall now dwell on matrix pseudodifferential operators of the form (6.45), where $A_0 = A_0^*$ and $\operatorname{ord} A_1 = m_1 < \operatorname{ord} A_0 = m$, $m > 0$. In this case A_0 and A_1 have Hermitian principal symbol $a_0(x, \xi)$. If $a_0 > 0$, i.e. $a_0(x, \xi)$ has on $T^*M \backslash 0$ positive eigenvalues only, then the results are analogous to those presented above. If $a_0(x, \xi)$ has eigenvalues of both signs, then A_0 has infinite sets of positive and negative eigenvalues and the eigenvalues of A lie in the union of parabolic neighborhoods of the rays $\mathbb{R}_-$ and $\mathbb{R}_+$. The corresponding abstract model of an operator close to selfadjoint is not much more complicated than the described above, and all the results can be easily generalized. It is convenient to consider the series (6.63), (6.64), and (6.67) as two-sided ($-\infty < l < \infty$). In (6.65) and (6.66) we replace λ^γ by $(-\lambda)^\gamma$ for $l < 0$; this function coincides with $|\lambda|^\gamma$ on $\mathbb{R}_-$. For details see, for example, (Markus 1986).

f. We shall briefly dwell on pseudodifferential operators elliptic in the sense of Douglis–Nirenberg. If such an A has Hermitian principal symbol, then we can assume that $l_j = m_j$ for all j. In this case it is convenient to consider the L_2-realization of A, i.e. its closure in the space $L_2(M) = H_0(M)$. The closure exists since there exists the formally adjoint operator A^*.

To get some idea of what effects one can expect in this case, we consider a trivial example of a block-diagonal operator

$$B = \operatorname{diag}(B_{11}, \ldots, B_{rr}) , \tag{6.79}$$

where B_{ll} is a pseudodifferential operator elliptic in the usual sense of order t_l, $t_1 > \ldots > t_r$. Suppose that these blocks are selfadjoint operators. Each block of nonzero order has its own asymptotics of eigenvalues. If we have t_j of different signs, then the eigenvalues condense at 0 and ∞. If there is a block of zero order, then it brings an essential spectrum consisting of eigenvalues of its principal symbol. If $t_j > 0$ for all j, then the principal term of the asymptotics of $N(\lambda; B)$ coincides with the principal term of the asymptotics of $N(\lambda, B_{rr})$ and the remaining blocks have no influence on this principal term.

In (Kozhevnikov 1981), as we have already mentioned in 3.2b, some assumptions are found that are sufficient for A to be similar to a weakly perturbed pseudodifferential operator of the form (6.79). We shall not dwell on the formulations of the corollaries for such A. In (Grubb 1978) the principal term of the asymptotics of $N(\lambda; A)$ is obtained for $A = A^* > 0$ with an estimate of the remainder without these additional assumptions. See also the references in (Grubb 1978) and (Kozhevnikov 1981).

6.3. Selfadjoint Elliptic Pseudodifferential Operators and Their Weak Perturbations on a Circle

a. In this section we consider an elliptic pseudodifferential operator A of order $m > 0$ on the circle S^1, scalar at first, with real principal symbol. We recall (see Sect. 2.4) that A has the form

$$Au(x) = \sum_{n=-\infty}^{\infty} \mathrm{e}^{\mathrm{i}nx} a(x,n) c_n(u) , \tag{6.80}$$

where $c_n(u)$ are Fourier coefficients of $u(x)$ and $a(x,\xi)$ is the complete symbol; $a(x,\xi) \sim a_0(x,\xi) + \ldots$, and all the functions here are 2π-periodic in x.

Some results, essentially stronger than in the previous section, can be obtained for A by means of *similarity transforms* of the form

$$A \to B = \Phi^{-1} A \Phi , \tag{6.81}$$

where Φ is a Fourier integral operator. The required Fourier integral operators on S can be easily globally defined by the Fourier type series:

$$\Phi u(x) = \sum_{n=-\infty}^{\infty} \mathrm{e}^{\mathrm{i}\varphi(x,n)} h(x,n) c_n(u) + Tu(x) . \tag{6.82}$$

Here $h(x,\xi)$ is a symbol from S^0_{ph}, 2π-periodic in x, and we impose on it the ellipticity condition

$$h(x,\xi) \neq 0 \quad \text{for} \quad \xi \neq 0 . \tag{6.83}$$

By $\varphi(x,\xi)$ we denote a phase function (not complete; cf. (5.111)); it is real, belongs to C^∞ for $\xi \neq 0$, and is positively homogeneous in ξ of the first degree. We rewrite it in the form

$$\varphi(x,\xi) = \varphi_\pm(x)\xi \qquad (\xi \neq 0) , \tag{6.84}$$

where the sign $\pm$ coincides with the sign of ξ. We assume that $\varphi_\pm(x)$ satisfies the conditions

$$\varphi'_\pm(x) > 0 , \quad \varphi_\pm(x + 2\pi) \equiv \varphi_\pm(x) + 2\pi . \tag{6.85}$$

By T we denote an operator of order $-\infty$ on S.

Theorem 6.3.1. *Under the above assumptions, Φ is an operator of zero order in the Sobolev scale $\{H_r(S)\}$ and Φ is Fredholm in any $H_r(S)$.*

For the proof we introduce the operators of change of variables

$$Z_{\pm}u(x) = u[\varphi_{\pm}(x)], \tag{6.86}$$

and we put $w_{\pm}(x,\xi) = Z_{\pm}^{-1}h(x,\xi)$, where the sign $\pm$ coincides with that of ξ. The operator Φ can be rewritten in the form

$$\Phi = Z_- W_- + Z_+ W_+ + T\,, \tag{6.87}$$

where $W_{\pm}$ are pseudodifferential operators of zero order of the form

$$W_- u(x) = \sum_{n<0} \mathrm{e}^{inx} w_-(x,n) c_n(u)\,, \quad W_+ u(x) = \sum_{n\geq 0} \mathrm{e}^{inx} w_+(x,n) c_n(u)\,.$$

It is clear from (6.87) that Φ is an operator of zero order. Next, $W = W_- + W_+$ is an elliptic pseudodifferential operator of zero order, and it has a parametrix R which is of the same kind. We write it in the form $R = R_- + R_+$ (R_- is the sum over $n < 0$ and R_+ is the sum over $n \geq 0$). We now put

$$\Psi = R_- Z_-^{-1} + R_+ Z_+^{-1}\,.$$

This is also a zero order operator, and one can easily verify (using the calculus of pseudodifferential operators) that it is a parametrix for Φ: $\Phi\Psi - I$ and $\Psi\Phi - I$ are operators of order $-\infty$. Consequently, Φ is a Fredholm operator in all $H_r(S)$.

Thus Fourier integral operators of the form (6.82) have the structure (6.87). We note that the parametrix Ψ is a Fourier integral operator of the analogous structure: $\Psi = Z_-^{-1}\widetilde{R}_- + Z_+^{-1}\widetilde{R}_+$, where $\widetilde{R}_{\pm} = Z_{\pm} R_{\pm} Z_{\pm}^{-1}$ are pseudodifferential operators of zero order (the sums over $n < 0$ and $n \geq 0$).

Additionally, one can establish that the indices $\kappa(\Phi)$ and $\kappa(W)$ coincide. If $\kappa(\Phi) = 0$, then choosing suitable T, we can get Φ invertible.

Theorem 6.3.2. *1) Suppose that A is an elliptic pseudodifferential operator from $\Psi_{\mathrm{ph}}^0(S)$ and Φ is a Fourier integral operator of the form* (6.87). *Then $A\Phi$ and ΦA are Fourier integral operators of the analogous form (with the same phase function).*

2) Suppose that A is a pseudodifferential operator from $\Psi_{\mathrm{ph}}^m(S)$ and Φ is an invertible Fourier integral operator of the form (6.87). *Then $\Phi^{-1}A\Phi$ is a pseudodifferential operator from $\Psi_{\mathrm{ph}}^m(S)$.*

Both assertions can be easily verified by means of the formulas $A = A_- + A_+$ and (6.87) and the theorems of the calculus of pseudodifferential operators. The second assertion is a variant for the circle of a well-known Egorov theorem (cf. (Egorov 1984, 1988), (Hörmander 1985b), (Taylor 1981), and (Tréves 1980)). We note that the symbol of $\Phi^{-1}A\Phi$ (more precisely, its full asymptotic expansion) can be computed by means of the theorems on composition and change of variables for pseudodifferential operators. The situation is similar for the symbols of $A\Phi$ and ΦA in the first assertion.

b. The following theorem is the main result in this section.

Theorem 6.3.3. *Suppose that A is a (scalar) elliptic pseudodifferential operator from $\Psi^m_{\rm ph}(S)$, $m \neq 0$, with real principal symbol. Then there exists an invertible Fourier integral operator Φ such that $B = \Phi^{-1}A\Phi$ is, up to a term of order $-\infty$, a pseudodifferential operator with the symbol independent of x.*

This theorem can be regarded as an analogue of the classical Lyapunov theorem on reducibility of a system of differential equations of the first order to a system with constant coefficients. We can also mention the transform operators in the theory of boundary problems for the Sturm–Liouville equation (see, for example, (Levitan and Sargsyan 1988) and (Marchenko 1977)).

The operator Φ is constructed in the form $\Phi = \Phi_1 G$, where

$$\Phi_1 u(x) = \sum_{n<0} e^{i\varphi_-(x)n} c_n(u) + \sum_{n\geq 0} e^{i\varphi_+(x)n} c_n(u) + T_1 u(x) \,,$$

T_1 is an operator of order $-\infty$, Φ_1 is an invertible Fourier integral operator, and G is an invertible elliptic pseudodifferential operator of zero order. We construct Φ_1 in such a way that $\Phi_1^{-1}A\Phi_1$ has principal symbol $b_0(\xi)$ independent of x. We outline this construction denoting $\Phi_1^{-1}A\Phi_1$ by B for the time being. The equality $A\Phi_1 = \Phi_1 B$ combined with the theorem on change of variables yields the following relation which allows us to define $\varphi_\pm(x)$ and $b_0(\xi)$:

$$a_0(x, \pm 1)[\varphi'_\pm(x)]^m = b_0(\pm 1) \,.$$

Here $a_0(x,\xi)$ is the principal symbol of A. We set

$$\arg b_0(\pm 1) = \arg a_0(x, \pm 1) \tag{6.88}$$

(for any value of $\xi = \pm 1$ this is 0 or π) and, further,

$$\varphi_\pm(x) = |b_0(\pm 1)|^{\frac{1}{m}} \int_{-\pi}^{x} |a_0^{-1}(y, \pm 1)|^{\frac{1}{m}} dy - \pi \,. \tag{6.89}$$

We now define $|b_0(\pm 1)|$ from the condition $\varphi_\pm(\pi) = \pi$. Next, we extend $b_0(\xi)$ to a function positively homogeneous of degree 0. The index $\kappa(\Phi_1)$ is equal to 0, and by choosing a suitable operator T_1 we can make Φ_1 invertible.

Suppose now that already A has the principal symbol $a_0(\xi) = b_0(\xi)$ independent of x; by B we understand the desired operator from the formulation of the theorem. The equality $AG = GB$ combined with the composition theorem allows us to construct recurrently the full asymptotic expansion $g_0(x,\xi) + g_1(x,\xi) + \ldots$ of the symbol of G and, simultaneously, the full asymptotic expansion $b_0(\xi) + b_1(\xi) + \ldots$ of the symbol of B. Here $b_j(\pm 1)$ are defined by the 2π-periodicity condition of $g_{j-1}(x, \pm 1)$ $(j = 1, 2, \ldots)$. We shall not dwell on the details.

The condition that the principal symbol $a_0(x,\xi)$ is real in Theorem 6.3.3 and in the next theorems can be replaced by the condition

$$\arg a_0(x, \pm 1) = c_\pm \ , \tag{6.90}$$

where $c_\pm$ do not depend on x. We cannot eliminate this condition since it is necessary even for the reduction of the principal symbol to that independent of x. Condition (6.90) is equivalent to the following one:

$$\mathrm{ord}(AA^* - A^*A) \leq 2m - 2 \ , \tag{6.91}$$

which is satisfied if $A = A_0 + A_1$, where A_0 is a normal operator and A_1 has order not greater than $m - 1$.

c. The pseudodifferential operator B obtained as the result of the similarity transform differs by a term of order $-\infty$ from the operator

$$B_0 u(x) = \sum_{-\infty}^{\infty} \mathrm{e}^{inx} b(n) c_n(u), \tag{6.92}$$

which is normal in all $H_r(S)$ since it has an orthogonal basis $\{\mathrm{e}^{inx}\}$ of eigenfunctions (if the scalar product corresponds to the norm (2.28)). The spectrum of B_0 clearly consists of eigenvalues $b(n)$, and they have a full asymptotic expansion in the powers of the number:

$$b(n) \sim b_0(\mathrm{sgn}\, n)|n|^m + b_1(\mathrm{sgn}\, n)|n|^{m-1} + \dots \ .$$

As a corollary we obtain

Theorem 6.3.4. *The eigenvalues of the operator A can be arranged, taking the multiplicities into account, in a sequence $\{\lambda_n(A)\}_{-\infty}^{\infty}$ in such a way that*

$$\lambda_n(A) \sim b_0(\mathrm{sgn}\, n)|n|^m + b_1(\mathrm{sgn}\, n)|n|^{m-1} + \dots \ . \tag{6.93}$$

For definiteness, we shall assume that $m > 0$ in what follows.

From Theorem 6.3.3 we can also deduce

Theorem 6.3.5. *There exists a system of subspaces $\mathcal{L}_l \subset H_\infty(S)$ ($j = 0, 1, \dots$) invariant with respect to A that form a Riesz basis in all $H_r(S)$, where $\mathcal{L}_0$ is finite-dimensional and $\mathcal{L}_l$ for $l = 1, 2, \dots$ are no more than two-dimensional.*

The essential improvement comparing with the analogous result in Sect. 6.2 (see Theorem 6.2.5: in the notation of Sect. 6.2 we have now $p(1 - q) \geq 1$) is precisely that the spaces $\mathcal{L}_1$, $\mathcal{L}_2, \dots$ are no more than two-dimensional. A proof of Theorem 6.3.5 can be given in the following way. The spectrum $\sigma(B)$ is contained in the union of the disks with centers at $b(n)$ and radii $C_\gamma |n|^{-\gamma}$, where γ is arbitrarily large. The union of these disks for a fixed sufficiently large γ is split into connected components, each containing the same number of eigenvalues of B and B_0, equal to 1 or 2 if the component is sufficiently far from the origin. We enumerate these components by natural numbers, denote them by $\Omega^{(1)}, \Omega^{(2)}, \ldots$, and denote the union of the remaining ones by $\Omega^{(0)}$. Let Γ_l be the boundary of the set $\Omega^{(l)}$ oriented positively with respect to $\Omega^{(l)}$. We define the Riesz projections P_l, Q_l by formulas of the form (6.62), replacing A, A_0 by B, B_0. One can prove that the inequality of the form (6.70) is satisfied (where $\|\cdot\|$ is the operator norm in any $H_r(S)$), i.e. $\{P_l H_\infty(S)\}$ is a Bary basis of subspaces in $H_r(S)$. Consequently, it is a Riesz basis of subspaces in $H_r(S)$, and by the similarity transform it again becomes a Riesz basis consisting of the spaces $\mathcal{L}_l$.

We note that if

$$\arg a_0(x, -1) \neq \arg a_0(x, +1) , \tag{6.94}$$

then the spaces $\mathcal{L}_l$ are one-dimensional starting from a certain number. Then all the eigenvalues $\lambda_n(A)$ with sufficiently large $|n|$ are simple, and the system of root functions of A is an *unconditional basis* in all $H_r(S)$, i.e. it is a basis after any rearrangement of its elements, and it can contain at most finitely many associated functions. In general case, if the subspace $\mathcal{L}_l$ is two-dimensional, it is spanned by two eigenfunctions corresponding to the same or to close eigenvalues, or by an eigenfunction and an associated function corresponding to the same eigenvalue.

It is interesting to mention that even if all eigenvalues of A are simple, the system of its eigenfunctions is not necessarily a basis. For example, let $A = A_0 + T$, where A_0 is an elliptic pseudodifferential operator of order $m > 0$,

$$A_0 \mathrm{e}^{\mathrm{i}nx} = a(n)\mathrm{e}^{\mathrm{i}nx} , \quad a(n) = a(-n) > 0 ,$$

$$T1 = 0 , \quad T(\mathrm{e}^{\mathrm{i}nx} \pm \alpha_n e^{-\mathrm{i}nx}) = \pm\beta_n(\mathrm{e}^{\mathrm{i}nx} \pm \alpha_n e^{-\mathrm{i}nx}) \quad (n \in \mathbb{N}) ,$$

and $\alpha_n \searrow 0$, $\beta_n \searrow 0$, $(\beta_n/\alpha_n)e^n \to 0$. Obviously, $\operatorname{ord} T = -\infty$ and A has the system of eigenfunctions

$$\{1, \mathrm{e}^{\mathrm{i}nx} \pm \alpha_n e^{-\mathrm{i}nx} \, (n \in \mathbb{N})\} , \tag{6.95}$$

complete in all $H_r(S)$, and the eigenvalues $a(0)$, $a(n) \pm \beta_n$ $(n \in \mathbb{N})$ can be made simple by choosing suitable β_n. The system biorthogonal to (6.95) has the form

$$\{(\sqrt{2\pi})^{-1}, (2\sqrt{\pi})^{-1}(\mathrm{e}^{\mathrm{i}nx} \pm \alpha_n^{-1} e^{-\mathrm{i}nx}) \, (n \in \mathbb{N})\} . \tag{6.96}$$

It is easy to see that condition (2.53) is not satisfied here; hence (6.95) is not a basis. However, it is an unconditional basis with parentheses: two-dimensional subspaces spanned by $\mathrm{e}^{\mathrm{i}nx} \pm \alpha_n e^{-\mathrm{i}nx}$, together with the one-dimensional space spanned by 1, form an orthogonal basis of subspaces in all $H_r(S)$.

On the other hand, if $\arg a_0(x, -1) = \arg a_0(x, +1)$, then it is not impossible that there exists even an orthogonal basis of root functions. For example, consider a pseudodifferential operator $A = A_0 + T$ with the same A_0 as above and

$$T\mathrm{e}^{\mathrm{i}nx} = e^{-n} e^{-\mathrm{i}nx} \; (n > 0) , \qquad T\mathrm{e}^{\mathrm{i}nx} = 0 \; (n \leq 0) .$$

The operator T has order $-\infty$, for any $n \in \mathbb{N}$ the exponential functions $\mathrm{e}^{\mathrm{i}nx}$ and $e^{-\mathrm{i}nx}$ form an orthogonal basis of the two-dimensional root subspace spanned by them, and $\{e^{inx}\}_{-\infty}^{\infty}$ is an orthogonal basis in all $H_r(S)$ consisting of root functions of A.

An additional useful result is that we can obtain a full asymptotic expansion for elements of a basis in $\mathcal{L}_l$. (Under condition (6.94), this is the asymptotic expansion of eigenfunctions.) For we notice that B_0 has eigenfunctions e^{inx} and B has a basis (probably with parentheses) with the asymptotics $e^{inx} + O(|n|^{-\gamma})$, where γ can be arbitrarily large. Suppose that the Fourier integral operator Φ of Theorem 6.3.3 is written in the form (6.82), where the symbol $h(x,\xi)$ has the asymptotic expansion $h_0(x,\xi) + h_1(x,\xi) + \dots$.

Theorem 6.3.6. *Under the above assumptions, there is a basis $u_n(x)$ in $\mathcal{L}_l$ with the asymptotics*

$$u_n(x) \sim e^{i\varphi(x,n)}[h_0(x, \operatorname{sgn} n) + \dots + h_k(x, \operatorname{sgn} n)|n|^{-k} + \dots] , \qquad (6.97)$$

where n takes the values such that $\lambda_n(A) \in \Omega^{(l)}$.

The precise meaning of the asymptotics (6.97) in the following: for any $k, r \in \mathbb{Z}_+$ we have

$$\|u_n(x) - e^{i\varphi(x,n)}[h_0(x, \operatorname{sgn} n) + \dots + h_k(x, \operatorname{sgn} n)|n|^{-k}]\|_r = O(|n|^{r-k-1}) .$$

Supplementary results can be obtained if we assume greater proximity of A to a selfadjoint operator. In particular, we have

Theorem 6.3.7. *Suppose that $\operatorname{Im} A = (A - A^*)/2i$ is a pseudodifferential operator of order $m - \nu - 1$, where $\nu \in \mathbb{N}$. Then all $b_j(\pm 1)$ $(j = 0, \dots, \nu)$ are real numbers and $\mathcal{L}_l$ form a Bary basis of subspaces in all $H_r(S)$.*

We could also give an analogue of Theorem 6.2.7 for pseudodifferential operators on S, but we shall not dwell on that.

As an illustration, we consider Example 2.4.8, where $\operatorname{Im} A$ has order $-\infty$. Using the expansion (2.45) of the complete symbol, one can easily verify that the asymptotic expansion of eigenvalues has the form

$$\lambda_n(A) = \frac{\alpha_0}{|n|} + \frac{\alpha_2}{|n|^3} + \frac{\alpha_4}{|n|^5} + O\left(\frac{1}{|n|^6}\right) \quad (n \to \pm\infty) , \qquad (6.98)$$

and the asymptotic expansion of the corresponding functions $u_n(x)$ has the form

$$u_{\pm n}(x) = e^{\pm inx}\left[1 \pm \frac{\gamma_3(x)}{|n|^3} + O\left(\frac{1}{|n|^4}\right)\right] \quad (n \to +\infty) , \qquad (6.99)$$

where

$$\alpha_4 = \frac{1}{2\pi}\int_{-\pi}^{\pi} \alpha_4(y)dy , \quad \gamma_3(x) = i\alpha_0^{-1}\int_{-\pi}^{x} [\alpha_4(y) - \alpha_4]dy . \qquad (6.100)$$

d. Now let A be an $N \times N$ matrix pseudodifferential operator of order $m \neq 0$ on S. The results analogous to those presented above can be obtained under the following assumptions on the principal symbol $a_0(x, \xi)$:

1) Any of the two matrices $a_0(x, \pm 1)$ is similar to a diagonal matrix.
2) The multiplicities $N_1^{\pm}, \ldots, N_{q\pm}^{\pm}$ of all its eigenvalues $\alpha_j(x, \pm 1)$ ($j = 1, \ldots, q_{\pm}$) do not depend on x.
3) The arguments of these eigenvalues do not depend on x.

The main result again is that under the above assumptions we can construct a matrix Fourier integral operator Φ such that $\Phi^{(-1)} A \Phi = B$ is a pseudodifferential operator of order m with symbol $b(\xi)$ independent of x, up to addition of an operator of order $-\infty$. The operators Φ and B are defined by the equation $A\Phi = \Phi B$. The operator Φ is Fredholm, but if A is not a differential operator, then we are not always able to make the index $\kappa(\Phi)$ equal to 0. In this case we choose Φ to be right invertible, and by $\Phi^{(-1)}$ we denote the operator right inverse to Φ. Thus, in a general case we have to consider *generalized* (in the sense just indicated) *similarity transforms* (Amosov 1988). (Another possibility is to make Φ invertible and the symbol of B 'almost constant', that is different from a symbol independent of x by inclusion of the factor e^{ikx} into some entries in the matrix principal symbol, say, for $\xi > 0$, see (Agranovich and Amosov 1981), (Agranovich 1984).) Nevertheless, this yields analogues to all theorems formulated above in the scalar case. In particular, all eigenvalues of A, except, possibly, finitely many, can be enumerated in such a way that with accuracy of $O(|n|^{\gamma})$, where γ is arbitrarily large, they coincide with eigenvalues of the matrix $b(n)$. Therefore, for eigenvalues of A we obtain full asymptotic expansions in fractional, in general, powers of $|n|$. Next, we construct a system $\{\mathcal{L}_l\}_0^{\infty}$ of finite-dimensional subspaces invariant with respect to A, which is a Riesz basis in all $H_r(S)$. In addition, $\dim \mathcal{L}_l$ do not exceed $2N$ for $l \geq 1$, and we obtain full asymptotics of some basis functions in $\mathcal{L}_l$. For details, see (Agranovich and Amosov 1981) and (Amosov 1988). Condition 2) is essential (see (Rozenblyum 1979)).

6.4. Elliptic Operators Far from Selfadjoint. We consider here elliptic operators of order $m > 0$ on M that have an angle or angles of ellipticity with parameter. As we have mentioned in Sect. 4.1, existence of such an angle can be regarded as a remote analogue of the selfadjointness.

a. Let first A be an operator with discrete spectrum in a separable Hilbert space H. The following completeness theorem is due to Dunford and Schwartz and is proved in their book (1963, Chap. XI).

Theorem 6.4.1. *Suppose that for $\lambda \in \varrho(A)$ the resolvent $R_A(\lambda)$ belongs to $\mathfrak{S}_\tau$ for some $\tau > 0$, and suppose that there exist rays $\Gamma_1, \ldots, \Gamma_r$ stemming out of the origin with the following properties.*

1) The points λ of each ray with sufficiently large $|\lambda|$ belong to $\varrho(A)$, and along these rays $\|R_A(\lambda)\| = O(|\lambda|^N)$ ($\lambda \to \infty$) for some $N \geq -1$.

2) The angles between neighboring rays are not greater than π/τ.
Then the system of root vectors of A is complete in H.

The proof is based on the representation of the resolvent $R_A(\lambda)$ as the ratio of two entire analytic functions, operator-valued and numerical, and the estimates of the moduli of these functions. If the vector f is orthogonal to all root vectors of A, then the function $\Phi(\lambda) = (R_A(\lambda)g, f)$ is an entire analytic function equal to the quotient of entire analytic functions $\Phi_1(\lambda)/\Phi_2(\lambda)$, where

$$|\Phi_j(\lambda)| \leq C_\varepsilon \exp(\varepsilon|\lambda|^\tau) \quad (j = 1, 2) \tag{6.101}$$

for any $\varepsilon > 0$. Estimating Φ_2 from below on some expanding contours, we can apply to $\Phi(\lambda)$ the Phragmen–Lindelöf theorem and show that $\Phi(\lambda) \equiv 0$. It follows that $f = 0$.

This theorem can be applied, in particular, to weak perturbations of normal or selfadjoint operators and contains some generalizations of the Keldysh theorem (Keldysh 1951). Namely, suppose that $A = A_0 + B$, where A_0 is a normal invertible operator with discrete spectrum lying in the union of close angles $\mathcal{L}_1, \dots, \mathcal{L}_r$ with common apex at 0 and without any other pairwise common points, and suppose that B is an operator compact relative to A, i.e. BA^{-1} is compact. If $A_0^{-1} \in \mathfrak{S}_\tau$ and if each angle is less than π/τ, then the system of root vectors of A is complete in H. In this case the N in the estimate of the norm $\|R_A(\lambda)\|$ is equal to -1.

Theorem 6.4.2. *Under the assumptions of Theorem 6.4.1 with $N = -1$, the Fourier series of any $f \in H$ with respect to the system of root vectors of A sums up to f by the Abel method with parentheses of order $\gamma = \tau + \varepsilon$ with sufficiently small $\varepsilon > 0$.*

More precisely, we have

$$f = \lim_{t \to +0} \sum_{j=1}^{r} \sum_{l=1}^{\infty} P_{jl\gamma}(t) f\,.$$

The operator $P_{jl\gamma}(t)$ here is defined by the formula of the form (6.65) with integration over the contours $\mathcal{T}_{jl}$ which split the angle between Γ_j and Γ_{j+1} (where Γ_{r+1} coincides with Γ_1) into large bounded domains, in general much broader that in 6.2a. The branch of the function λ^γ in the definition of $P_{jl\gamma}(t)$ is chosen such that for a fixed j the real part of this function increases in modulus as l grows. This result is weaker than that of Theorem 6.2.4, but it is obtained under more general assumptions.

In application to elliptic operators we obtain the following theorem:

Theorem 6.4.3. *Suppose that A is a pseudodifferential operator of order $m > 0$ on M, elliptic with parameter along the rays $\Gamma_1, \dots, \Gamma_r$ stemming out of the origin, where the angle between any two neighboring rays does not exceed $\pi m/n$. Then the system of root functions of A is complete in any $H_r(M)$.*

Moreover, in this case the Fourier series of any function $f(x) \in H_s(M)$ with respect to the root functions of A sums up to $f(x)$ by the Abel method with parentheses of order $\gamma = (n/m) + \varepsilon$ with sufficiently small $\varepsilon > 0$.

b. Theorem 6.4.3 can be applied to matrix pseudodifferential operators, elliptic with parameter along Γ_j in the usual sense or in the sense of Douglis–Nirenberg (see 4.3a,b).

To obtain corollaries for pseudodifferential operators (4.1) we need some notions of the theory of polynomial pencils (see, for example (Gohberg and Krejn 1965), (Keldysh 1971), and (Markus 1986)). Suppose that $A_0, \ldots, A_n$ are linear operators in a Hilbert space H, not necessarily bounded, where $A_0 = -I$. We put

$$A(\lambda) = \sum_{l=0}^{p} \lambda^{p-l} A_l \,, \quad A(\partial_t) = \sum_{l=0}^{p} A_l \partial_t^{p-l} \,. \tag{6.102}$$

One can easily prove that the function

$$u(t) = \mathrm{e}^{\lambda_0 t} \left(\frac{t^s}{s!} u_0 + \ldots + u_s \right) \tag{6.103}$$

is a solution of the equation $A(\partial_t)u(t) = 0$ if and only if

$$A(\lambda_0)u_k + \ldots + \frac{1}{k!} A^{(k)}(\lambda_0)u_0 = 0 \quad (k = 0, \ldots, s) \tag{6.104}$$

(and, in particular, $A(\lambda_0)u_0 = 0$; here $A^{(k)}(\lambda)$ is the k-th derivative of $A(\lambda)$ with respect to λ). In this case we call (6.103) an *elementary solution* of the equation $A(\partial_t)u = 0$, and we call the set $\{u_0, \ldots, u_p\}$ a chain of an *eigenvector* u_0 and *associated* to it *vectors* $u_1, \ldots, u_p$ of the pencil $A(\lambda)$ corresponding to the *eigenvalue* λ_0.

Suppose that $H^{(1)}, \ldots, H^{(p)}$ are Hilbert spaces densely embedded in H. We put $\mathcal{H} = H^{(1)} \times \ldots \times H^{(p)}$. If the initial data $(u(0), \ldots, u^{(p-1)}(0))$ for all possible elementary solutions (6.103) of $A(\partial_t)u(t) = 0$ form a system complete in $\mathcal{H}$ (i.e. their finite linear combinations are dense in $\mathcal{H}$), then we say that the system of root vectors (eigenvectors and associated vectors) of the pencil $A(\lambda)$ is *p-fold complete* in $\mathcal{H}$ (cf. (Keldysh 1951, 1971)). The sense of the word 'fold' is connected with the possibility of simultaneous approximation of any set $(f_1, \ldots, f_p) \in \mathcal{H}$ by linear combinations of the initial data of elementary solutions. For $p = 1$ this is the usual completeness of the root vectors of the operator A_1.

Under some additional assumptions, which we are not giving here, the p-fold completeness for $A(\lambda)$ is equivalent to the usual completeness of the linearization of this pencil, analogous to that considered in 4.3c. In application to (4.1) we obtain the following result.

Theorem 6.4.4. *Suppose that the pseudodifferential operator* (4.1) *is elliptic with parameter along the rays* $\Gamma_1, \dots, \Gamma_r$ *and the angles between the neighboring rays does not exceed* $\pi m/n$. *Then for any* s *the system of root functions of the pencil* (4.1) *is* p*-fold complete in*

$$H_{s+m(p-1)}(M) \times \dots \times H_s(M) . \tag{6.105}$$

In fact, we obtain here 'p-fold summability' by the Abel method. In addition we can obtain generalizations to the case of matrix A_j. We shall not dwell on that.

c. We now discuss the question about rough asymptotics of the distribution function $N_1(\lambda) = N_1(\lambda; A)$ for the moduli of eigenvalues of a pseudodifferential operator A of order $m > 0$ on M, under the assumption that there is an angle of ellipticity with parameter. We suppose that A is scalar and that the bisector of the angle is $\mathbb{R}_-$, and we put

$$\alpha_1 = c_{n/m,q}^{-1} \int\int_{T^*M} [a_0(x,\xi) + 1]^{-q} dx đ\xi , \tag{6.106}$$

$$\alpha_2 = c_{n/m,q}^{-1} \int\int_{T^*M} [|a_0(x,\xi)| + 1]^{-q} dx đ\xi , \tag{6.107}$$

where $c_{\delta,q} = \delta B(\delta, q-\delta)$ and $qm > n$. It is easy to verify that these quantities do not depend on q. If $a_0(x,\xi) > 0$, then $\alpha_1 = \alpha_2 = \alpha_0$ (see (6.7)). One can also verify that $|\alpha_1| \le \alpha_2$. We give two theorems from the paper of Agranovich and Markus (1989).

Theorem 6.4.5. *The following inequalities hold:*

$$\liminf_{\lambda\to+\infty} \lambda^{-\frac{n}{m}} N_1(\lambda) > 0 \quad \text{if} \quad \alpha_1 \ne 0 , \tag{6.108}$$

$$\liminf_{\lambda\to+\infty} \lambda^{-\frac{n}{m}} N_1(\lambda) \le \alpha_2 . \tag{6.109}$$

The inequality (6.109) is true for any elliptic pseudodifferential operator A with discrete spectrum. In the proof of relation (6.108) we employ the beginning of the expansion (5.96) of the trace of the resolvent (if $m < n$, of the trace of a power of the resolvent) and combine it with a certain rough Tauberian theorem of the Hardy–Littlewood type.

The quantity α_1 is certainly different from 0 if

$$\frac{n}{m} |\arg a_0(x,\xi)| < \frac{\pi}{2} . \tag{6.110}$$

For $n > 1$ we can replace $<$ by $\le$. As is seen from Theorem 6.4.3, condition (6.110) also guarantees the completeness of the system of root functions.

Theorem 6.4.6. *We have the inequalities*

$$\limsup_{\lambda\to\infty} \lambda^{-\frac{n}{m}} N_1(\lambda) \geq |\alpha_1| \,, \tag{6.111}$$

$$\limsup_{\lambda\to\infty} \lambda^{-\frac{n}{m}} N_1(\lambda) \leq e\alpha_2 \,. \tag{6.112}$$

The inequality (6.112) is true for any elliptic A with discrete spectrum.

Corollary 6.4.7. *If the limit* $\lim_{\lambda\to\infty} \lambda^{-\frac{n}{m}} N_1(\lambda)$ *exists, then it belongs to the segment* $[|\alpha_1|, \alpha_2]$.

We do not know whether such a limit always exists. On the example of a differential operator $A = [a(x_1)D_1]^2 + [b(x_2)D_2]^2$ on the two-dimensional torus one can show, by changing a and b, that this limit can coincide with $|\alpha_1|$, coincide with α_2, and take any intermediate value.

If A is a matrix pseudodifferential operator (or scalar, but $n = 1$), then it is possible that the eigenvalues of the principal symbol of A are distributed in two closed angles $\mathcal{L}_0$ and $\mathcal{L}_1$ having the unique common point, their apex at the origin, and separated by angles of ellipticity with parameter. In this case we can obtain inequalities analogous to the above for $\liminf_{\lambda\to\infty} \lambda^{-\frac{n}{m}} N_{1,0}(\lambda)$ and $\limsup_{\lambda\to\infty} \lambda^{-\frac{n}{m}} N_{1,0}(\lambda)$, where $N_{1,0}(\lambda)$ is the distribution function of the moduli of eigenvalues in a neighborhood of the angle $\mathcal{L}_0$, see (Agranovich and Markus 1989); we do not give the formulation for the shortage of space. In particular, an interesting case is when $\mathcal{L}_0$ is a ray. Then there exists the asymptotics of the eigenvalues of A close to this ray ((Agranovich 1987), cf. (Kozhevnikov 1977)) even with an estimate of the remainder (Bojmatov and Kostyuchenko 1990).

d. As we have already mentioned, the condition of ellipticity with parameter can disappear when we take a power of a pseudodifferential operator, though completeness of the root functions and the asymptotics of eigenvalues remain if they took place before. In this connection we mention that in the case of a differential operator of the first order on a circle (see Example 2.6.4) a proper spectral asymptotics takes place with unusually defined leading coefficient (it was observed by Seeley (1986)), and the criterion for the completeness of the system of root vectors shown in Theorem 2.6.5 also has unusual global character. This brings a thought that possibly criteria of the completeness and of the presence of proper asymptotics of eigenvalues, more general that we have discussed, have not been discovered yet.

Remarks and Bibliographical Notes

Here we supplement the bibliographical notes made in the text of the article and give some additional remarks. As regards the list of references, we mention that for the lack of space we have not included, as a rule, brief communications preceding the detailed papers, and in some cases we had to restrict ourselves to references to books and surveys containing more detailed bibliography.

To § 1 and § 2. The notions of (principal) symbol, ellipticity, parametrix (or regularizator) appeared independently in the theory of singular integral equations (see (Mikhlin 1962), (Muskhelishvili 1962), and the survey (Prößdorf 1987)) and in the theory of partial differential equations. Although between these theories some links were established (in some papers elliptic boundary problems were reduced to singular integral equations), relations between the notions and methods apparently remained unclear to the end of the fifties. Even the terminology was different: in the former theory the term 'ellipticity' was absent (for example in (Muskhelishvili 1962) the term 'normal type operator' is used), and in the latter we would not find the term 'symbol' (the notion of characteristic polynomial was used). In this respect see the survey (Seeley 1967b) and the notes in (Hörmander 1985a).

The construction of the calculus of pseudodifferential operators on the level of principal symbols was prepared by the papers of Calderón and Zygmund (1957) and Dynin (1961) (see also Agranovich 1965). In (Calderón and Zygmund 1957) differential and singular integral operators were studied jointly; in (Dynin 1961) an algebra generated by differential and singular integral operators was considered and elliptic operators in this algebra were studied. This was stimulated by the necessity to ensure the freedom of the homotopy of the symbols in the study of the index of elliptic differential operators.

With the calculus of pseudodifferential operators we overcome most of the technical difficulties in the proofs of main theorems about elliptic equations.

A complete calculus of pseudodifferential operators in $\mathbb{R}^n$ was constructed in the paper of Kohn and Nirenberg (1965). In this paper it does not refer to the theory of singular integral operators, except the theorem on change of variables for pseudodifferential operators which was proved by Hörmander in 1965. Close results were obtained by Vishik and Eskin (see the supplement in (Vishik and Eskin 1965)) and Unterberger and Bokobza (1964).

The calculus of pseudodifferential operators on the torus, given by means of the Fourier type series, was considered in the late sixties by Volevich (unpublished). The presentation of Sect. 2.4 follows the author's paper (1984) and (Amosov 1988).

To § 3. Not all variants of the theory of elliptic operators in $\mathbb{R}^n$ and on a closed manifold were touched upon. Some other variants were mentioned in the surveys (Rozenblyum et al.1989) and (Fedosov 1990).

To § 4. The notion of ellipticity with parameter arose first in the theory of differential boundary problems: see (Agmon 1962), (Agmon and Nirenberg 1963), (Agranovich and Vishik 1964).

Linearization with respect to a parameter used in Sect. 4.3 is a traditional method in the theory of abstract nonselfadjoint operators, see, for example, (Gohberg and Krejn 1965), (Keldysh 1971), and (Markus 1986).

To § 5. The literature on functions of abstract operators is very extensive; see, in particular, the references in (Dunford and Schwartz 1958), (Krasnosel'skij et al. 1966), and (Reed and Simon 1978). The formulas of the type (5.28) and (5.29) are contained in (Dikij 1968) and (Duistermaat and Guillemin 1975), respectively.

First precise estimates of s-numbers of operators of finite order appeared in (Paraska 1965). The kernels of operators of order $m < -n/2$ were investigated before Agmon by Browder and Maurin, see the references in (Agmon 1965).

The results of Sects. 5.3 and 5.5 related to the powers of elliptic operators are in principal due to Seeley (1967a), see also (Burak 1968) for the case of differential operators. The results of Sect. 5.6 in the case of differential operators are due to Greiner (1971) and Seeley (1969b), where differential elliptic problems were also considered. The classical paper of Minakshisundaram and Pleijel (1949) on the Laplace–Beltrami operator served as a model. The case of pseudodifferential operators was considered by Duistermaat and Guillemin (1975) (a scalar $A = A^* > 0$), who first noticed the possibility of the presence of logarithmic terms in the asymptotics, and Widom (1980). The approach we have outlined in the case of nonselfadjoint pseudodifferential operators was mentioned in (Agranovich 1987), and it is close to that used in (Duistermaat and Guillemin 1975). For one more approach see the book (Grubb 1986).

The coefficients of the asymptotic expansion of the function $\operatorname{tr} e^{-tA}$ are, of course, spectral invariants, since they are defined by the spectrum of the operator A. These coefficients constitute a traditional subject of study in various problems of spectral geometry (see (Bérard 1977), (Gilkey 1974) and the references therein). The reader may notice that equivalent collections of spectral invariants appear in the theorems of Sects. 5.5, 5.7, and 5.8; see also Theorems 6.1.6 and 6.3.4.

The results of Sect. 5.7 for differential operators are known from the paper of Burak (1968). Similar information is contained in the papers on boundary problems for elliptic differential operators selfadjoint or close to selfadjoint of Agmon, Pham The Lai, Robert (see the references in (Agranovich 1993)), Métivier (1983) and others. In the last papers the asymptotics of the kernel of the resolvent was investigated in the complement to the domain (5.98) (in (Métivier 1983) even for $\theta = 1$). Full expansions in the case of pseudodifferential operators were obtained in the author's note (1987).

In 5.8a we give well-known results on Cauchy problem for strictly hyperbolic equation (cf. (Taylor 1981), (Ivrii 1988)) under simplifying assumptions (A is elliptic and M is closed).

The theory of Fourier integral operators was initiated by Hörmander (1968, 1971). A theory of canonical operator of Maslov (1965), which appeared earlier, is conceptually close to it; see also the survey (Fedoryuk 1988) and the references therein.

To Sect. 6.1. The 'Tauberian' approach to finding spectral asymptotics goes back to Carleman papers of the thirties, see (Carleman 1935). The method of hyperbolic equation, which permits to obtain the best progress, was first applied by Avakumovich (1956) and Levitan (1956) to differential operators of the second order. Competing with this method, a method of resolvent allowed Métivier to give in (1983) a new proof of Hörmander's theorem (in (Métivier 1983) elliptic boundary problems were also considered). The conjecture that for $N(\lambda)$ (in the cases of simple elliptic problems) two-terms asymptotic formulas may be true was formulated by H. Weyl in 1912.

Outside our presentation remained the variational approach to the problem of finding spectral asymptotics, especially useful in problems with irregularities; see the surveys (Birman and Solomyak 1977) and (Rozenblyum et al. 1989).

In the formulation of Theorem 6.1.4 we have taken account of a result of Safarov on the coincidence of conditions that the set of periodic points and the set of 'absolutely periodic' points have measure 0, given in (Vasil'ev and Safarov 1988).

In (Colin de Verdière 1979), (Duistermaat and Guillemin 1975), and (Hörmander 1985b) the case when all points on S^*M are periodic is considered. Characteristic in this case is the 'cluster' (with condensations) asymptotics of eigenvalues. More general situation was investigated by Safarov (1988).

For elliptic in the sense of (3.8) (and hypoelliptic) pseudodifferential operators $A = A^* > 0$ in $\mathbb{R}^n$ the asymptotics of $N(\lambda)$ with some estimate of the remainder was obtained by Shubin and Tulovskij, see (Shubin 1978). Helffer and Robert obtained analogues of the Hörmander theorem (Theorem 6.1.2) and some theorems of (Colin de Verdière 1979) for globally elliptic pseudodifferential operators, see (Helffer 1984). Spectral asymptotics was also investigated for 'anisotropic-elliptic' pseudodifferential operators (with various 'weights' for various variables), see (Helffer and Robert 1982a) and (Aramaki 1987). In the papers of Boutet de Monvel and Guillemin and Sternberg the relations between spectral properties of operators on $\mathbb{R}^n$ and operators on compact manifolds were studied. For references, see (Helffer 1984).

For generalized Toeplitz operators the analogues of the Hörmander theorem and of some theorems from (Colin de Verdière 1979) are presented in (Boutet de Monvel and Guillemin 1981).

To Sect. 6.2. The material, to considerable extent, is a development of the results of a classical note of Keldysh (1951) (see also (Keldysh 1971)) under additional assumptions on subordination of the operator A_1 to the operator A_0 and on the spectral asymptotics of the latter. Theorems 6.2.4, 6.2.5, and 6.2.6 are variants of the Katsnel'son theorems (Katsnel'son 1967a,b) (in these papers $q \geq 0$) and the Markus theorem (Markus 1962) (see also (Vizitej and Markus 1965)), respectively. We take into account the presence of asymptotics of eigenvalues $\lambda_j(A_0)$ which allows us to put parentheses more often. In (1981) Markus and Matsaev gave new, more constructive proof of the Katsnel'son theorem of type 6.2.5. It employs the method of creating an 'artificial gap' in the spectrum of A_0. The same method was used in (Markus and Matsaev 1982) in the proof of an abstract theorem which implies Theorem 6.2.8 and covers many special results for elliptic operators; see also (Markus 1986). Unfortunately, we could not touch upon details concerned with application of subtle theorems from the function theory. In general, the role played by the methods of the function theory in the spectral theory is very considerable, see the survey (Nikol'skij and Khrushchev 1987).

The importance of the case $q < 0$, considered in (Agranovich 1976), becomes clear from the problems posed by physicists in (Vojtovich et al. 1977). (Previously, the main example for theorems of the type given in Sect. 6.2 were elliptic differential problems close to selfadjoint, and in such problems $q \geq 0$.) These problems lead to pseudodifferential operators of Examples 2.2.3 and 2.4.8 and to related interesting operators. Completeness of the root functions of the operator (2.9) was proved by Ramm (1973). The analysis of these operators by the methods of the theory of elliptic pseudodifferential operators (Agranovich 1977) yields essentially more than the application of well-known tools of the operator theory (cf. (Colton and Kress 1983)).

In the papers of Rozenblyum (1982, 1983, 1986) pseudodifferential operators close to normal operators were considered. Friedlander considered in (1980) analytic perturbations of analytic selfadjoint pseudodifferential operators and in (1977) the perturbations of selfadjoint operators in the metrics of $C^{(\alpha)}$ and L_p.

To Sect. 6.3. The idea of using the similarity transform to find the full asymptotics of eigenvalues of selfadjoint pseudodifferential operators on a circle S is due to Rozenblyum (1977, 1978). The papers of Agranovich and Amosov (1981, 1984) and (Amosov 1988) are mainly devoted to nonselfadjoint and matrix pseudodifferential operators on S. We also mention that the spectral properties of differential operators on an interval with periodic and other boundary conditions were investigated long ago and quite completely (see, for example (Marchenko 1977), (Najmark 1969)). In the paper of Helffer and Robert (1982b) a full asymptotic expansion of the eigenvalues of selfadjoint pseudodifferential operators on the line $\mathbb{R}^1$ was obtained. The relations between spectral properties of operators on the line and on the circle were considered in the dissertation of Milevskij (1989).

To Sect. 6.4. The assertion on completeness in Theorem 6.4.3 is contained in (Burak 1968). This is an analogue of the theorem from (Agmon 1962) on problems elliptic with parameter. The 'Abel method' was introduced by Lidskij, see (1962). In his theorem of 1960 on completeness and summability by the Abel method the conditions were imposed not on the resolvent but on the numerical domain of the operator. Later, the Abel method was considered by many authors, see, in particular (Kostyuchenko and Radzievskij 1974). A simple variant of Theorem 6.4.2 was shown in (Agranovich 1977).

A series of formulations in the present survey were taken from the author's book 'Nonselfadjoint Elliptic Operators' which is being prepared in English for Springer-Verlag.

L.S.Dzhanlatyan and Ya.A.Rojtberg have read the manuscript of this article; the author is grateful to them for a series of remarks.

References*

Agmon, S. (1962): On the eigenfunctions and on the eigenvalues of general elliptic boundary value problems. Commun. Pure Appl. Math. **15**, 119–147. Zbl.109,327

Agmon, S. (1965): On kernels, eigenvalues, and eigenfunctions of operators related to elliptic problems. Commun. Pure Appl. Math. **18**, 627–663. Zbl.151,202

Agmon, S., Douglis, A., Nirenberg, L. (1964): Estimates near the boundary for solutions of elliptic partial differential equations satisfying general boundary conditions. II. Commun. Pure Appl. Math. **17**, No. 1, 35–92. Zbl.123,287

Agmon, S., Nirenberg, L. (1963): Properties of solutions of ordinary differential equations in Banach spaces. Commun. Pure Appl. Math. **16**, No. 2, 121–239. Zbl.117,110

Agranovich, M.S. (1965): Elliptic singular integro-differential operators. Usp. Mat. Nauk **20**, No. 5, 3–120. English transl.: Russ. Math. Surv. **20**, No. 5, 1–121 (1965). Zbl.149,361

Agranovich, M.S. (1976): On summability of series in root vectors of non-selfadjoint elliptic operators. Funkts. Anal. Prilozh. **10**, No. 3, 1–12. English transl.: Funct. Anal. Appl. **10**, 165–174 (1977). Zbl.355.35068

Agranovich, M.S. (1977): Spectral properties of difraction problems. Supplement to (Vojtovich et al. 1977), 289–416 (Russian). Zbl.384.35002

Agranovich, M.S. (1980): Convergence of the series in root vectors of almost selfadjoint operators. Tr. Mosk. Mat. O-va **41**, 163–180. English transl.: Trans. Moscow Math. Soc. No. 1, 167–183 (1982). Zbl.479.47012

* For the convenience of the reader, references to reviews in *Zentralblatt für Mathematik* (Zbl.), compiled using the MATH database, have, as far as possible, been included in this bibliography.

Agranovich, M.S. (1984): Elliptic pseudodifferential operators on a closed curve. Tr. Mosk. Mat. O-va **47**, 22–67. English transl.: Trans. Moscow Math. Soc. 23–74. (1985). Zbl.573.35071

Agranovich, M.S. (1987): Some asymptotic formulas for elliptic pseudodifferential operators. Funkts. Anal. Prilozh. **21**, No. 1, 63–65. English transl.: Funct. Anal. Appl. **21**, 53–56 (1987). Zbl.631.35074

Agranovich, M.S. (1993): Elliptic boundary problems. In: Itogi Nauki Tekh. Ser. Sovrem. Probl. Mat., Fundam. Napravleniya **79** (in print). English transl. in: Encyclopaedia of Mathematical Sciences **79**. Springer–Verlag: Berlin–Heidelberg–New York (in preparation).

Agranovich, M.S., Amosov, B.A. (1981): Elliptic pseudodifferential matrix operators on a closed curve. Funkts. Anal. Prilozh. **15**, No. 3, 79–81. English transl.: Funct. Anal. Appl. **15**, 217–219 (1981). Zbl.475.35083

Agranovich, M.S., Markus, A.S. (1989): On spectral properties of elliptic pseudodifferential operators far from self-adjoint ones. Z. Anal. Anwend. **8**, No. 3, 237–260. Zbl.696.35123

Agranovich, M.S., Vishik, M.I. (1964): Elliptic problems with parameter and parabolic problems of general form. Usp. Mat. Nauk **19**, No. 3, 53–161. English transl.: Russ. Math. Surv. **19**, No. 3, 53–157 (1964). Zbl.137.296

Alekseevskij, D.V., Vinogradov, A.M., Lychagin, V.V. (1988): Basic ideas and notions of differential geometry. Itogi Nauki Tekh. Ser. Sovrem. Probl. Mat., Fundam. Napravleniya **28**. English transl.: Encyclopaedia of Mathematical Sciences 28. Springer–Verlag, Berlin–Heidelberg–New York, 1991. Zbl.675.53001

Alimov, Sh.A., Il'in, V.A., Nikishin, E.M. (1976, 1977): Convergence problems of multiple trigonometric series and spectral decompositions. I, II. Usp. Mat. Nauk **31**, No. 6, 28–83, **32**, No. 1, 107–130. English transl.: Russ. Math. Surv. **31**, No. 6, 29–86 (1976), **32**, No. 1, 115–139 (1977). Zbl.345.42005, Zbl.353.42005

Amosov, B.A. (1988): On the theory of pseudodifferential operators on the circle. Usp. Mat. Nauk **43**, No. 3, 169–170. English transl.: Russ. Math. Surv. **43**, No. 3, 197–198 (1988). Zbl.671.35083

Aramaki, J. (1987): On the asymptotic behaviour of the spectrum of quasi-elliptic pseudodifferential operators on $\mathbb{R}^n$. Tokyo J.Math. **10**, No. 2, 481–505. Zbl.655.35063

Atiyah, M.F., Bott, R. (1967, 1968): A Lefschetz fixed point formula for elliptic complexes. I, II. Ann. Math. II. Ser. **86**, 374–407, **88**, 451–491. Zbl.161,432, Zbl.167,217

Atiyah, M.F., Singer, I.M. (1963): The index of elliptic operators on compact manifolds. Bull. Am. Math. Soc. **69**, 422–433. Zbl.118,312

Avakumović, V.G. (1956): Über die Eigenfunktionen auf geschlossenen Riemannschen Mannigfaltigkeiten. Math. Z. **65**, 324–344. Zbl.70,326

Beals, M., Fefferman, Ch., Grossman, R. (1983): Strictly pseudocomplex domains in $\mathbb{C}^n$. Bull. Am. Math. Soc., New. Ser. **8**, No. 2, 125–326. Zbl.546.32008

Bérard, P.H. (1977): On the wave equations on a compact Riemannian manifold without conjugate points. Math.Z. **155**, 249–276. Zbl.341.35052

Bérard, P.H. (1986): Spectral Geometry: Direct and Inverse Problems. Lect. Notes Math. **1207**. Zbl.608.58001

Birman, M.Sh., Solomyak, M.Z. (1977): Asymptotics of the spectrum of differential equations. Itogi Nauki Tekh. Ser. Mat. Anal. **14**, 5–58. English transl.: J. Sov. Math. **12**, 247–283 (1979). Zbl.417.35061

Bitsadze, A.V., Vinogradov, V.S., Dezin, A.A., Il'in, V.A. (1987): Partial differential equations. Tr. Mat. Inst. AN SSSR **176**, 259–299. English transl.: Proc. Steklov Inst. Math. **176**, 263–300 (1987). Zbl.636.35003

Bojmatov, K.Kh., Kostyuchenko, A.G. (1990): Spectral asymptotics of nonselfadjoint elliptic systems of differential operators in bounded domains. Funkts. Anal. Prilozh. **24**, No. 1, 64–66. English transl.: Funct. Anal. Appl. **24**, No. 1, 54–57 (1990). Zbl.726.35093

Boutet de Monvel, L., Guillemin, V. (1981): The Spectral Theory of Toeplitz Operators. Ann. Math. Stud. **99**. Zbl.469.47021

Burak, T. (1968): Fractional powers of elliptic differential operators. Ann. Sc. Norm. Super. Pisa, III. Ser. **22**, 113–132. Zbl.159,156

Calderón, A.P., Zygmund, A. (1957): Singular integral operators and differential equations. Amer. J. Math. **79**, 901–921. Zbl.81,335

Carleman, T. (1935): Properiétés asymptotiques des functions fondamentales des membranes vibrantes. 8. Congr. Math. Scand., Stockholm (1934), 34–44. Zbl.12,70

Colin de Verdière, Y. (1979): Sur le spectre des opérateurs elliptiques à bicaractéristiques toutes périodiques. Comment. Math. Helv. **54**, 508–522. Zbl.459.58014

Colton, D., Kress, R. (1983): Integral Equation Methods in Scattering Theory. Wiley, Interscience Publishers. Zbl.522.35001

Dikij, L.A. (1958): Trace formulas for Sturm–Liouville differential operators. Usp. Mat. Nauk **13**, No. 3, 111–143. English transl.: Am. Math. Soc., Transl., II. Ser. **18**, 81–115 (1961). Zbl.80,67

Douglis, A., Nirenberg, L. (1955): Interior estimates for elliptic systems of partial differential equations. Commun. Pure Appl. Math. **8**, 503–538. Zbl.66,80

Dubrovin, B.A., Novikov, S.P., Fomenko, A.P. (1986): Modern Geometry - Methods and Applications. 2nd ed. Nauka: Moscow. English transl.: Springer–Verlag: New York, Berlin, Heidelberg, Grad. Texts Math. **93**, **104**, 1984, 1985. Zbl.433.53001, Zbl.529,53002

Duistermaat, J.J. (1981): On operators of trace class in $L^2(X, \mu)$. Proc. Indian Acad. Sci., Math. Sci. **90**, No. 1, 29–32. Zbl.482.47012

Duistermaat, J.J., Guillemin V.W. (1975): The spectrum of positive elliptic operators and periodic bicharacteristics. Invent. Math. **29**, No. 1, 39–79. Zbl.307.35071

Dunau, J. (1977): Fonctions d'un opérateur elliptique sur une variété compacte. J. Math. Pures et Appl., IX.Ser. **56**, No. 4, 367–391. Zbl.323.58017

Dunford, N., Schwartz, J.T. (1958): Linear Operators. Part I: General Theory. New York, London: Interscience Publishers. Zbl.84,104

Dunford, N., Schwartz, J.T. (1963): Linear Operators. Part II: Spectral Theory. New York, London: Interscience Publishers. Zbl.128,348

Dynin, A.S. (1961): Singular operators of arbitrary order on a manifold. Dokl. Akad. Nauk SSSR **141**, No. 1, 21–23. English transl.: Sov. Math., Dokl. **2**, No. 6, 1375–1377 (1961). Zbl.124,311

Egorov, Yu.V. (1984): Linear Differential Equations of Principal Type. Nauka: Moscow. English transl.: Consultants Bureau: New York 1986. Zbl.574.35001

Egorov, Yu.V. (1988): Microlocal analysis. In Itogi Nauki Tekh. Ser. Sovrem. Probl. Mat., Fundam. Napravleniya **33**, 5–156. English transl. in: Encyclopaedia of Mathematical Sciences **33**, 1–147, 1993.

Egorov, Yu.V., Shubin, M.A. (1988a): Linear partial differential equations. Foundations of the classical theory. Itogi Nauki Tekh. Ser. Sovrem. Probl. Mat., Fundam. Napravleniya **30**. English transl.: Encyclopaedia of Mathematical Sciences **30**, 1992. Zbl.657.35002

Egorov, Yu.V., Shubin, M.A. (1988b): Linear partial differential equations. Elements of modern theory. In Itogi Nauki Tekh., Ser. Sovrem. Probl. Mat., Fundam. Napravleniya **31**, 5–125. English transl. in Encyclopaedia of Mathematical Sciences **31** (in preparation). Zbl.686.35005

Elschner, J. (1985): Singular Ordinary Differential Operators and Pseudodifferential Equations. Akademie–Verlag: Berlin. Zbl.571.47038. Also Lecture Notes Math. **1128**.

Fedoryuk, M.V. (1988): Equations with rapidly oscillating solutions. In Itogi Nauki Tekh. Ser. Sovrem. Probl. Mat., Fundam. Napravleniya **34**, 6–56. English transl. in Encyclopaedia of Mathematical Sciences **34** (in preparation). Zbl.657.35118

Fedosov, B.V. (1990): Index theorems. In Itogi Nauki Tekh., Ser. Sovrem. Probl. Mat., Fundam. Napravleniya **65**, 165–268. English transl. in Encyclopaedia of Mathematical Sciences **65** (in preparation).

Friedlander, L.F. (1977): Convergence of series in root functions of operators that are nearly selfadjoint in the spaces C and L_p. Funkts. Anal. Prilozh. **11**, No. 4, 94–95. English transl.: Funct. Anal. Appl. **11**, 325–327 (1977). Zbl.375.47011

Friedlander, L.F. (1980): On certain spectral properties of very weak nonselfadjoint perturbations of selfadjoint operators. Tr. Mosk. Mat. O-va **41**, 181–216. English transl.: Trans. Moscow Math. Soc. 1982, **1**, 185–218. Zbl.473.47010

Friedrichs, K.O. (1968): Pseudo-Differential Operators. An Introduction. Courant Institute of Math. Sci., New York Univ. Zbl.226.47028

Gel'fand, I.M. (1960): On elliptic equations. Usp. Mat. Nauk **15**, No. 3, 121-132. English transl.: Russ. Math. Surv. **15**, No. 3, 113–123 (1960). Zbl.95,78

Gilkey, P.B. (1974): The index theorem and the heat equation. Princeton Univ., Math. Lecture Series **4**. Publish or Perish. Zbl.287.58006

Gohberg, I.Ts., Krejn, M.G. (1965): Introduction to the Theory of Linear Nonselfadjoint Operators in Hilbert Spaces. Nauka: Moscow. English transl.: Am. Math. Soc.: Providence 1969. Zbl.138,78

Greiner, P. (1971): An asymptotic expansion for the heat equation. Arch. Ration. Mech. Anal. **41**, No. 1, 163–218. Zbl.238.35038

Grubb, G. (1978): Remainder estimates for eigenvalues and kernels of pseudodifferential elliptic systems. Math. Scand. **43**, 275–307. Zbl.407.35075

Grubb, G. (1986): Functional Calculus of Pseudo-Differential Boundary Problems. Birkhäuser: Boston–Basel–Stuttgart. Zbl.622.35001

Guillemin, V., Sternberg, Sh. (1979): On the spectra of commuting pseudo-differential operators; recent work of Kac–Spencer, Weinstein and others. In: Partial Differ. Equations and Geometry. Proc. Park City Conf. (1977). New York–Basel, Lect. Notes Pure Appl. Math. **48**, 149–165. Zbl.502.58036

Helffer, B. (1984): Théorie Spectrale pour des Opérateurs Globalement Elliptiques. Astérisque **112**. Zbl.541.35002

Helffer, B., Robert, D. (1982a): Properiétés asymptotiques du spectre d'opérateurs pseudodifférentiels sur $\mathbb{R}^n$. Commun. Partial Differ. Equations **7**, No. 7, 795–882. Zbl.501.35081

Helffer, B., Robert, D. (1982b): Asymptotique des niveaux d'énergie pour des hamiltoniens à un degré de liberté. Duke Math. J. **49**, No. 4, 853–868. Zbl.519.35063

Hörmander, L. (1966): On the Riesz means of spectral functions and eigenfunction expansions for elliptic differential operators. Recent Advances in the Basic Sciences, Yechiva Univ. Conf. (1966), 155–202.

Hörmander, L. (1968): The spectral function of an elliptic operator. Acta Math. **121**, 193–218. Zbl.164,132

Hörmander, L. (1971): Fourier integral operators. Acta Math. **127**, 79–183. Zbl.212,466

Hörmander, L. (1983a,b, 1985a,b): The Analysis of Linear Partial Differential Operators. Vols. 1, 2, 3, 4. Springer–Verlag: Berlin–Heidelberg–New York. Zbl.521.35001, Zbl.521.35002, Zbl.601.35001, Zbl.612.35001

Ivrii, V.Ya (=Ivrij, V.Ya) (1982): Precise spectral asymptotics for elliptic operators that act in vector bundles. Funkts. Anal. Prilozh. **16**, No. 2, 30–38. English transl.: Funct. Anal. Appl. **16**, 101–108 (1982). Zbl.505.35066

Ivrii, V.Ya. (=Ivrij, V.Ya.) (1984): Precise spectral asymptotics for elliptic operators acting in fiberings over manifolds with boundary. Lect. Notes Math. **1100**. Zbl.565.35002

Ivrii, V.Ya. (=Ivrij, V.Ya.) (1988): Linear hyperbolic equations. In: Itogi Nauki Tekh. Ser. Sovrem. Probl. Mat., Fundam. Napravleniya **33**, 157–247. English transl. in: Encyclopaedia of Mathematical Sciences **33**, 1–147, 1993.
Katsnel'son, V.E. (1967a): Convergence and summability of series in root vectors for some classes of non-selfadjoint operators. Candidate Dissertation, Khar'kov (Russian).
Katsnel'son, V.E. (1967b): Conditions under which systems of eigenvectors form a basis. Funkts. Anal. Prilozh. **1**, No. 2, 39–51. English transl.: Funct. Anal. Appl. **1**, 122–131 (1967). Zbl.172,174
Keldysh, M.V. (1951): Eigenvalues and eigenfunctions of some classes of non-selfadjoint equations. Dokl. Akad. Nauk SSSR **77**, No. 1, 11–14 (Russian). Zbl.45,394
Keldysh, M.V. (1971): On the completeness of the eigenfunctions of some classes of non-selfadjoint operators. Usp. Mat. Nauk **26**, No. 4, 15-41. English transl.: Russ. Math. Surv. **26**, No. 4, 15–44 (1971). Zbl.225.47008
Kohn, J.J., Nirenberg, L. (1965): An algebra of pseudodifferential operators. Commun. Pure Appl. Math. **18**, No. 1/2, 269–305. Zbl.171,351
Kostyuchenko, A.G., Radzievskij, G.O. (1974): On summation of n-tuple expansions by the Abel method. Sib. Math. J. **15**, No. 4, 855–870. English transl.: Sib. Mat. Zh. **15**, 607–617 (1974). Zbl.295.40009
Kozhevnikov, A.N. (1977): Asymptotic behaviour of the eigenvalues of elliptic systems. Funkts. Anal. Prilozh. **11**, No. 4, 82–83. English transl.: Funct. Anal. Appl. **11**, 314–315 (1978). Zbl.384.35046
Kozhevnikov, A.N. (1981): Remainder estimates for eigenvalues and complex powers of the Douglis–Nirenberg elliptic systems. Commun. Partial Differ. Equations **6**, 1111–1136. Zbl.469.58016
Krasnosel'skij, M.A., Zabrejko, P.P., Pustyl'nik, E.I., Sobolevskij, P.E. (1966): Integral Operators in Spaces of Summable Functions. Nauka: Moscow. English transl.: Leyden: Noordhoff 1976. Zbl.145,397
Kumano-go, H. (1981): Pseudodifferential Operators. Massachusetts Institute of Technology. Zbl.489.35003
Lax, P.D. (1957): Asymptotic solutions of oscillatory initial value problems. Duke Math. J. **24**, No. 4, 627–646. Zbl.83,318
Levitan, B.M. (1956): Some questions of spectral theory of selfadjoint differential operators. Usp. Mat. Nauk **11**, No. 6, 117–144. English transl.: Am. Math. Soc., Transl., II. Ser. **18**, 49–79 (1961). Zbl.73,320
Levitan, B.M., Sargsyan, I.S. (1988): Sturm–Liouville and Dirac Operators. Nauka: Moscow. English transl.: Dordrecht; Kluwer 1990. Zbl.657.34002
Lidskij V.B. (1962): On summability of series in principal vectors of non-selfadjoint operators. Tr. Mosk. Mat. O-va **11**, 3–35. English transl.: Am. Math. Soc., Transl., II. Ser. **40**, 193–228 (1964). Zbl.117,331
Marchenko, V.A. (1977): Sturm–Liouville Operators and Applications. Naukova Dumka: Kiev. English transl.: Birkhäuser: Basel 1986. Zbl.399.34022
Markus, A.S. (1962): Expansion in root vectors of a slightly perturbed selfadjoint operator. Dokl. Akad. Nauk SSSR **142**, No. 3, 538–541. English transl.: Sov. Math., Dokl. **3**, No. 1, 104–109 (1962). Zbl.122,116
Markus, A.S. (1986): Introduction to the Spectral Theory of Polynomial Operator Pencils. Shtiintsa: Kishinev. English transl.: Am. Math. Soc.: Providence 1988. Zbl.678.47006
Markus, A.S., Matsaev, V.I. (1981): Convergence of expansions in eigenvectors of an operator close to selfadjoint. Mat. Issled., **61**, 104–129 (Russian). Zbl.477.47016
Markus, A.S., Matsaev, V.I. (1982): Comparison theorems for spectra of linear operators and spectral asymptotics. Tr. Mosk. Mat. O-va **45**, 133–181. English transl.: Trans. Mosc. Math. Soc., No. 1, 139–188 (1984). Zbl.557.47009

Maslov, V.P. (1965): Perturbation Theory and Asymptotic Methods. Moscow University Press: Moscow. French transl.: Dunod: Paris 1972. Zbl.247.47010
Métivier, G. (1983): Estimation du reste en théorie spectrale. Rend. Semin. Mat., Fasc. Spec., Convegno su linear partial and pseudo-differential operators, Torino (1982), 157–180. Zbl.542.35055
Mikhlin, S.G. (1962): Multidimensional Singular Integrals and Integral Equations. Fizmatgiz: Moscow. English transl.: Pergamon Press: Oxford 1965. Zbl.105,303
Milevskij, A.S. (1989): Similarity transformations and spectral properties of hypoelliptic pseudodifferential equations on a circle. Funkts. Anal. Prilozh. **23**, No, 3, 71–72. English transl.: Funct. Anal. Appl. **23**, No. 3 (1989).
Minakshisundaram, S., Pleijel A. (1949): Some properties of the eigenfunctions of the Laplace operator on Riemannian manifolds. Can. J. Math. **1**, 242–256. Zbl.41,427
Muskhelishvili, N.I. (1962): Singular Integral Equations. Fizmatgiz: Moscow. German transl.: Akademie–Verlag: Berlin 1965. Zbl.103,75
Najmark, M.A. (1969): Linear Differential Operators. 2nd ed. Nauka: Moscow. English transl.: Harrop: London 1967. Zbl.193,41; Zbl.57,71
Nikol'skij, N.K., Khrushchev, S.V. (1987): A functional model and some problems in the spectral theory of function. Tr. Mat. Inst. Steklova **176**, 97–210. English transl.: Proc. Steklov Inst. Math. **176**, 101–214 (1988). Zbl.649.47010
Paraska, V.I. (1965): Asymptotics of eigenvalues and singular numbers of linear smoothing operators. Mat. Sb., Nov. Ser. **68**, No. 4, 621–631. English transl.: Am. Math. Soc. Transl., II. Ser. **79**, 87–95 (1969).
Pattisier, G. (1977): Sur les fonctions d'un opérateur pseudo-differentiel elliptique. Publ. Dép. Math., Lyon **14**, No. 3, 21–58. Zbl.423.35088
Petrovskij, I.G. (1939): Sur l'analyticité des solutions des systèmes d'equations différentielles. Mat. Sb. **5(47)**, 3–68. Zbl.22,226
Pleijel, A. (1963): On a theorem by P.Malliavin. Isr. J. Math. **1**, 166–168. Zbl.126,115
Prößdorf, S. (1987): Linear integral equations. In: Itogi Nauki Tekh. Ser. Sovrem. Probl. Mat., Fundam. Napravleniya **27**, 5–130. English transl. in: Encyclopaedia of Mathematical Sciences **27**. 1–126. 1991. Zbl.677.45001
Ramm, A.G. (1973): Expansion in eigenfunctions of discrete spectrum in diffraction problems. Radiotekhnika i Elektronika **18**, No. 3, 496–501 (Russian).
Reed, M., Simon, B. (1978): Methods of Modern Mathematical Physics. Vol. 4. Academic Press: New York. Zbl.401.47001
Rozenblyum, G.V. (1977): Almost-similarity of pseudodifferential systems on a circle. Probl. Mat. Anal. **6**. Leningrad University Press: Leningrad, pp. 143–157 (Russian). Zbl.447.35069
Rozenblyum, G.V. (1978): Almost-similarity of operators and spectral asymptotics of pseudodifferential operators on a circle. Tr. Mosk. Mat. O-va **36**, 59–84. English transl.: Trans. Moscow Math. Soc. **36**, No. 2, 57–82. Zbl.434.47040
Rozenblyum, G.V. (1979): Fundamental solution and spectral asymptotics of systems with turning points. Probl. Mat. Fiz.**9**, 122–128 (Russian). Zbl.493.34045
Rozenblyum, G.V. (1982): Spectral asymptotics of normal operators. Funkts. Anal. Prilozh. **16**, No. 2, 82–83. English transl.: Funct. Anal. Appl. **16**, 145–146 (1982). Zbl.496.47024
Rozenblyum, G.V. (1983): On an operator almost commuting with its adjoint. Teor. Oper. Teor. Funkts. **1**, 111–115 (Russian). Zbl.583.47041
Rozenblyum, G.V. (1986): Angular asymptotics of the spectra of operators close to normal. Probl. Mat. Anal. **10**, 180–195. English transl.: J. Sov. Math. **45**, 1250–1261 (1989). Zbl.691.47003
Rozenblyum, G.V., Solomyak, M.Z., Shubin, M.A. (1989): Spectral theory of differential operators. Itogi Nauki Tekh., Ser. Sovrem. Probl. Mat., Fundam. Napravleniya **64**, 1–243. English transl.: Encyclopaedia of Mathematical Sciences **64** (in preparation). Zbl.715.35057

Sadov, S.Yu. (1987): Two analogues of the Pleijel inequality. Funkts. Anal. **27**, 156–164 (Russian). Zbl.649.44005

Sadovnichij, V.A. (1986): Operator Theory. 2sd ed. Moscow University Press: Moscow. English transl,: New York, Consultants Bureau 1991. Zbl.517.47001, Zbl.732.47002

Safarov, Yu. G. (1988): Asymptotic of the spectral function of a positive elliptic operator without the nontrap condition. Funkts. Anal. Prilozh. **22**, No. 3, 53–65. English transl.: Funct. Anal. Appl. **22**, 213–223 (1988). Zbl.679.35074

Seeley, R. T. (1963): The index of elliptic systems of singular integral operators. J. Math. Anal. Appl. **7**, 289–309. Errato ibid. **14**, 552 (1966). Zbl.133,376

Seeley, R.T. (1965): Refinement of the functional calculus of Calderón and Zygmund. Nederl. Akad. Wet., Proc., Ser. A. **68**, No. 3, 521–531. Zbl.141,133

Seeley, R.T. (1967a): Complex powers of an eliptic operator. Proc. Symp. Pure Math., **10**, 288–307. Zbl.159,155

Seeley, R.T. (1967b): Elliptic singular integral equations. Proc. Symp. Pure Math., **10**, 308–315. Zbl.177,386

Seeley, R.T. (1969a): The resolvent of an elliptic boundary problem. Am. J. Math. **91**, No. 4, 889–920. Zbl.191,118

Seeley, R.T. (1969b): Analytic extension of the trace associated with elliptic boundary problems. Am. J. Math. **91**, No. 4, 963–983. Zbl.191,118

Seeley, R.T. (1986): A simple example of spectral pathology for differential operators. Commun. Partial Differ. Equations **11**, No. 6, 595–598. Zbl.598.35013

Shilov, G.E. (1984): Mathematical Analysis. Second Special Course. 2-nd edition, Moscow University Press: Moscow. English transl.: Generalized Functions and Partial Differential Equations. Gordon/Breach, New York etc. 1968. Zbl.177,363; Zbl.591.46030

Shubin, M.A. (1978): Pseudodifferential Operators and Spectral Theory. Nauka: Moscow. English transl.: Springer–Verlag: Berlin–Heidelberg–New York 1987. Zbl.451.47064

Sobolev, S.L. (1950): Some Application of Mathematical Analysis to Mathematical Physics. Leningrad University Press: Leningrad (Russian).

Strichartz, R.S. (1972): A functional calculus for elliptic pseudo-differential operators. Am. J. Math. **94**, 711–722. Zbl.246.35082

Subkhankulov, M.A. (1976): Tauberian Theorems with Remainder. Nauka: Moscow (Russian).

Taylor, M.E. (1981): Pseudo-Differential Operators. Princeton Univ. Press. Zbl.453.47046

Trèves, F. (1980a,b): Introduction to Pseudo-Differential and Fourier Integral Operators. Vol. 1,2. Plenum: New York–London. Zbl.453.47027

Unterberger, A., Bokobza, J. (1964): Opérateurs de Calderón-Zygmund précisés. C. R. Acad. Sci., Paris **259**, 1612–1614. Zbl.143,369

Vasil'ev, D.G., Safarov, Yu.G. (1988): Branching Hamiltonian billiards. Dokl. Akad. Nauk. SSSR **301**, No. 2, 271–275. English transl.: Sov. Math., Dokl. **38**, No. 1, 64–68 (1989). Zbl.671.58012

Vishik, M.I., Eskin, G.I. (1965): Convolution equations in a bounded domain. Usp. Mat. Nauk **20**, No. 3, 89–152. English transl.: Russ. Math. Surv. **20**, No. 3, 85–151 (1965). Zbl.152,342

Vizitej, V.N., Markus, A.S. (1965): Convergence of multiple expansions in eigenvectors and adjoint vectors of operator penčil. Mat. Sb. **66**, No. 2, 287–320. English transl.: Am. Math. Soc., Transl., II. Ser. **87**, 187–227 (1970). Zbl.148,129

Vladimirov, V.C. (1979): Generalized Functions in Mathematical Physics. 2nd ed. Nauka: Moscow. English transl.: Moscow: Mir 1979. Zbl.515.46033, Zbl.403.46036

Vojtovich, N.N., Katsenelenbaum, B.Z., Sivov, A.N. (1977): Generalized Method of Eigen-Oscillations in the Diffraction Theory. Nauka: Moscow (Russian). Zbl.384.35002

Volovoj, A.V. (1987): Improved two-term asymptotics of distribution function of eigenvalues of an elliptic operator on a compact manifold. Dokl. Akad. Nauk SSSR, **294**, No. 5, 1037–1041. English transl.: Sov. Math., Dokl. **35**, No. 3, 604–607 (1987). Zbl.701.35122

Widder, D.V. (1941): The Laplace Transform. Princeton Univ. Press.

Widom, H. (1978): Families of pseudodifferential operators. Topics in Functional Analysis. Adv. Math., Suppl. Stud. **3**, 345–395. Zbl.477.58035

Widom, H. (1980): A complete symbolic calculus for pseudodifferential operators. Bull. Sci. Math., II Sér. **104**, 19–63. Zbl.434.35092

II. Degenerate Elliptic Equations and Boundary Problems

S. Z. Levendorskij and B. Paneah

Translated from the Russian
by Roger Cooke

Contents

Preface

A differential operator A on a manifold Ω is called *elliptic* if its principal symbol $A_0(x,\xi)$ does not vanish on $T_0^*(\Omega)$ (see Egorov, Shubin 1988, Sect. 1). If Ω is a closed manifold and the order of the operator is m, then for any $s \in \mathbb{R}^1$ it is a *Fredholm* $H^{s+m}(\Omega) \to H^s(\Omega)$ *operator* and the following *a priori* estimate holds:

$$\|u\|_s \leq c(s)(\|Au\|_{s-m} + \|u\|_{s-1}), \quad u \in H^\infty(\Omega). \tag{0.1}$$

What has just been said applies also to elliptic pseudodifferential operators.

Now suppose that Ω is a manifold with boundary $\partial\Omega = M$ and A is an elliptic differential operator on Ω of order $2m$. We assign to it the boundary value problem

$$Au = f \text{ in } \Omega, \quad B_j u = g_j \text{ on } M, \quad 1 \leq j \leq m, \tag{0.2}$$

where B_j are differential operators of orders $m_j < 2m$. The condition that this problem be elliptic is also equivalent to the nonvanishing of the symbol of some pseudodifferential operator $B(x',\xi')$ on $T_0^*(M)$ (Egorov, Shubin 1987, Chap. 2, Sect. 2). The operator of the problem $\mathfrak{A} : H^s(\Omega) \ni u \mapsto (Au, B_1u|_M, \ldots, B_mu|_M) \in H^{s-m}(\Omega) \times \prod_{j=1}^m H^{s-m_j-\frac{1}{2}}(M)$ is a Fredholm operator for $s > m - \frac{1}{2}$ and the following a priori estimate holds:

$$\|u\|_s \leq c(s)\Big(\|Au\|_{s-m} + \sum \|B_ju, M\|_{s-m_j-\frac{1}{2}} + \|u\|\Big) \tag{0.3}$$

(Egorov, Shubin 1987, Chap. 2, Sect. 2). If the ellipticity condition $A_0(x,\xi) \neq 0$ is violated only at the points of the set $\Gamma \subset \Omega$, the operator A is said to be *degenerate* on Γ. Similarly, if the condition $B(x',\xi') \neq 0$ is violated only for $x' \in \Gamma' \subset M$, then the boundary value problem (0.2) is called *degenerate* on Γ'. When the operator A and the problem (0.2) are degenerate, the a priori estimates (0.1) and (0.3) fail, as do the theorems on the Fredholm property. However, many properties of elliptic operators and boundary value problems carry over to the cases $\Gamma \neq \varnothing$, $\Gamma' \neq \varnothing$, while the Fredholm property and the a priori estimates hold in other pairs of spaces. A survey of the corresponding results constitutes the subject matter of the present article.

Chapter 1
Elliptic Differential Operators that Are Degenerate on the Boundary and Hypoelliptic Pseudodifferential Operators

By S. Z. Levendorskij

In this chapter we study mostly operators that degenerate into operators of lower order on the boundary Γ of a bounded domain $\Omega \subset \mathbb{R}^n$, $n \geq 2$; the nature of the degeneracy is described by functions that depend only on the distance to the boundary or part of the boundary. For large classes of such operators one can construct scales of weighted spaces of Sobolev type and sets of symbols whose invertibility implies that the operators under consideration are Fredholm operators. We give a natural classification of the operators both according to the methods used to study them and according to the properties and forms of the symbols. We show how to determine the number and form of the boundary conditions for each class and which calculus of pseudodifferential operators is connected with it. We also study certain classes of hypoelliptic pseudodifferential operators with principal symbols that vanish on a subset of the cotangent bundle.

For simplicity in this chapter we study only scalar-valued operators, and the coefficients of the differential operators are assumed infinitely smooth in the domain. Almost all results carry over naturally to systems, and the smoothness conditions can be weakened.

§1. Description of the Classes of Operators that Are Degenerate on the Entire Boundary

1.1. Classes of Operators and Weighted Spaces of Sobolev Type. In a bounded domain Ω with boundary $\Gamma \in C^\infty$ consider the equation

$$Au(x) = f(x), \quad x \in \Omega, \tag{1.1}$$

where A is an elliptic operator in Ω. To describe the nature of the degeneracy of the operator A we regard a small ε-neighborhood V_ε of the boundary Γ in Ω as $\Gamma \times (0, \varepsilon)$; let y_n be a coordinate along the normal to Γ and y' local coordinates on Γ. We assume that the operator A has the form

$$A = \sum_{|\alpha| \leq m} a_\alpha(y) y_n^{\hat{k}_\alpha} D_y^\alpha, \tag{1.2}$$

in V_ε, where $D^\alpha = (-i)^{|\alpha|}\partial^\alpha$, $a_\alpha \in C^\infty(\Gamma \times \overline{\mathbb{R}}_+)$, $\hat{k}_\alpha = k_{|\alpha'|,\alpha_n}$, and the points (l, j, k_{lj}) lie on the lower surface Q' of a convex polytope A with vertices at some points of the form (l, j, k_{lj}) with $lj = 0$ and $k_{lj} \geq k_{l'j'}$ if $(l, j) \geq (l', j')$ (i.e., if $l \geq l'$ and $j \geq j'$). We shall write $A \in L^m_{(k)}(\Omega)$.

We fix the norms in the Sobolev spaces $H^l(\Gamma)$ and $H^m(\Omega)$, and a function $\varphi_2 \in C_0^\infty(\Omega)$ such that $\operatorname{supp}(1-\varphi_2) \subset \overline{V}_\varepsilon$ and set

$$\varphi_1 = 1 - \varphi_2, \quad \tilde{\varphi}_1 u(y) = (\varphi_1 u)(x(y)),$$
$$\|u\|_{(k),m} = \Big(\|\varphi_2 u\|^2_{H^m(\Omega)} + \sum_{l+j\le m} \int_0^\varepsilon \|y_n^{k_{lj}} D_n^j \tilde{\varphi}_1 u\|^2_{H^l(\Gamma)}\, dy_n\Big)^{\frac{1}{2}},$$

and we denote by $H^m_{(k)}(\Omega)$ the space of distributions on Ω with finite norm $\|\cdot\|_{(k),m}$. It is not difficult to show that $H^m_{(k)}(\Omega)$ is a complete space and that it remains invariant as a topological space under a change in (a sufficiently small) $\varepsilon > 0$, the function φ_2, and the norms in $H^l(\Gamma)$ and $H^m(\Omega)$. It is not difficult to show that if $\hat{k}_\alpha - \alpha_n \ge \hat{k}_\beta - \beta_n$ for all $\alpha \ge \beta$, then $C_0^\infty(\Omega)$ is dense in $H^m_{(k)}(\Omega)$ and if $\hat{k}_\alpha \ge 0$ for all α, then $C^\infty(\overline{\Omega})$ is dense in $H^m_{(k)}(\Omega)$. For simplicity we shall consider only this kind of function $\alpha \mapsto \hat{k}_\alpha$, although it is known that one of the spaces $C_0^\infty(\Omega)$ and $C^\infty(\overline{\Omega})$ is dense in $H^m_{(k)}(\Omega)$ for other functions $\alpha \mapsto \hat{k}_\alpha$ (Olejnik, Radkevich 1971; Triebel 1978).

We shall be interested in learning the conditions under which an operator $A: H^m_{(k)}(\Omega) \to L_2(\Omega)$ is a Fredholm operator and

$$\|u\|_{(k),m} \le C\Big(\|Au\|_{L_2(\Omega)} + \|u\|_{L_2(K)}\Big), \tag{1.3}$$

where $K \subset \Omega$ is a compact set. If Eq. (1.1) requires additional conditions, estimate (1.3) is modified in the usual way.

1.2. Model Classes and Essentially Different Types of Operators. The simplest classes of operators $L^m_{(k)}(\Omega)$ and spaces $H^m_{(k)}(\Omega)$ are defined by functions $(l,j) \mapsto k_{lj}$ of the form

$$k_{lj} = s^0 + \gamma' l + \gamma_n j \ \text{ or } \ k_{lj} = \max\{\bar{s}, s^0 + \gamma' l + \gamma_n j\}, \tag{1.4}$$

where $\gamma', \gamma_n \ge 0$ and $0 < \bar{s} - s^0 \le \min\{\gamma' m, \gamma_n m\}$. Using Hardy's inequality (cf., for example, Triebel 1978, p. 321), it is not difficult to show that the norm in $H^m_{(k)}(\Omega)$ defined by a function of the first form in (1.4) with $\gamma_n < 1$ is equivalent to some norm defined by a function $(l,j) \mapsto k'_{lj}$ of the second form in (1.4) with $\gamma_n = 1$ and satisfying the condition $k'_{lj} \le k_{lj}$. Therefore estimate (1.3) implies the same kind of estimate with k' in place of k, and $L^m_{(k)}(\Omega) \subset L^m_{(k')}(\Omega)$. Consequently we can replace k by k' throughout and assume in what follows that $\gamma_n \ge 1$.

Under this assumption the following four essentially different types of operator classes and spaces defined by functions of the form (1.4) can be distinguished: *type I* is defined by a function of the first form in (1.4) with $\gamma' > 0$ and $\gamma_n > 0$; *type II* by a function of the first form in (1.4) with $\gamma' = 0$ and $\gamma_n > 1$; *type III* by a function of the first form in (1.4) with $\gamma' = 0$ and $\gamma_n = 1$; *type IV* by any function (1.4) with $\gamma' > 0$ and $\gamma_n = 1$. The study of operators of a general form can be reduced to the study of operators defined by the functions (1.4) by following the general outline of Sect. 3, and the study of operators of the second form in (1.4) with $\gamma_n > 1$ can be reduced to the case $\gamma_n = 1$.

1.3. Methods of Study, Symbols, and Boundary Conditions for Operators of the Various Types. Operators of type I are so strongly degenerate on the boundary that they can be studied using a suitable infinite partition of unity on Ω (cf., for example, Triebel 1978). Therefore the presence of the boundary is inessential (it can be regarded as the set of infinitely distant points) and boundary conditions are not needed. As in the case of elliptic operators with weight on the entire space, the conditions for the operator to be a Fredholm operator can be stated in terms of the invertibility of scalar symbols.

The study of operators of type II in a neighborhood of the boundary reduces to studying, in a neighborhood of zero, strongly degenerate operators on a half-line with operator-valued symbols. Hence they too can be studied using an infinite partition of unity on a half-line. Another method of studying model operators of type II in a neighborhood of Γ is by a power change of variable y_n taking the operator A to an elliptic operator in a half-cylinder with coefficients that stabilize at infinity. Replacing the coefficients by their limits and taking the Fourier transform on the radial variable, we obtain an operator-valued symbol whose invertibility together with the invertibility of certain scalar symbols is equivalent to the condition that the operator be a Fredholm operator.

This last method of investigation, only with a logarithmic change of variable instead of a power, also makes it possible to study model operators of type III, which are operators of Euler type in y_n. Here, as in the case of operators of type II, there arises an operator-valued symbol that assumes values in the class of elliptic operators on Γ and is parameterized by the points of the line.

Like operators of type I, operators of types II and II do not require boundary conditions.

The operators closest to elliptic operators in terms of their properties are operators of type IV. Thus the class of elliptic operators is one of the model classes to whose study the study of a variety of classes of degenerate operators of type IV reduces following the general outline of Sect. 3. Like elliptic operators, operators of type IV as a rule require additional boundary conditions, and their regularizers can be pieced together from local regularizers at the points of Ω and at the points of Γ (we note that the regularizers of operators of types II and III cannot be pieced together from local regularizers). The local regularizers at the points of Γ can be constructed by rectifying Γ locally and freezing the coefficients a_α. The operators so obtained in a half-space can be studied following the usual outline for elliptic operators – applying the Fourier transform along directions tangent to the boundary and studying the *operator-value symbol* that arises, a family of operators on a half-line parameterized by the points of $T^*\Gamma$. These operators on a half-line, as a rule, are also degenerate, so that their index and the number of additional conditions that must be adjoined to Eq. (1.1) in general depend on more than the leading coefficients and in some cases are independent of the leading coefficients. They may also depend on the weighted space of Sobolev type in which the solution of Eq. (1.1) is sought.

Thus operators of type I can be described by scalar symbols alone, while an operator-valued symbol (specific to each type) is needed to describe operators of types II–IV.

1.4. Hypoellipticity of Degenerate Operators. We shall say that A is *hypoelliptic* on the set $K \subset \overline{\Omega}$ if there exists a neighborhood $U \subset \mathbb{R}^n$ of the set K such that if $u \in H^m_{(k)}(\Omega)$ and $Au \in C^\infty(U \cap \overline{\Omega})$, then $u \in C^\infty(U \cap \overline{\Omega})$. In the case of integer $k_{lj} \geq 0$ it also makes sense to talk about hypoellipticity on a subset of the boundary for an operator (1.2) on $\overline{\Omega}$. Under this condition only Fredholm operators of types I and IV are hypoelliptic on every subset of the boundary, and for the majority of classes of type IV u must have a priori a high degree of smoothness.

Fredholm operators of type II are hypoelliptic on Γ, but not on every subset of Γ, while operators of type III are not hypoelliptic on any subset of the boundary. In particular, operators of types I, II, and IV are hypoelliptic on $\overline{\Omega}$, but operators of type III are not.

§2. Model Classes of Operators

2.1. Operators of Type I (Strongly Degenerate Operators). A model class of type I is defined by a function $(l,j) \mapsto k_{lj} = s^0 + \gamma' l + \gamma_n j$, where $\gamma' > 0$ and $\gamma_n > 1$. For $r \in \mathbb{Z}_+$ and $s \in \mathbb{R}$ we set $k'_{lj} = s + \gamma' l + \gamma_n j$, $H^{r,s}(\Omega) = H^r_{(k')}(\Omega)$, $\| \cdot \|_{r,s} = \| \cdot \|_{(k'),r}$, and we construct the symbol

$$a_{(s^0)}(y', \eta) = \sum_{|\alpha| \leq m} a_\alpha(y', 0)\eta^\alpha. \tag{2.1}$$

The symbol $a_{(s^0)}$ is well-defined as a function on $T^*\Gamma \times \mathbb{R}$ if and only if A is an operator of type I.

Theorem 2.1. *The following conditions are equivalent:*

a) *the principal symbol $a_{(-\infty)}$ and the symbol $a_{(s^0)}$ are nondegenerate, i.e.,*

$$a_{(-\infty)}(x, \xi) \neq 0 \quad \forall\, (x, \xi) \in T^*\Omega \setminus 0, \tag{2.2}$$

$$a_{(s^0)}(y', \eta) \neq 0 \quad \forall\, (y', \eta) \in T^*\Gamma \times \mathbb{R}; \tag{2.3}$$

b) *$A : H^{r+m, s^0+s}(\Omega) \to H^{r,s}(\Omega)$ is a Fredholm operator;*

c) *there exist $c_{r,s} > 0$ and a compact set $K \subset \Omega$ such that*

$$\|u\|_{m+r,s^0+s} \leq c_{r,s}\big(\|Au\|_{r,s} + \|u\|_{L_2(K)}\big) \tag{2.4}$$

for any $u \in \cup H^{0,s'}(\Omega)$.

Example 2.1. Let $\Omega = \{x \,|\, |x| < 1\}$, and let A be an elliptic operator of order 2 in Ω admitting the representation

$$A = a_0(r,\varphi) - (1-r)^{2\gamma'} a_1(r,\varphi)\Delta_\Gamma - (1-r)^{2\gamma_n} a_2(r,\varphi)\partial_\gamma^2, \tag{2.5}$$

in some neighborhood of $\Gamma = S_{n-1}$, where Δ_Γ is the Laplace-Beltrami operator on Γ and $a_j \in C^\infty([0,\infty) \times \Gamma)$.

Then A is an operator of type I if $\gamma' > 0$ and $\gamma_n > 1$; moreover condition (2.3) holds if and only if

$$a_0(1,\varphi) + a_1(1,\varphi)|\eta'|^2 + a_2(1,\varphi)\eta_n^2 \neq 0 \quad \forall\, (\eta,\varphi) \in \mathbb{R}^n \times \Gamma. \tag{2.6}$$

If condition (2.6) does not hold, for example, if

$$a_0(1,\varphi) = 0 \quad \forall \varphi \in \Gamma, \tag{2.7}$$

then the operator A is not a Fredholm operator of type I; but under condition (2.7) it can be a Fredholm operator of type III (if $\gamma' = \gamma_n - 1$) or type IV (if $\gamma' > \gamma_n - 1$). In the first case one must set $k_{lj} = \gamma' l + j$ and in the second case $k_{lj} = 2(\gamma_n - 1) + l(\gamma' - \gamma_n + 1) + j$.

We set

$$S_{d(x)}(\Omega) = \{u \in C^\infty(\Omega) \,|\, \sup d^l(x)|u^{(\alpha)}(x)| < \infty \ \forall\, l, |\alpha|\},$$

where $d(x) = \operatorname{dist}(x,\Gamma)$.

Corollary 2.1. *Suppose the hypotheses of Theorem* 2.1 *are satisfied.*

a) *If* $u \in \bigcup_s H^{0,s}(\Omega)$ *and* $Au \in S_{d(x)}(\Omega)$, *then* $u \in S_{d(x)}(\Omega)$.

b) *If* $s^0 \leq 0$, γ' *and* γ_n *are integers,* $u \in H^{0,s}(\Omega)$, *and* $Au \in C^\infty(\overline{\Omega})$, *then* $u \in C^\infty(\overline{\Omega})$.

Remark. Local analogs of the assertions in Corollary 2.1 hold. In particular A is hypoelliptic at any point $x^0 \in \overline{\Omega}$.

The following theorem can be deduced in the usual manner from Theorem 2.1 and Corollary 2.1.

Theorem 2.2. *Suppose* $s^0 < 0$ *and conditions* (2.2) *and* (2.3) *hold.*

a) *The operator* A, *regarded as an unbounded operator on* $H^{r,s}(\Omega)$ *with domain* $H^{r+m,s+s^0}(\Omega)$, *is closed.*
b) *If* A *is formally self-adjoint, then* A *is self-adjoint.*
c) *The spectrum of the operator* A *is independent of* r *and* s *and is either discrete or fills up the entire complex plane.*

d) *Each point of the spectrum is an eigenvalue of finite multiplicity and all the eigenfunctions and associated functions belong to* $S_{d(x)}(\Omega)$.

2.2. Operators of Type II (Elliptic along Directions Tangent to the Boundary and Strongly Degenerate along the Normal). A model class of type II is defined by a function $(l,j) \mapsto k_{lj} = s^0 + \gamma_n j$, where $\gamma_n > 1$. From this γ_n and $\gamma' = 0$ we define the space $H^{r,s}(\Omega)$ as in Sect. 2.1 and construct the operator-valued symbol

$$\hat{A}_{\mathrm{II}}(\lambda) = \sum_{|\alpha|\le m} \lambda^{\alpha_n} a_\alpha(y',0) D_{y'}^{\alpha'}, \tag{2.8}$$

parameterized by the points $\lambda \in \mathbb{C}$ and assuming values in an algebra of operators on Γ. Replacing the summation over $|\alpha| \le m$ in (2.1) by summation over $|\alpha| = m$, we obtain the symbol $a_{(s^0_+)}$. If

$$a_{(s^0_+)}(y',\eta) \neq 0 \quad \forall\,(y',\eta) \in T^*\Gamma \times \mathbb{R}, \quad |\eta| \neq 0, \tag{2.9}$$

then $\hat{A}_{\mathrm{II}}(\lambda)$ is an elliptic operator with parameter on the closed manifold Γ. Consequently the inverse operator $\hat{A}_{\mathrm{II}}(\lambda)^{-1} : H^r(\Gamma) \to H^{m+r}(\Gamma)$ is a meromorphic function of $\lambda \in \mathbb{C}$, and there exist N and $\delta > 0$ such that in the region where $|\lambda| > N$ and $|\arg\lambda| < \delta$ or $|\arg\lambda - \pi| < \delta$ and $|\lambda| > N$ there are no poles of the operator $\hat{A}_{\mathrm{II}}(\lambda)^{-1}$. The poles themselves are independent of r.

Theorem 2.3. *The following conditions are equivalent:*

a) *conditions* (2.2) *and* (2.9) *hold, and on the line* $\operatorname{Im}\lambda = 0$ *the operator* $\hat{A}_{\mathrm{II}}(\lambda)^{-1}$ *has no poles;*
b) *the operator* $A : H^{r+m,s^0+s}(\Omega) \to H^{r,s}(\Omega)$ *is a Fredholm operator;* (2.10)
c) *the a priori estimate* (2.4) *holds.*

Example 2.2. The operator (2.5) is an operator of type II if $\gamma' = 0$ and $\gamma_n > 1$. For this operator condition (2.9) assumes the form

$$a_1(1,\varphi)|\eta'|^2 + a_2(1,\varphi)\eta_n^2 \neq 0 \quad \forall\,(\varphi,\eta) \in \Gamma \times (\mathbb{R}^n \setminus 0), \tag{2.11}$$

while condition (2.10) holds if and only if none of the eigenvalues μ_j of the operator $a_2(1,\varphi)^{-1}(a_0(1,\varphi) - a_1(1,\varphi)\Delta_\Gamma)$ lie on the negative real axis.

If conditions (2.2), (2.9), and (2.10) hold, then Corollary 2.1 and Theorem 2.2 hold for operators of type II, but Remark 2.1 does not (it may happen that $Au \in C^\infty(U_{x^0} \cap \overline{\Omega})$ but $u \notin C^\infty(U_{x^0} \cap \overline{\Omega})$ for any sufficiently small neighborhood U_{x^0} of some point $x^0 \in \Gamma$); however A is hypoelliptic on the entire boundary Γ.

2.3. Operators of Type III (Elliptic along Directions Tangent to the Boundary and Euler Operators along the Normal). The model class of type III is defined by the function $(l,j) \mapsto k_{lj} = s^0 + j$. From $\gamma' = 0$, $\gamma_n = 1$ we define the space $H^{r,x}(\Omega)$ as in Sect. 2.1 and construct the operator-valued symbol

$$\hat{A}_{\mathrm{III}}(\lambda) = \sum_{|\alpha| \le m} \lambda(\lambda + i) \cdots (\lambda + i(\alpha_n - 1)) a_\alpha(y', 0) D^{\alpha'} . \tag{2.12}$$

Under condition (2.9) $\hat{A}_{\mathrm{III}}(\lambda)$ possesses all the properties of the operator $\hat{A}_{\mathrm{II}}(\lambda)$ enumerated in Sect. 2.2. In particular $\hat{A}_{\mathrm{III}}(\lambda)^{-1}$ is a meromorphic function, the set of numbers s for which there are poles on the line $\operatorname{Im} \lambda = s$ is discrete, and between any two lines $\operatorname{Im} \lambda = s$ and $\operatorname{Im} \lambda = s'$ there is only a finite set of poles.

Theorem 2.4. *The following conditions are equivalent:*

a) *conditions* (2.2) *and* (2.9) *hold, and there are no poles of the operator* $\hat{A}_{\mathrm{III}}(\lambda)^{-1}$ *on the line* $\operatorname{Im} \lambda = s^0 + s + \frac{1}{2}$;
b) *the operator* $A : H^{r+m, s+s^0}(\Omega) \to H^{r,s}(\Omega)$ *is a Fredholm operator;*
c) *the estimate* (2.4) *holds for any* $u \in H^{r+m, s+s^0}(\Omega)$.

Example 2.3. The operator (2.5) is an operator of type III if $\gamma' = 0$ and $\gamma_n = 1$. It satisfies the hypotheses of Theorem 2.4 if condition (2.11) holds and for all j the inequality $\lambda(\lambda + i) + \mu_j \ne 0$ holds when $\operatorname{Im} \lambda = s + \frac{1}{2}$.

We note that Corollary 2.1 and Theorem 2.2 fail for operators of type III, but if $f \in H^{r,s'}(\Omega)$ in Eq. (1.1) with $s' < s$ and there are no poles of the operator $\hat{A}_{\mathrm{III}}(\lambda)^{-1}$ on the lines $\operatorname{Im} \lambda = s + s^0 + \frac{1}{2}$ and $\operatorname{Im} \lambda = s' + s^0 + \frac{1}{2}$, one can write out the asymptotics of the solution from the poles lying in the strip bounded by these lines.

2.4. Operators of Type IV (Equations Requiring Boundary or Coboundary Conditions). The model classes of type IV are distinguished by the condition $k_{lj} = \max\{\bar{s}, s^0 + \gamma' l + j\}$, where $0 \le \bar{s} - s^0 \le \min\{\gamma' m, m\}$, $\gamma' > 0$. For the sake of simplicity we shall limit ourselves to the case $\bar{s} = 0$ in what follows.

The study of operators of type IV in a neighborhood of the boundary reduces to the study of the operator-valued symbol

$$\hat{A}_{\mathrm{IV}}(y', \eta') = \sum_{s^0 + \gamma'|\alpha'| + \alpha_n \ge \bar{s}} a_\alpha(y', 0) \eta'^{\alpha'} y_n^{\hat{k}_\alpha} D_n^{\alpha_n} : \widetilde{H}^m_{(k)}(\mathbb{R}_+) \to L_2(\mathbb{R}_+) \tag{2.13}$$

for $(y', \eta') \in T^*\Gamma \setminus 0$, where $\widetilde{H}^m_{(k)}(\mathbb{R}_+)$ is the space of distributions on $\mathbb{R}_+$ with finite norm

$$\|u\|^{\sim}_{m,(k)} = \left(\sum_{|\alpha| \le m} \left\| y_n^{\hat{k}_\alpha} D_n^{\alpha_n} u \right\|^2_{L_2(\mathbb{R}_+)} \right)^{\frac{1}{2}} . \tag{2.14}$$

The operators $\hat{A}_{\mathrm{IV}}(y', \eta')$ are described by the symbol $a_{(s^0_+)}$ and the symbol

$$M_R(y',\lambda) = \sum_{0\le j\le m-r} a_{(0,r+j)}(y',0)\lambda(\lambda+i)\cdots(\lambda+i(j-1)),$$

where $r = \min\{j \in \mathbb{Z} |\, k_{0j} \ge \bar{s}\}$.

Theorem 2.5. *Suppose condition* (2.9) *holds and that for all* $y' \in \Gamma$

$$M_r(y',\lambda) \ne 0 \ \textit{for}\ \operatorname{Im}\lambda = s^0 + r + \frac{1}{2}. \tag{2.15}$$

Then the operator (2.13) *is a Fredholm operator and its index* $\varkappa$ *equals* $\nu + \nu' + r - m$, *where* ν *is the number of roots of the equation* $M_r(y',\lambda) = 0$ *with* $\operatorname{Im}\lambda < s^0 + r + \frac{1}{2}$ *and* ν' *is the number of roots* $\eta_n = \eta_n(y',\eta')$ *of the equation*

$$a_{(s^0_+)}(y',\eta',\eta_n) = 0 \ \textit{with}\ \operatorname{Im}\eta_n > 0.$$

Theorem 2.5 will be proved at the end of this section.

We shall assume that $\varkappa$ is independent of $(y',\eta') \in T^*\Gamma \setminus 0$; if the boundary is connected and $n > 2$, then this condition is demonstrably satisfied. The difference between the number of boundary conditions and the number of co-boundary conditions that must be adjoined to Eq. (1.1) is $\varkappa$. The question of the form of the boundary conditions for Eq. (1.1) and the conditions under which it suffices to adjoin only $\varkappa$ boundary conditions when $\varkappa > 0$ and introduce $|\varkappa|$ operators of potential type into (1.1) in the case $\varkappa < 0$ has been studied by Vishik and Grushin (1969b, Sect. 5). In what follows we assume that $\varkappa > 0$ and that it suffices to adjoin $\varkappa$ boundary conditions.

Since $\bar{s} = 0$, $s^0 \le \bar{s} \le 0$, $k_{m0} \ge 0$, and $k_{0m} \ge 0$, it follows that $H^m_{(k)}(\mathbb{R}^n_+)$ is continuously imbedded in the space $\widehat{H}^{-s^0}(\gamma',\mathbb{R}^n_+)$ of functions on $\mathbb{R}^n_+$ admitting an extension lu to $\mathbb{R}^n$ with finite norm

$$\|lu\|^{\wedge}_{-s^0,\gamma'} = \|(|\xi'|^{1/\gamma'} + |\xi_n|^{-s^0}\widehat{lu}(\xi)\|_{L_2(\mathbb{R}^n)},$$

where $\widehat{lu}$ is the Fourier transform of the function lu. The restriction operator $r' : \widehat{H}^{\mu}(\gamma',\mathbb{R}^n_+) \to H^{(\mu-\frac{1}{2})/\gamma'}(\mathbb{R}^{n-1})$ to the boundary of the half-space is bounded if and only if $\mu > \frac{1}{2}$, and so the boundary conditions adjoined to Eq. (1.1) in the case $\varkappa \ge -s^0 + \frac{1}{2}$ cannot be merely differential equations in y_n. For the form of the admissible pseudodifferential boundary conditions see Sect. 5 of Vishik and Grushin 1969b; we assume for simplicity that $\varkappa < -s^0 + \frac{1}{2}$ and we adjoin to Eq. (1.1) the boundary conditions

$$\sum_{l\le j} B_{lj}(y',D')r'_\Gamma D^l_n u = g_j, \quad j = 0,1,\ldots,\varkappa-1, \tag{2.16}$$

which are differential expressions in y_n, where r'_Γ is the operation of restriction to Γ and B_{lj} are classical pseudodifferential operators on Γ of degree $(j-l)/\gamma'$ with principal symbols b_{lj}.

With the boundary problem (1.1) and (2.16) we connect the operator

$$\mathfrak{A} : H^m_{(k)}(\Omega) \to L_2(\Omega) \oplus \mathcal{H}'(\Gamma),$$

where $\mathcal{H}'(\Gamma) = \bigoplus_{a\le j\le \varkappa-1} H^{\left(-s^0-j-\frac{1}{2}\right)/\gamma'}(\Gamma)$ and boundary symbol $\widehat{\mathfrak{A}}(y',\eta') :$ $\widetilde{H}^m_{(k)}(\mathbb{R}_+) \to L_2(\mathbb{R}_+) \oplus \mathbb{C}^\varkappa$ acting according to the rule

$$\widehat{\mathfrak{A}}(y',\eta')u = \left(\hat{A}_{\mathrm{IV}}(y',\eta')u; \sum_{l\le j} b_{lj}(y',\eta')u^{(l)}(+0),\ j = 0,1,\dots,\varkappa-1\right).$$

Theorem 2.6. *Suppose conditions* (2.2), (2.9), *and* (2.15) *hold and that the boundary symbol is invertible on* $T^*\Gamma \setminus 0$.

Then $\mathfrak{A}$ *is a Fredholm operator and for all* $u \in H^m_{(k)}(\Omega)$

$$\|u\|_{m,(k)} \le c\|\mathfrak{A}u\|_{L_2(\Omega)\oplus\mathcal{H}'(\Gamma)} + c_1\|u\|_{L_2(\Omega)}. \tag{2.17}$$

Example 2.4. Let $0 \le m_1 < m_2$ be even, $0 \le k_1 < k_2$, and

$$A = bA_1 + A_2, \ \text{ where } A_j = d(x)^{k_j}(-\Delta)^{m_j/2}. \tag{2.18}$$

Then it is natural to set $\bar{s} = k_1$, $\gamma' = \gamma_n = \max\{1,\gamma\}$, where $\gamma = (k_2 - k_1)/(m_2 - m_1)$ and distinguish the following cases.

a) If $k_2 - k_1 < m_2 - m_1$, then the first term in (2.18) is compact with respect to the second, and therefore the number and form of the boundary conditions for the operator A are determined by the operator A_2, for which $m = r = m_2$, $\nu' = m_2/2$, $\nu = 0$, and $\varkappa = m_2/2$. Since $s^0 = k_2 - m_2$, it is possible to impose differential boundary conditions alone if and only if $k_2 < m_2/2$.
b) If $m_1 = 0$, $\operatorname{Im} b \notin \overline{\mathbb{R}}_-$, and $\gamma > 1$, then A is a Fredholm operator of type I.
c) If $m_1 > 0$, $\operatorname{Im} b \notin \overline{\mathbb{R}}_-$, and $\gamma > 1$, then by the general outline of Sect. 3 the study of the operator A reduces to the study of the operator bA_1; the number and form of the boundary conditions for the operator A is determined by the operator A_1 (cf. case a)).
d) If $m_1 \ge 0$ and $\gamma = 1$, then A is an operator of type IV for which $s^0 = m_1 - k_1$, $r = m_1$, $a_{(s^0_+)}(y',\eta) = |\eta|^{m_2}$, and

$$\hat{A}_{\mathrm{IV}}(y',\eta') = by_n^{k_1}(|\eta'|^2 + D_n^2)^{\frac{m_1}{2}} + y_n^{k_2}(|\eta'|^2 + D_n^2)^{m_2/2},$$
$$M_r(y',\lambda) = b + \lambda(\lambda+i)\cdots(\lambda + i(m_2 - m_1 - 1)).$$

Consequently condition (2.9) holds, and $\nu' = m_2/2$ is independent of b while (2.15) holds for some b but not others. The number ν also depends on b. For example, if $|b|$ is small, then $\nu = m_2 - m_1$, while if it is large, then $\nu = (m_2 - m_1)/2$ (and condition (2.15) holds in both cases). Therefore $\varkappa = (m_2 - m_1) + m_2/2 + m_1 - m_2 = m_2/2$ in the first case and $\varkappa = \frac{m_1}{2}$ in the second.

Example 2.5. Let $\varphi \in C^\infty(\overline{\Omega})$ be a function equal to 1 in a small neighborhood of Γ and to 0 outside a slightly larger neighborhood; let ∂_n be the derivative along the normal, $b, c \in \mathbb{C}$, and $A = d(x)^2(-\Delta) - ic\varphi(x)d(x)\partial_n + b$. Then $\bar{s} = s_n = 0$, $\gamma' = \gamma_n = 1$, $\nu' = 1$, $r = 0$, $m = 2$, and ν is the number of roots of the polynomial $\lambda(\lambda + i) + c\lambda + b$ in the half-plane $\operatorname{Im}\lambda < \frac{1}{2}$ (on the line $\operatorname{Im}\lambda = \frac{1}{2}$ there can be no roots). With various b and c the cases $\nu = 2$, $\nu = 1$, and $\nu = 0$ are possible. In the first case $\varkappa = 1$, in the second $\varkappa = 0$, and in the third $\varkappa = -1$. Consequently in the first case one must adjoin one boundary condition to Eq. (1.1), which must be a pseudodifferential condition, since $s^0 = 0$. In the second case one may omit the boundary conditions, while in the third case one must introduce an operator of potential type into (1.1).

Now let us consider boundary value problems with data possessing a high degree of smoothness. For simplicity we shall restrict ourselves to the case of integral $k_{lj} \geq 0$, i.e., to the case of coefficients of class $C^\infty(\overline{\Omega})$ and *isotropic degeneracy* $k_{lj} = \max\{0, l + j - r\}$, where $r \in [0, m)$ is an integer. We note that under this condition $s^0 = -r$ and $\gamma' = \gamma_n = 1$.

For an integer $\tilde{r} \geq 0$ we define the space $H^m_{(k,\tilde{r})}(\Omega)$, replacing m by $m + \tilde{r}$ and k_{lj} by $\max\{0, l + j - r - \tilde{r}\}$ in the definition of the space $H^m_{(k)}(\Omega)$, and we consider the operator $A : H^m_{(k,\tilde{r})}(\Omega) \to H^{\tilde{r}}(\Omega)$. As in the case $\tilde{r} = 0$, the study of this operator reduces to the study of the family of operators

$$\hat{A}_{\mathrm{IV}}(y', \eta') : \widetilde{H}^m_{(k,\tilde{r})}(\mathbb{R}_+) \to H^{\tilde{r}}(\mathbb{R}_+), \quad (y', \eta') \in T^*\Gamma \setminus 0, \tag{2.19}$$

where $\widetilde{H}^m_{(k,\tilde{r})}(\mathbb{R}_+)$ is defined like $\widetilde{H}^m_{(k)}(\mathbb{R}_+)$: it is only necessary to carry out the replacement of m and k_{lj} indicated above. Using the isomorphism $(D_n + i|\eta'|)^{\tilde{r}} : H^{\tilde{r}}(\mathbb{R}_+) \to L_2(\mathbb{R}_+)$, we reduce the study of the family (2.19) to the study of the family

$$\hat{A}_{\mathrm{IV},\tilde{r}}(y', \eta') = (D_n + i|\eta'|)^{\tilde{r}} \hat{A}_{\mathrm{IV}}(y', \eta') : \widetilde{H}^m_{(k,\tilde{r})}(\mathbb{R}_+) \to L_2(\mathbb{R}_+).$$

Applying Theorem 2.5 to this family, we obtain the following theorem:

Theorem 2.7. *If condition* (2.9) *holds and for all* $y' \in \Gamma$

$$M_r(y', \lambda) \neq 0 \ \textit{for} \ \operatorname{Im}\lambda = \tilde{r} + \frac{1}{2}, \tag{2.20}$$

then the operator (2.19) *is a Fredholm operator and its index is* $\varkappa(\tilde{r}) = \nu(\tilde{r}) + \nu' + r - m$, *where* $\nu(\tilde{r})$ *is the number of roots of the equation* $M_r(y', \lambda) = 0$ *in the half-plane* $\operatorname{Im}\lambda < -\tilde{r} + \frac{1}{2}$.

Thus, if there are roots of the equation $M_r(y', \lambda) = 0$ in the half-plane $\operatorname{Im}\lambda < \frac{1}{2}$, then the index of the operators (2.19), and hence also the number of boundary and/or co-boundary conditions that must be adjoined to Eq. (1.1) depends on $\tilde{r}$; moreover condition (2.20) may fail for certain values of

$\tilde{r} \in \mathbb{Z}_+$. At the same time, under condition (2.9) there exists r_0 (depending on A, or more precisely, on $M_r(y',\lambda)$) such that condition (2.9) holds for $\tilde{r} \geq r_0$, and $\nu(\tilde{r}) = m - r$ and $\varkappa(\tilde{r}) = \nu'$ are independent of $\tilde{r}$. Furthermore $\operatorname{Ker} \hat{A}_{\mathrm{IV}}(y',\eta') \subset S(\mathbb{R}_+)$ when $r \geq r_0$, and consequently the kernel is independent of r. Hence one can construct a boundary value problem for which the corresponding operator-valued symbol has a left inverse for all $\tilde{r} \geq r_0$. Therefore for all $\tilde{r} \geq r_0$ the following *a priori estimate* holds:

$$\|u\|_{m,(k,\tilde{r})} \leq c_{\tilde{r}}\big(\|\mathfrak{A}u\|_{H^{\tilde{r}}(\Omega)\oplus\mathcal{H}'_{\tilde{r}}(\Gamma)} + \|u\|_{L_2(\Omega)}\big),$$

where $\mathcal{H}'_{\tilde{r}} = \oplus H^{r+\tilde{r}-j-\frac{1}{2}}(\Gamma)$. In particular if $u \in H^{r_0}(\Omega)$ and $\mathfrak{A}u \in C^\infty(\overline{\Omega}) \oplus C^\infty(\Gamma)^\varkappa$, then $u \in C^\infty(\overline{\Omega})$.

Of course this last assertion and the a priori estimate admit the usual localization.

We conclude this section by proving Theorem 2.5. We fix $(y',\eta') \in T^*\Gamma \setminus 0$ and set $\hat{A} = \hat{A}_{\mathrm{IV}}(y',\eta')$, $I_1 = (0,1]$, $I_2 = [1,\infty)$. We define the spaces $H_j = \widetilde{H}^m_{(k)}(I_j)$ in analogy with the space $H = \widetilde{H}^m_{(k)}(\mathbb{R}_+)$, and we denote by $\hat{A}_j$ the operator $\hat{A} : H_j \to L_2(I_j)$. It will be shown below that

$$\hat{A}_1 \text{ is a Fredholm operator with index } r+\nu \tag{2.21}$$

$$\hat{A}_2 \text{ is a Fredholm operator with index } \nu'. \tag{2.22}$$

Since H is a subspace in $H_1 \oplus H_2$ of codimension m, it follows from (2.21) and (2.22) that $\hat{A}$ is a Fredholm operator whose index is $\varkappa = r + \nu + \nu' - m$.

To prove (2.21) we denote by $\widetilde{H}_{r,s}$ the space of functions on $(0,1]$ with finite norm

$$\|u\|^{\sim}_{r,s} = \Big(\sum_{0\leq j\leq r} \|D^j_n y_n^{j+s} u(y_n)\|^2_{L_2((0,1])}\Big)^{\frac{1}{2}}.$$

We set

$$\widetilde{H}_r = \widetilde{H}_{r,0}, \quad B_1 = h_n^{s^0+r} D^r_n : H_1 \to \widetilde{H}_r,$$

$$B_2 = \sum a_{(0,j)}(y',0) y_n^{s^0+j} D_n^{j-r} y_n^{-s^0-r} : \widetilde{H}_r \to L_2(I_1), \quad \hat{A}'_1 = B_2 B_1,$$

and we remark that $\hat{A}_1 - \hat{A}'_1 : H_1 \to L_2(I_1)$ is a compact operator, while the norm in H_1 is equivalent to the norm

$$\|u\|_{m,(k)} = \Big(\sum_{r\leq j\leq m} \|y_n^{s^0+j} D^j_n u\|^2_{L_2} + \sum_{0\leq j\leq r} \|y_n^{s^0+r} D^j_n u\|^2_{L_2}\Big)^{\frac{1}{2}}$$

(both assertions follow from the condition $\gamma' > 0$ and Hardy's inequality). Therefore it suffices to prove (2.21) for the operator $\hat{A}'_1$.

Making the change of variable $y_n = e^t$, we show easily that B_2 is similar to the operator $M_r(y', D_t + i(s^0 + r + \frac{1}{2})) : H^{m-r}(\mathbb{R}_+) \to L_2(\mathbb{R}_-)$, which is surjective when condition (2.15) holds and has a kernel of dimension ν. Consequently the same is true of B_2.

We begin our study of the operator B_1 by proving the a priori estimate

$$\|u\|_{H_1} \le C\big(\|B_1u\|_{\tilde{H}_r} + \|u\|_{L_2([c,1])}\big) \quad \forall u \in H_1, \tag{2.23}$$

where $C > 0$ and $c > 0$ are independent of u. Since B_1 is elliptic on $(0,1]$, it suffices to prove (2.23) for all $u \in C_0^\infty([0,1))$. But then $\|y_n^{s^0+r}D_n^j u\|_{L_2} \le c\|B_1u\|_{L_2}$, $j = 0,1,\dots,r$, by Hardy's inequality; and, applying Leibniz' formula, we have successively

$$\|y_n^{s^0+r+1}D_n^{r+1}u\|_{L_2} \le c\|y_n^{s^0+r}D_n^r u\|_{L_2} + c_1\|D_n y_n y_n^{s^0+r}D_n^r u\|_{L_2} \le c_2\|B_1u\|_{\tilde{H}_1},$$
$$\|y_n^{s^0+r+j}D_n^{r+j}u\|_{L_2} \le c_j\|B_1u\|_{\tilde{H}_j}, \quad j = 2,3,\dots .$$

These estimates give estimate (2.23), and from this estimate and the compactness of the form $\|u\|^2_{L_2([c,1])}$ with respect to the form $\|u\|^2_{\tilde{H}_r}$, it follows that $\operatorname{Im} B_1$ is closed and $\operatorname{Ker} B_1$ is finite-dimensional. Obviously any vector v from the dense subspace $C_0^\infty((0,1])$ in $\tilde{H}_r$ has a pre-image u in $C_0^\infty((0,1])$, and so B_1 is surjective and a Fredholm operator. It is also obvious that $\operatorname{Ker} B_1 = \mathcal{L}(1, y_n, \dots, y_n^{r-1})$, so that $\operatorname{Ind} B_1 = r$.

Property (2.21) for $\hat{A}' = B_2B_1$, and hence also for $\hat{A}_1$, follows from the properties just established for the operators B_1 and B_2.

To prove (2.22) we set $\mu = \frac{1}{\gamma'}$ and make the change of variable $y_n = t^\mu$, $u(y_n) = t^\sigma v(t)$, $f(y_n) = t^\omega g(t)$ in the equation $\hat{A}_2 u = f$, where $\sigma = -(s^0 + m)\mu - (1-\mu)/2$ and $\omega = -(1-\mu)/2$. This change of variable shows that $\hat{A}_2$ is similar to the operator $\hat{A}_2' + \hat{A}_2'' : H^m([1,\infty)) \to L_2([1,\infty))$, where $\hat{A}_2' = a_{(s^0_+)}(y', \eta', \frac{1}{\mu}D_t)$ and $\hat{A}_2''$ has the form

$$\hat{A}_2'' = \sum_{0\le j<m} b_j(y',\eta',t)D_t^j$$

with $|b_j(y',\eta',t)| \le Ct^{-\delta_j}$ for some $\delta_j > 0$. Consequently $\hat{A}_2''$ is compact, and it suffices to prove (2.22) for the operator $\hat{A}_2'$. But this is a well-known fact.

§3. General Classes of Operators

3.1. Definition of the Types of Operators and Their Symbols. Let A be an operator of the form (1.2). We set $s' = \min\{k_{l0} \,|\, 0 < l \le m\}$, $s_n = \min\{k_{0j} - j \,|\, 0 < j \le m\}$ and assign the operator A to *type I* if $s' > k_{00}$ and $s_n > k_{00}$, to *type II* if $s_n > s' = k_{00}$, to *type III* if $s_n = s' = k_{00}$, and to *type IV* if $s' > s_n$ and $s_n \le k_{00}$. Let the polytope Q be the same as in Sect. 1.1. For $s \le s^0 = \min\{k_{00}, s_n, s'\}$ we denote by Q_s the plane of support of Q passing through the point $(0,0,s)$ and through at least two points of the form $(l,0,k_{l0})$ and $(0,j,k_{0j})$ with $lj \ne 0$, and for $s \le s^0$ we set $I(s) = \{(l,j) \,|\, (l,j,k_{lj}) \in Q_s\}$,

$\hat{I}(s) = \{\alpha | (|\alpha'|, \alpha_n) \in I(s)\}$, $\tilde{a}_{(s)}(y,\eta) = \sum_{\alpha\in\hat{I}(s)} a_\alpha(y',0) y_n^{\hat{k}_\alpha}\eta^\alpha$, $a_{(s)}(y',\eta) = \tilde{a}_{(s)}(y',1,\eta)$. For $s < s^0$ the symbols $\tilde{a}_{(s)}$ and $a_{(s)}$ are well defined as functions on $T^*\Gamma \times \mathbb{R}_+ \times \mathbb{R}$ and $T^*\Gamma \times \mathbb{R}$ respectively, and for operators of type I the symbol $a_{(s^0)}$ is also well defined. The number of distinct symbols $\tilde{a}_{(s)}$ is finite, since $Q_s \cap Q'$ is either a face or an edge of the surface Q'. Let $s^1 > \cdots > s^{\hat{n}}$ be all of the $s < s^0$ for which $Q_s \cap Q'$ is a face. We set $s^{\widehat{n+1}} = -\infty$ and choose $s^i_+ \in (s^{i+1}, s^i)$; then $a_{(s^0)}$, $\tilde{a}_{(s^i)}$, and $a_{(s^i_+)}$, $i = 1, 2, \ldots, \hat{n}$, is a complete set of symbols.

Example 3.1. The following figure shows a surface Q' of type IV for which $Q_{s_0} \cap Q'$ is the triangle ABC, $Q_{s^1} \cap Q'$ is the quadrilateral $ACDE$, and $Q_{s^0_+} \cap Q'$ and $Q_{s^1_+} \cap Q'$ are the edges AC and DE.

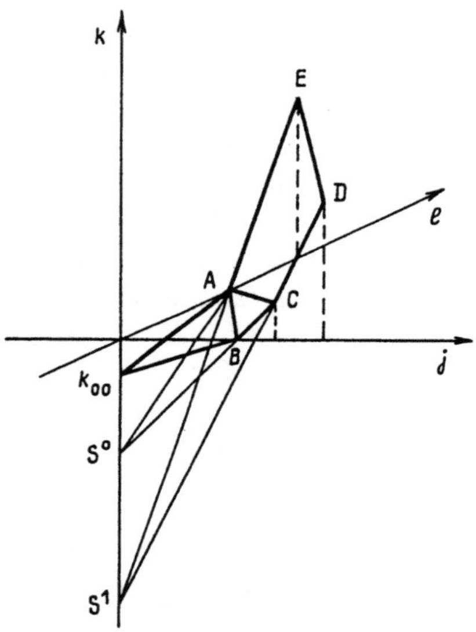

Fig. 1

For operators of types II–IV we define the symbols $\hat{A}_{\mathrm{II}}$, $\hat{A}_{\mathrm{III}}$, and $\hat{A}_{\mathrm{IV}}$ by Eqs. (2.8), (2.12), and (2.13) respectively, replacing summations over $|\alpha| \leq m$ by summations over $\alpha \in \hat{I}(s^0)$.

3.2. Conditions for an Operator to Be a Fredholm Operator.

Theorem 3.1. *Let $A : H^m_{(k)}(\Omega) \to L_2(\Omega)$ be an operator of type I (resp. type II; type III). Then the following conditions are equivalent:*

a) *Condition* (2.2) *holds along with the conditions*

$$\forall i > 0\, \tilde{a}_{(s^i)}(y,\eta) \neq 0 \quad \forall (y,\eta) \in T^*\Gamma \times \dot{\mathbb{R}}_+ \times \mathbb{R}, \quad |\eta| \neq 0, \tag{3.1}$$

$$\forall i > 0\, a_{(s^i_+)}(y',\eta) \neq 0 \quad \forall (y',\eta) \in T^*\Gamma \times \mathbb{R}, \quad |\eta| \neq 0, \tag{3.2}$$

as well as conditions (2.3) *(resp. condition* (2.10)*; the condition that the operator* $\hat{A}_{\mathrm{III}}(\lambda)^{-1}$ *has no poles on the line* $\operatorname{Im}\lambda = s^0 + \frac{1}{2}$*).*

b) *The operator* $A : H^m_{(k)}(\Omega) \to L_2(\Omega)$ *is a Fredholm operator.*

c) *The a priori estimate* (1.3) *holds.*

As in Sect. 2 one can introduce scales of weighted spaces and state an analog of Theorem 3.1 for Eq. (1.1) with right-hand side having a high degree of smoothness. For operators of types I and II there are analogs of Corollary 2.1, Remark 2.1, and Theorem 2.2.

Now let A be an operator of type IV, and suppose for simplicity that all $k_{lj} \geq 0$. From the symbol $\hat{A}_{\mathrm{IV}}(y',\eta')$ we determine the number and form of the boundary and co-boundary conditions adjoined to Eq. (1.1), as was done in Sect. 2.4, and we denote by $\mathfrak{A}$ the operator of the corresponding boundary value problem and by $\hat{\mathfrak{A}}(y',\eta')$ its operator-valued boundary symbol.

Theorem 3.2 *Suppose conditions* (2.2), (3.1), *and* (3.2) *hold and that for all* $(y',\eta') \in T^*\Gamma \setminus 0$ *the operators* $\hat{\mathfrak{A}}(y',\eta')$ *are invertible.*

Then $\mathfrak{A}$ *is a Fredholm operator, and the a priori estimate analogous to* (2.17) *holds.*

As in Sect. 2.4 we can consider only problems with data having a high degree of smoothness. The number of boundary conditions generally depends on the smoothness of the data, i.e., on the spaces in which the operator is considered, but there are classes of operators for which the number of boundary conditions is independent of the smoothness. These are operators for which the operator-valued symbol $\hat{\mathfrak{A}}(y',\eta')$ is a family of elliptic operators with parameter on a half-line, i.e., operators for which $I(s^0)$ has the form

$$\{(l,j) \,|\, l/m_0' + j/m_{n0} = 1\}, \quad \text{where } m_0', m_{n0} \in \mathbb{N}, \tag{3.3}$$

and $k_{lj} = 0$ for all $(l,j) \in I(s^0)$.

An example of such an operator is the operator (2.18) under the hypotheses of part c) of Example 2.4 when $\hat{A}_{\mathrm{IV}}(y',\eta') = (|\eta'|^2 - \partial_n^2)^{\frac{m_1}{2}}$.

We now sketch the proofs of Theorems 3.1 and 3.2. Operators of type I are hypoelliptic operators in the sense of the calculus of pseudodifferential operators with double symbols (cf. Sect. 6.2). Therefore Theorem 3.1 for operators of type I follows from the general theorems of this calculus. Operators of type II can be studied in exactly the same way after they have been realized as operators on $\mathbb{R}_+$ with operator-valued symbols. In addition, in both the first and second cases one can use an infinite partition of unity (on Ω and $\mathbb{R}_+$ respectively).

Theorem 3.2 reduces to Theorem 2.6 in the following way. Operators of type IV admit a localization in a neighborhood of any point of the boundary and a local rectification of the boundary, so that it suffices to construct a local regularizer for the operator in a half-space. To construct a local regularizer at zero, we partition $\mathbb{R}^n_+ \times \mathbb{R}^n$ into two zones such that A is hypoelliptic in the inner zone and differs by only a small amount from the operator A^0 defined like A, only with summation over $\alpha \in \hat{I}(s^0)$, in the zone adjacent to the boundary. In the second zone A is close to A^0, so that a microlocal regularizer can be constructed in this zone starting from the regularizer of the boundary value problem for the operator A^0, while in the first zone a microlocal regularizer can be constructed using the calculus of pseudodifferential operators with double symbols. The usual calculus can be used only in the case of coefficients that are smooth all the way to the boundary, i.e., in the case $k_{lj} \in \mathbb{Z}_+$. The microlocal regularizers are pieced together using the same calculus.

Theorem 3.1 can be proved for operators of types II and III following approximately the same outline. Here the microlocalization in a neighborhood of $T^*\Gamma$ is slightly more complicated to carry out, since regularizers cannot be pieced together from local regularizers in the usual way.

3.3. Non-power Degeneracy. Let $\rho \in C^\infty(\mathbb{R}_+)$, $\rho > 0$, $\rho(t) = 1$ for $t \geq 1$, $\rho(+0) = 0$, and $|\rho^{(s)}| \leq c_s \rho^{1-s}$ for all $s \in \mathbb{Z}_+$. In (1.2) and below we replace y_n by $\rho(y_n)$; then for operators of types I and II all the results (Theorems 2.1–2.3 and 3.1, Corollary 2.1, and Remark 2.1 for operators of type I) remain valid, except that the space $S_{d(x)}(\Omega)$ must be replaced by $S_{\rho(d(x))}(\Omega)$.

Theorem 3.2 remains valid if condition (3.3) holds and

$$k_{lj} = 0 \text{ for } l/m_0' + j/m_{n0} \leq 1. \tag{3.4}$$

All the proofs are modified in the obvious way.

3.4. L_p-Coercivity ($1 \leq p < \infty$). We define the analog of the space $H^m_{(k)}(\Omega)$ starting from $L_p(\Omega)$ rather than from $L_2(\Omega)$. Then the assertions of Theorem 2.1 and Theorem 3.1 for operators of type I and Theorem 3.2 under conditions (3.3) and (3.4) remain valid (even in the case of non-power degeneracy).

§4. Degeneracy on Part of the Boundary and the Case of a Nonsmooth Boundary

4.1. Isotropic Degeneracy on a Nonsmooth Boundary. We shall call the degeneracy *isotropic* if the operator has the form

$$A = \sum_{|\alpha| \leq m} a_\alpha(x) \rho_{|\alpha|}(x) D^\alpha,$$

where ρ_j are functions that determine the character of the degeneracy. Only degeneracy of types I and IV can be isotropic.

If A is strongly degenerate on the whole boundary, it can be included in the class of pseudodifferential operators with double symbols and studied using the general theorems of Sect. 6.2 (cf. Example 6.2); it can also be studied using an infinite partition of unity (cf. Triebel 1978, Ch. VI). When this is done, no restrictions are imposed on the smoothness of the boundary.

If A is an isotropic operator of type IV that degenerates strongly on the boundary into an elliptic operator A^0 of lower order, then the conditions on the smoothness of Γ are determined by the operator A^0 alone, since in the case of isotropic degeneracy we can reduce the study of the operator A in a neighborhood of $T^*\Gamma$ to the study of the operator A^0 without rectifying the boundary locally (cf. Levendorskij 1988a).

If the limiting operator A^0 is nonelliptic, the smoothness requirement can be reduced by considering the operator in divergence form

$$L = \sum_{|\alpha|,|\beta|\leq m} D^\alpha a_{\alpha\beta}(x) d(x)^{-2m-2r+|\alpha|+|\beta|} D^\beta. \tag{4.1}$$

Lizorkin (1985) and Lizorkin and Nikol'skij (1981) (cf. also the bibliographies in the second of these papers) have studied the first boundary problem for the operator (4.1) in generalized and variational formulations; they have obtained conditions for solvability and unique solvability and for the operator to be a Fredholm operator, and have also studied the smoothness of the solutions as functions of the smoothness of the data of the problem. It was assumed that $0 < m + r - \frac{1}{2} < m$ and $\Gamma \in C^{s+1}$, where $s \in \mathbb{N}$ satisfies the conditions $m+r-\frac{1}{2} \leq s < m+r+\frac{1}{2}$, $2s \geq m$; s conditions are imposed on the boundary. We note that the condition $\Gamma \in C^{s+1}$ is in general weaker than the condition $\Gamma \in C^m$, which makes it possible to rectify Γ locally in studying the bilinear form corresponding to the operator (4.1).

4.2. Regular Degeneracy on Part of the Boundary and Different Kinds of Degeneracy on Different Subsets of the Boundary. Let $\partial\Omega = \Gamma_1 \cup \Gamma_2$, where Γ_1 and Γ_2 intersect at a nonzero angle, and suppose that either A is degenerate only on Γ_1 or A is degenerate on Γ_1 and Γ_2 independently, i.e., the functions defining the degeneracy are products of functions that depend either only on the distance from Γ_1 or only on the distance from Γ_2. The analysis of the methods of studying the different types of operators carried out in Sect. 1.3 shows that if there is degeneracy on Γ_1 of one of the types I, II, or III, then analogs of the theorems of Sects. 2 and 3 can be obtained. In doing so, if there is no degeneracy or degeneracy of type IV on Γ_2 and it is necessary to impose boundary conditions on Γ_2, then weight functions depending on the distance from Γ_1 must be introduced into the conditions.

Regular degeneracy on a submanifold Γ' of the boundary $\partial\Omega$ or at isolated points of the boundary reduces to the cases studied above (it suffices to perform a change of variable in a neighborhood of Γ').

Certain cases of degeneracy considered in this section were studied earlier (Bagirov 1987; Fursikov 1971ac; cf. also the bibliography in Glushko and Savchenko 1985).

§5. Some Classes of Hypoelliptic Pseudodifferential Operators

5.1. Slowly Varying Hypoelliptic Operators. We recall that the operator A is *hypoelliptic* if

$$\operatorname{sing\,supp} u = \operatorname{sing\,supp} Au, \quad u \in \mathcal{D}'.$$

Here $\operatorname{sing\,supp} u$ is the complement of the largest open subset on which $u \in C^\infty$. An elliptic classical pseudodifferential operator (with singular support, like all the operators below) is hypoelliptic. To prove this one can construct a parametrix using the calculus of classical pseudodifferential operators. One can also prove that certain classes of nonelliptic pseudodifferential operators are hypoelliptic following this scheme, but it is necessary to enlarge the class of operators and use more general calculi.

The simplest extension of the class of classical symbols is the class of symbols $S^m_{\rho,\delta}(\Omega)$, where $\Omega \subset \mathrm{R}^n$ is an open set, $0 \le \delta < \rho \le 1$, $m \in \mathbb{R}$, satisfying the following condition: for any compact set $K \subset \Omega$ and $(x, \xi) \in K \times \mathbb{R}^n$

$$|a^{(\alpha)}_{(\beta)}(x,\xi)| \le c_{\alpha\beta}\langle\xi\rangle^{m-\rho|\alpha|+\delta|\beta|} \quad \forall\alpha,\beta,$$

where $a^{(\alpha)}_{(\beta)} = \partial^\alpha_\xi D^\beta_x$ and $\langle\xi\rangle = (1+|\xi|^2)^{\frac{1}{2}}$. If in addition there exists m_0 such that

$$c|\xi|^{m_0} \le |a(x,\xi)|, \quad |a^{(\alpha)}_{(\beta)}(x,\xi)a(x,\xi)^{-1}| \le c_{\alpha\beta}|\xi|^{-\rho|\alpha|+\delta|\beta|},$$

for any compact set $K \subset \Omega$ when $x \in K$ and $|\xi| \ge C$, where $c > 0$, C, and $c_{\alpha\beta}$ ($|\alpha|+|\beta| = 0, 1, \dots$) depend only on K and a, then we write $a \in HS^{m,m_0}_{\rho,\delta}(\Omega)$. If all the conditions hold with $K = \Omega = \mathbb{R}^n$, we write $a \in S^m_{\rho,\delta}$, $a \in HS^{m,m_0}_{\rho,\delta}$.

The corresponding classes of operators will be denoted by the symbols $L^m_{\rho,\delta}(\Omega)$, $HL^{m,m_0}_{\rho,\delta}(\Omega)$, $L^m_{\rho,\delta}$, and $HL^{m,m_0}_{\rho,\delta}$.

Theorem 5.1. *The operator $A \in HL^{m,m_0}_{\rho,\delta}(\Omega)$ is hypoelliptic in Ω. If $\omega \subset \Omega$ is an open subset, then*

$$Au\big|_\omega \in H^s_{\mathrm{loc}}(\omega) \Rightarrow u\big|_\omega \in H^{s+m-\varepsilon}_{\mathrm{loc}}(\omega), \tag{5.1}$$

where $\varepsilon = m - m_0$.

Proof W e construct a proper pseudodifferential operator B with symbol equal to $a(x,\xi)^{-1}$ for $|\xi| \ge C$. Then

$$AB = I - T_1, \quad BA = I - T_2, \text{ where } T_j \in L^{-\Delta}_{\rho,\delta}(\Omega) \tag{5.2}$$

and $\Delta = \rho - \delta > 0$, so that the same procedure as in the case of classical pseudodifferential operators enables us to construct a *parametrix* $R \in L^{-m_0}_{\rho,\delta}(\Omega)$ of the operator A, i.e., an operator such that $AR - I$ and $RA - I$ are operators with C^∞ kernels. This proves hypoellipticity; to prove (5.1) we need to remark that a proper pseudodifferential operator of class $L^{-m_0}_{\rho,\delta}(\Omega)$ is bounded as a mapping from $H^s_{\text{loc}}(\Omega)$ into $H^{s+m_0}_{\text{loc}}(\Omega)$ for all s.

Example 5.1. Let $\nu, \mu \in \mathbb{N}$, $\mu < \nu$, and $c > 0$. Then $A = c + |x|^{2\nu}(-\Delta)^\mu$ satisfies the hypotheses of Theorem 5.1 with $\Omega = \mathbb{R}^n$, $m = 2\mu$, $m_0 = 0$, $\rho = 1$, and $\delta = \mu/\nu$.

Remark.

a) If $\varepsilon > 0$ in (5.1), we say that A is *hypoelliptic with the loss of* ε *derivatives* (for an elliptic operator $\varepsilon = 0$).
b) If $1 - \rho \leq \delta$, the classes of operators just introduced are invariant with respect to changes of variables, and so one can state an analog of Theorem 5.1 for operators on a manifold. If the manifold is closed, one can construct weighted spaces of Sobolev type in which the operator will be a Fredholm operator.
c) We have obtained the key equalities (5.2) by an elementary method because of the condition $\rho > \delta$. One can construct large classes of operators that do not belong to the class $HL^{m,m_0}_{\rho,\delta}$ with $\rho > \delta$, but belong to slightly more general algebras of pseudodifferential operators in which analogs of Eqs. (5.1) can be obtained equally simply. We shall discuss the corresponding calculus in Sect. 6.1. In the next section we consider a class of operators for which (5.2) can be obtained using a more difficult technique.

5.2. Hypoelliptic Operators with Smooth Multiple Characteristics. Let Ω be a closed manifold, $\Sigma \subseteq T^*\Omega \setminus 0$ a smooth conic submanifold, and $m \in \mathbb{R}$. Let $k \geq 1$ be an integer, and A a classical pseudodifferential operator of order m on Ω such that in any local coordinate system its symbol $a \sim \sum a_{m-j}$ satisfies the following conditions:

the functions a_{m-j} *have zeros of order at least* $2(k-j)$ *on* Σ, $j = 0, 1, \ldots, k-1$, and

$$a_m \text{ vanishes with order precisely } 2k. \tag{5.3}$$

With the operator A we associate a family of differential operators in $\mathbb{R}^n$ with polynomial coefficients

$$\sigma_\rho(A) = \sum_{|\alpha+\beta|+2j=2k} \frac{i^{|\beta|}}{\alpha!\beta!}(a_{m-j})^{(\alpha)}_{(\beta)}(x,\xi) y^\alpha D^\beta_y,$$

parameterized by the points $\rho = (x, \xi) \in \Sigma$.

Theorem 5.2. *Suppose condition* (5.3) *holds,*

$$a_m(x,\xi) \neq 0 \quad \forall\, (x,\xi) \in (T^*\Omega \setminus 0) \setminus \Sigma \tag{5.4}$$

and

$$\sigma_\rho(A) \text{ is invertible on } S(\mathbb{R}^n) \quad \forall \rho \in \Sigma. \tag{5.5}$$

Then A is hypoelliptic and for all s and N

$$\|u\|_{s+m-k} \le C_s\|Au\|_s + c'_{s,N}\|u\|_{-N} \tag{5.6}$$

for any $u \in \mathcal{D}'(\Omega)$ for which the right-hand side is finite.

Corollary 5.1 *Let $m > k$, and let A satisfy the hypotheses of Theorem 5.2 and be formally self-adjoint with respect to some positive smooth density on Ω. Then A is a self-adjoint operator with discrete spectrum.*

We note that conditions (5.3)–(5.5) are satisfied simultaneously in all local coordinate systems.

Outline of the Proof of Theorem 5.2. It suffices to construct a parametrix, which must be a bounded operator from $H^s(\Omega)$ into $H^{s+m-k}(\Omega)$ for all s. Under the conditions (5.3) and (5.4) the operator A can belong locally to the class $HL^{m,m-k}_{\rho,\delta}$ only for $\rho = \delta = \frac{1}{2}$ and therefore it is not possible to construct the parametrix by the elementary method of Sect. 5.1.

Since the construction of the parametrix is a microlocal problem, we can assume that $\Omega = \mathbb{R}^n$ and $\Sigma \subset \mathbb{R}^n \times (\mathbb{R}^n \setminus 0)$. We set $d(x,\xi) = \inf\{|x-y| + |\eta - \xi|/|\xi| \mid (y,\eta) \in \Sigma\}$ and distinguish the set $V_{c,\varepsilon}$ by the condition $d(x,\xi) \le c\langle\xi\rangle^{-\frac{1}{2}+\varepsilon}$. Under the conditions (5.3) and (5.4) a belongs to $HS^{m,m-k}_{\rho,\delta}$ on the set $V_{-c,\varepsilon} = \mathbb{R}^{2n} \setminus U_{c,\varepsilon}$ with $\rho = \frac{1}{2}+\varepsilon$, $\delta = \frac{1}{2} - \varepsilon$ for any $c, \varepsilon > 0$, and therefore we can construct a microlocal parametrix on $V_{-c,\varepsilon}$ as in Sect. 5.1. On the set $V_{c,\varepsilon}$ we can carry out a further microlocalization and reduce the study to the construction of microlocal parametrices on the ellipsoids

$$V^\varepsilon_\rho = \{(y,\eta) \mid \langle\xi\rangle^{1-2\varepsilon}|x-y|^2 + \langle\xi\rangle^{-1-2\varepsilon}|\xi-\eta|^2 < 1\}$$

with centers at the points $\rho = (x,\xi) \in \Sigma$. If $\varepsilon > 0$ is sufficiently small, the operator A differs little from $\sigma_\rho(A)$ on V^ε_ρ (the two differ by an operator of order less than $m-k$), and therefore the construction of microlocal parametrices on V^ε_ρ can be begun with the construction of the inverses of the operators $\sigma_\rho(A)$. To study $\sigma_\rho(A)$ we choose canonical coordinates in $\mathbb{R}^{2n}$ so that $T_\rho\Sigma = \{(x,\xi) \mid x^2 = 0,\ \xi^3 = 0,\ x^4 = \xi^4 = 0\}$, where $x = (x^1, x^2, x^3, x^4)$ is a partition of the coordinates into groups (some groups may be absent). It can be shown that conditions (5.3)–(5.5) are symplectically invariant. By (5.3) the symbol $a_\rho(y,\eta)$ of the operator $\sigma_\rho(A)$ is a polynomial in $(y^2, \eta^3, y^4, \eta^4)$, and therefore to describe $\sigma_\rho(A)$ it suffices to describe the inverse of the operator-valued function $(y^2,\eta^3) \mapsto a_\rho(y^2,\eta^3,y^4,D^4_y)$. Using the global calculus of pseudodifferential operators on the entire space one can show that $a_\rho(y^2,\eta^3,y^4 D^4_y)^{-1}$ is a pseudodifferential operator and its symbol, as a function of (y^4,η^4) and the parameters (y^2,η^3), satisfies the necessary estimates. Consequently $\sigma_\rho(A)^{-1}$ is also a pseudodifferential operator with the required properties.

Remark.

a) Along with the proof of Theorem 5.2 Boutet de Monvel, Grigis, and Helffer (1976) also describe the class of pseudodifferential operators to which the parametrix belongs. The model operators were first studied by Grushin (1970; 1971; 1972).

b) It is possible to exhibit pairs of spaces in which A determines a Fredholm operator.

c) For hypoellipticity and the estimate (5.6) it suffices to require the existence of a left inverse for the operators $\sigma_\rho(A)$ (under assumption (5.3) the existence of a left inverse is equivalent to the assumption that the kernels are trivial).

d) The symbol $\sigma_\rho(A)$ can be replaced by other symbols (cf. Levendorskij 1988a; Helffer 1976). For example if Σ is involutive one can use a scalar symbol defined on $T(T^*\Omega \setminus 0)|_\Sigma / T\Sigma$; if Σ is symplectic, one can use an operator-valued symbol defined on Σ and assuming values in $\mathcal{L}(S(\mathbb{R}^d))$, where $2d = \operatorname{codim} \Sigma$.

e) If $\Sigma = T^*\Gamma$, where $\Gamma \subset \Omega$ is a submanifold, and condition (5.5) does not hold, one can adjoin to the equation $Au = f$ the *interior boundary and coboundary conditions* on Γ in analogy with the formulation of the boundary value problems for operators of type IV on the boundary. For another approach to the statement of the problems for such operators see Chapter 2.

f) If Σ is the same as in d), and condition (5.5) does not hold, then the hypoellipticity of the operator A can be studied by constructing an auxiliary operator on Γ (Grushin 1971), the analog of reducing the study of a boundary value problem to the study of an auxiliary operator on the boundary.

g) A (micro)local variant of Theorem 5.2 can also be stated.

The following example illustrates parts d)–g) of Remark 5.3.

Example 5.2. The operator $A = D_2^2 + x_2^2 D_1^2 + \lambda(x) D_1$ satisfies condition (5.3) with $\Sigma = \{(x,\xi) |\, x_2 = \xi_2 = 0\}$, $m = 2$, $k = 1$, and the operator is hypoelliptic at 0 if the operators $D_2^2 + x_2^2\xi_1^2 + \lambda(0)\xi_1$ are invertible in $S(\mathbb{R}^1)$ when $\xi_1 \neq 0$. Since these operators are unitarily equivalent to the operators $|\xi_1|(\Lambda + \lambda(0)\operatorname{sgn}\xi_1)$, where $\Lambda = D_2^2 + x_2^2$, and the eigenvalues of the operator Λ are $2n+1$, $n \in \mathbb{Z}_+$, the condition of hypoellipticity at 0 assumes the form

$$\lambda(0) \neq \pm(2n+1) \quad \forall n \in \mathbb{Z}_+. \tag{5.7}$$

If the condition (5.7) is not satisfied, we can adjoin the interior boundary condition $u|\Gamma = g$, where $\Gamma = \{x |\, x_2 = 0\}$. The corresponding *interior boundary problem* will be hypoelliptic, since the eigenspaces of the operator Λ are one-dimensional and the eigenfunctions do not vanish at zero.

It can be shown that the condition (5.7) is a necessary condition for hypoellipticity of the operator A; however the nonhypoelliptic operator $A =$

$x_2^2D_1^2 + D_2^2 + D_1$ can be made hypoelliptic by the weak perturbation $iax_2^2D_1$, where $a \neq 0$ (Grushin 1971). To prove this we construct an auxiliary operator on the line $x_2 = 0$ and prove that it is hypoelliptic.

Remark. If $k = 1$ and the principal symbol satisfies the condition

$$a_m(x, \xi) \in U = \{z \in \mathbb{C} |\, |\mathrm{Im}\, z| \leq \gamma \mathrm{Re}\, z\}, \tag{5.8}$$

then the eigenvalues of the operator $\sigma_{x,\xi}(A)$ are easily computed, making it possible to restate condition (5.5). Let ω be a symplectic form on $T^*\Omega$ and $Q = Q_{x,\xi}$ the Hessian of the function $a_m(x,\xi)/2$ at the point $(x,\xi) \in \Sigma$. This is a quadratic form on $T(T^*\Omega)$; we denote the corresponding bilinear form by $Q(u,v)$. We now introduce the Hamiltonian mapping F associated with Q/i by means of the equality $\omega(u, Fv) = Q(u,v)/i$, and we denote the eigenvalues of F belonging to $V \setminus \{0\}$ by $\mu_1, \mu_2, \dots$. Then under the condition (5.3) condition (5.5) is equivalent to

$$a_{m-1}^s(x,\xi) + Q(\bar{v}, v) + \sum (2\alpha_j + 1)\mu_j \neq 0, \tag{5.9}$$

for all $\alpha_j \in \mathbb{Z}_+$, $|\xi| = 1$, where v is an associated vector corresponding to the eigenvalue zero and $a_{m-1}^s = a_{m-1} + \frac{i}{2} \sum \partial^2 a_m / \partial x_j \partial \xi_j$ is the subprincipal symbol of the operator A.

When $k = 1$, estimate (5.6) means that A is hypoelliptic with the loss of one derivative; if $\Sigma \in C^\infty$ and conditions (5.3) and (5.8) hold, then (5.9) is necessary for hypoellipticity with the loss of one derivative. Under condition (5.8) there also exists a necessary and sufficient condition for hypoellipticity with the loss of one derivative in the case of an arbitrary Σ (cf. Hörmander 1979, Sect. 22.3).

More refined conditions for hypoellipticity are studied in Chapter 2.

5.3. The Sharp Gårding Inequality. Energy estimates are used in the proofs of existence, uniqueness, and finiteness of the rate of propagation for hyperbolic equations. Many of these reduce to *Gårding's inequality* (cf. Egorov, Shubin 1988), which is proved using the construction of an approximate square root of a positive classical pseudodifferential operator with positive principal symbol. This is easily done, since the square root of a positive elliptic symbol is also an elliptic symbol. By the same scheme one can prove a more refined result (Theorem 5.3), but in doing so it is necessary to go not only outside the class of classical pseudodifferential operators, but even outside the classes $L^m_{\rho,\delta}$, i.e., to use more complicated calculi (cf. Sect. 6.1).

In the following theorems A is a classical proper self-adjoint pseudodifferential operator of order m in the domain $\Omega \subset \mathbb{R}^n$ with symbol a, principal symbol a_m, and subprincipal symbol a_{m-1}^s, $K \subset \Omega$ is an arbitrary compact set, and Σ is the set of zeros of the principal symbol.

Theorem 5.3 (the sharp Gårding inequality). *If*

$$a_m(x,\xi) \geq 0 \quad \forall\,(x,\xi) \in T^*\Omega, \tag{5.10}$$

then the estimate

$$(Au,u) \geq -c_K\|u\|^2_{(m-\alpha)/2} \quad \forall\, u \in C_0^\infty(K) \tag{5.11}$$

holds when $\alpha = 1$.

The following theorem is significantly more difficult to prove than Theorem 5.3.

Theorem 5.4 (the Fefferman-Phong inequality). *If* $a(x,\xi) \geq 0$ *for all* $(x,\xi) \in T^*\Omega$, *then estimate* (5.11) *holds for* $\alpha = 2$.

Analogs of Theorems 5.3 and 5.4 also hold for symbols of significantly more general classes (cf. Hörmander 1979, Sect. 18.6). Theorem 5.3 also holds in the vector-valued case, but Theorem 5.4 does not.

Theorem 5.4 is very precise in some respects, but it requires the symbol to be nonnegative. We give below some results that hold without this hypothesis, but Σ must be a manifold.

Theorem 5.5 (Hörmander's inequality). *Suppose condition* (5.10) *holds and* $a^s_{m-1}(x,\xi) + \mathrm{Tr}^+ Q_{x,\xi} \geq 0$ *for all* $(x,\xi) \in \Sigma$, *where* $\mathrm{Tr}^+ Q_{x,\xi} = \Sigma\mu_j$. *Suppose also that the symplectic form has constant rank on* Σ *and* $T_{(x,\xi)}\Sigma$ *coincides with the radical of the form* $Q_{x,\xi}$ *for* $(x,\xi) \in \Sigma$. *Then the estimate* (5.11) *holds for* $\alpha = 2$.

Theorem 5.6 (Melin's inequality). *Let* a_m *and* a^s_{m-1} *satisfy condition* (5.10) *and* $a^s_{m-1}(x,\xi) + \mathrm{Tr}^+ Q_{x,\xi} > 0$ *for all* $(x,\xi) \in \Sigma$. *Then for any compact set* $K \subset \Omega$ *there exists* $c_K > 0$ *and* C_K *such that for* $k = 1$

$$(Au,u) \geq c_K\|u\|^2_{(m-k)/2} - C_K\|u\|^2_{(m-k-1)/2} \quad \forall\, u \in C_0^\infty(K). \tag{5.12}$$

The estimate in Theorem 5.6 is weaker than in Theorem 5.5, but additional conditions were imposed on Σ in Theorem 5.5.

The following theorem generalizes Melin's inequality and can be proved in analogy with Theorem 5.2.

Theorem 5.7. *Let the symbol* a *satisfy condition* (5.3), *let* a_m *satisfy condition* (5.10), *and let* $\sigma_p(A)$ *be positive definite for all* $\rho \in \Sigma$.

Then estimate (5.12) holds.

§6. Algebras of Pseudodifferential Operators Connected with Various Classes of Degenerate and Hypoelliptic Operators

6.1. General Calculi of Pseudodifferential Operators in $\mathbb{R}^n$. Beals and Fefferman (1974) and Beals (1975) have proposed describing classes of symbols using the estimates

$$|a^{(\alpha)}_{(\beta)}(x,\xi)p(x,\xi)^{-1}| \leq c_{\alpha\beta}\Phi(x,\xi)^{-\alpha}\varphi(x,\xi)^{-\beta}, \quad \forall\, \alpha,\beta \tag{6.1}$$

where $\Phi^\alpha = \Phi_1^{\alpha_1}\cdots\Phi_n^{\alpha_n}$ and the weight vectors Φ and φ and the function p must satisfy certain conditions. In other words, the derivatives of the symbols are described using the metric

$$g_{x,\xi}(y,\eta) = \Sigma(\Phi_j(x,\xi)^{-2}\eta_j^2 + \varphi_j(x,\xi)^{-2}y_j^2). \tag{6.2}$$

Hörmander (1985) used more general metrics; we state Hörmander's conditions below for a metric of the form (6.2).

We write $a \in S(p,g)$ if condition (6.1) holds and there exist $c > 0$ and C such that if $|x_j - y_j| \leq c\varphi_j(x,\xi)$ and $|\xi_j - \eta_j| \leq c\Phi_j(x,\xi)$ for all j, then

$$\sum(\varphi_k(x,\xi)\varphi_k(y,\eta)^{-1} + \Phi_k(x,\xi)\Phi_k(y,\eta)^{-1}) + p(x,\xi)p(y,\eta)^{-1} \leq C; \tag{6.3}$$

$$\sum(\varphi_k(x,\xi)\varphi_k(y,\eta)^{-1} + \Phi_k(x,\xi)\Phi_k(y,\eta)^{-1}) + p(x,\xi)p(y,\eta)^{-1} \leq$$
$$\leq C\Big(1 + \sum(\varphi_j(y,\eta)|\xi_j - \eta_j| + \Phi_j(y,\eta)|x_j - y_j|)\Big)^C; \tag{6.4}$$

$$h(x,\xi) = \max(\Phi_j\varphi_j)(x,\xi)^{-1} \leq C. \tag{6.5}$$

Example 6.1. Let $0 \leq \delta \leq \rho \leq 1$, $\delta < 1$, and

$$g_{x,\xi}(y,\eta) = \langle\xi\rangle^{-2\rho}|\eta|^2 + \langle\xi\rangle^{2\delta}|y|^2, \quad p(x,\xi) = \langle\xi\rangle^m.$$

Then g and p satisfy conditions (6.3)–(6.5) and $S(p,g) = S^m_{\rho,\delta}$.

We denote by $\mathcal{L}(p,g)$ the class of pseudodifferential operators on $\mathbb{R}^n$ with symbols in $S(p,g)$. The pseudodifferential operators of $\mathcal{L}(p,g)$ are continuous in $S(\mathbb{R}^n)$ and in $S'(\mathbb{R}^n)$, and the pseudodifferential operators of the class $\mathcal{L}(1,g)$ are bounded in $L_2(\mathbb{R}^n)$.

If $A = a(x,D) \in \mathcal{L}(p,g)$ and $B = b(x,D) \in \mathcal{L}(q,g)$, then AB is a pseudodifferential operator with symbol $c \in S(pq,g)$, and for all N

$$c - \sum_{|\alpha|<N}\frac{1}{\alpha!}a^{(\alpha)}b_{(\alpha)} \in S(h^N pq, g); \tag{6.6}$$

A^* is a pseudodifferential operator with symbol $a^* \in S(p,g)$, and for all N

$$a^* - \sum_{|\alpha|<N}\frac{1}{\alpha!}\bar{a}^{(\alpha)}_{(\alpha)} \in S(ph^N, g). \tag{6.7}$$

We shall say that $a \in S(p,g)$ is a *hypoelliptic symbol* if $|a(x,\xi)| \geq cp(x,\xi)$ for $|\xi| \geq C$, where $c > 0$. If $h(x,\xi) \leq C\langle\xi\rangle^{-c}$ for some C and $c > 0$, then the same reasoning as in Sect. 5.1 makes it possible to construct the parametrix of the operator $A = a(x,D)$ and thereby prove that it is hypoelliptic in $\mathbb{R}^n$.

Thus to prove hypoellipticity it suffices to include the symbol of the operator as a hypoelliptic symbol in some algebra of symbols satisfying the condition $h(x,\xi) \leq C\langle\xi\rangle^{-c}$.

The composition theorem makes it possible to prove that certain operators are hypoelliptic even though they have nonhypoelliptic symbols (Beals 1975). In this case it is necessary to invert the auxiliary operator-valued symbols (compare with Sect. 5.2).

The theorems just stated also make it possible to construct a scale of weighted spaces in which operators of class $\mathcal{L}(p,g)$ act and to study whether they are Fredholm operators, their spectral properties, etc. (cf. Beals 1975, 1981).

We note that a general calculus can be used to simplify the proofs of many theorems on classical pseudodifferential operators and differential operators and obtain new results, cf. Beals 1975, 1981; Hörmander 1979, 1985, Sect. 18. As an example we give the outline of the proof of Gårding's inequality (5.10) in the case $m = 1$, to which the general case is easily reduced. We must prove that the self-adjoint pseudodifferential operator $A \in \mathcal{L}(h^{-1},g)$ with symbol $a = a_1 + a_0$ is bounded below, where $a_1 \geq 0$ and $a_0 \in S(1,g)$. We set $G = (a_1+1)^{-1}h^{-1}g$, $B = (a_1+1)^{\frac{1}{2}}$, $B = b(x,D)$. It can be shown that the metric G and the function b satisfy the conditions (6.3)–(6.5), and since $a+1 \in S(b^2,G)$ and $b \in S(b,G)$, applying (6.6) and (6.7), we obtain $A+I = B^*B + C + a_0(x,D)$, where $C \in \mathcal{L}((a+1)^{-1},G) \subset \mathcal{L}(1,G)$. Applying the theorem on boundedness in L^2, we obtain (5.10).

6.2. Pseudodifferential Operators with Double Symbols. Let $\Omega \subset \mathbb{R}^n$ be a fixed open set. If we wish to construct a calculus of operators on Ω analogous to Sect. 6.1, we must impose an additional condition on the supports of their kernels so as to have $Au \in C_0^\infty(\Omega)$ for $u \in C_0^\infty(\Omega)$. This is conveniently done by considering instead of the usual pseudodifferential operators the so-called *pseudodifferential operators with double symbols*, which act in $C_0^\infty(\Omega)$ according to the rule

$$((\operatorname{Op} a)u)(x) = (2\pi)^{-n} \int d\xi \int dy\, e^{i\langle x-y,\xi\rangle} a(x,y,\xi)u(y).$$

Let g be a metric on $\Omega \times \mathbb{R}^n$ of the form (6.2). We shall write $a \in \hat{S}(p,g)$ and $A = \operatorname{Op} a \in \hat{\mathcal{L}}(p,g)$ if

1) $\varphi_j(x,0) \leq C\inf\{|x-y| \mid y \in \partial\Omega,\ x_i = y_i\, \forall i \neq j\}$ for all j;

2) $\varphi_j(x,\xi) \leq C\varphi_j(x,0)$, $C^{-1}h(x,\xi) \leq h(x,0) \leq C$ for all j;

3) condition (6.3) holds with $\varphi_j(x,0)$ in place of $\varphi_j(x,\xi)$ and condition (6.4) with $x = y$; a function p satisfying these conditions will be called a *weight*;

4) for all multi-indices α, β, and γ

$$|\partial_\xi^\alpha D_x^\beta D_y^\gamma a(x,y,\xi)|p(x,\xi)^{-1} \le c_{\alpha\beta\gamma}\Phi(x,\xi)^{-\alpha}\varphi(x,\xi)^{-\beta-\gamma};$$

5) for any compact set $K \subset \Omega$ there exists a compact set $K' \subset \Omega$ such that if $x \in K$ (resp. $y \in K$) and $(x,y,\xi) \in \operatorname{supp} a$, then $y \in K'$ (resp. $x \in K'$);

6) if $(x,y,\xi) \in \operatorname{supp} a$, then for any j and any weight q we have

$$\varphi_j(x,\eta)^{\pm 1}\varphi_j(y,\eta)^{\mp 1} + \Phi_j(x,\eta)^{\pm 1}\Phi_j(y,\eta)^{\mp 1} + q(x,\eta)^{\pm 1}q(y,\eta)^{\mp 1} \le C,$$

where C depends on q (and, naturally, on $\hat{a}$).

Operators of class $\hat{\mathcal{L}}(p,g)$ are continuous mappings on $C_0^\infty(\Omega)$ and $\mathcal{D}'(\Omega)$, and operators of class $\hat{\mathcal{L}}(1,g)$ are continuous mappings on $L_2(\Omega)$. If $a \in S(p,g)$ is the usual symbol of a differential operator A, we can represent A in the form $A = \operatorname{Op}\hat{a} \in \hat{\mathcal{L}}(p,g)$. To do this we construct functions $\psi_j \in S(\varphi_j, g)$ such that $\varphi_j \le C\psi_j$, we fix $\theta \in C_0^\infty(\mathbb{R})$, $0 \le \theta \le 1$, $\theta\big|_{|t|>1} = 0$, $\theta\big|_{|t|<\frac{1}{2}} = 1$, and we set

$$\theta_c(x,y,\xi) = \theta\Big(c^{-1}\sum(\psi_j(x,0)^{-2} + \psi_j(y,0)^{-2})\big|x_j - y_j\big|^2\Big).$$

Then for small $c > 0$ we have $\hat{a} = \theta_c a \in \hat{S}(p,g)$ and $A = \operatorname{Op}\hat{a}$. In particular $I = \operatorname{Op}\theta_c \in \hat{\mathcal{L}}(1,g)$.

On the other hand each operator $A = \operatorname{Op} a \in \hat{\mathcal{L}}(p,g)$ can be represented in the form $A = \operatorname{Op}\theta_c\tilde{a} + A'$, where $A' \in \hat{\mathcal{L}}^{-\infty}(p,g) = \cap\hat{\mathcal{L}}(h^N p, g)$ and $\tilde{a} \in S(p,g)$ satisfies the condition that for all N

$$\tilde{a}(x,\xi) - \sum_{|\alpha|<N} \frac{1}{\alpha!}\partial_\xi^\alpha D_y^\alpha a(x,y,\xi)\big|_{y=x} \in S(h^N,p,g). \tag{6.8}$$

If $A = \operatorname{Op} a \in \hat{\mathcal{L}}(p,g)$ and $B = \operatorname{Op} b \in \hat{\mathcal{L}}(q,g)$, then $AB = \operatorname{Op} c$, where $c \in \hat{S}(pq,g)$ satisfies the condition that for all N

$$c(x,z,\xi) - \sum_{|\alpha|>N} (\alpha!)^{-1} D_y^\alpha(a^{(\alpha)}(x,y,\xi)b(y,z,\xi))\big|_{y=x} \in \hat{S}(pqh^N, g). \tag{6.9}$$

The *hypoelliptic symbols* $a \in \hat{S}(p,g)$ are distinguished by the condition

$$|a(x,x,\xi)^{-1}p(x,\xi)| \le c \;\text{ for }\; |x| + |\xi| + \operatorname{dist}(x,\partial\Omega)^{-1} > c.$$

Under this condition one can construct a function $b \in S(p^{-1},g)$ equal to $a(x,x,\xi)^{-1}$ outside a certain compact set and, setting $B = \operatorname{Op}\theta_c b \in \hat{\mathcal{L}}(p^{-1},g)$ and applying (6.8) and (6.9), one can obtain

$$AB = I + T_1, \quad BA = I + T_2, \tag{6.10}$$

where $T_j \in \mathcal{L}(h,g)$. We then modify B in the usual way and obtain (6.10) with $T_j \in \hat{\mathcal{L}}^{-\infty}(1,g)$. The second of relations (6.10) implies an a priori estimate for solutions of the equation $Au = f$. If $h(x,\xi) \to 0$ as $|x| + |\xi| + \operatorname{dist}(x,\partial\Omega)^{-1} \to$

∞, then the operators of the class $\hat{\mathcal{L}}(h,g)$ are compact in $L_2(\Omega)$ (and in the weighted spaces of distributions on Ω connected with the metric g), so that (6.10) implies that the operator A is compact in these spaces. We shall not give the general definitions and theorems here (cf. Levendorskij 1982), but rather limit ourselves to an example.

Example 6.2. Let $q \geq 1$, $q(x) \to \infty$ as $x \to \partial\Omega$ and as $x \to \infty$, $|q^{(\alpha)}| \leq c_\alpha q^{1+|\alpha|}$ for all α, and let $k_1 - k_2 > 2m$ for $m \in \mathbb{N}$. We set

$$g_{x,\xi}(y,\eta) = q(x)^2|y|^2 + (|\xi| + q(x)^\gamma)^{-2}|\eta|^2,$$
$$p(x,\xi) = q(x)^{k_2}|\xi|^{2m} + q(x)^{k_1}, \quad A = \nabla^{*m} q(x)^{k_2}\nabla^m + q(x)^{k_1},$$

where $\gamma = (k_1 - k_2)/2m$, and for $r \in \mathbb{Z}_+$, $s \in \mathbb{R}$ we introduce the spaces

$$H_q^{r,s}(\Omega) = \{u |\, q^{s+(r-|\alpha|)\gamma}u^{(\alpha)} \in L_2(\Omega),\ |\alpha| \leq r\}.$$

We have $H_q^{-r,-s}(\Omega) = H_q^{r,s}(\Omega)^*$ with the natural topology.

Then $A \in \hat{\mathcal{L}}(p,g)$ and $A : H_q^{r+m,s+k_2}(\Omega) \to H_q^{r,s}(\Omega)$ is a Fredholm operator. As an unbounded operator on $H_q^{r,s}(\Omega)$ it is the closure of its restriction to $C_0^\infty(\Omega)$ (and a self-adjoint operator if $r = s = 0$); its spectrum is discrete, and the eigenfunctions are independent of r and s and belong to $S_{q(x)^{-1}}(\Omega)$.

6.3. Algebras of Pseudodifferential Operators Connected with Operators of Type III. We note that the study of equations of type III is important, since the study of elliptic equations on manifolds with conic points or edges reduces to it.

In local coordinates model operators of type III have the form $A = \sum a_\alpha(y) y_n^{s^0+\alpha_n} D_y^\alpha$, and hence can be regarded as the pseudodifferential operators with double symbols that can be described by the metric

$$g_{x,\xi}(y,\eta) = |y'|^2 + x_n^{-2}y_n^2 + (\langle\xi'\rangle + |x_n\xi_n|)^{-2}|\eta'|^2 + x_n^2(\langle\xi'\rangle + |x_n\xi_n|)^{-2}\eta_n^2, \quad (6.11)$$

but the calculus of Sect. 6.2 does not enable us to obtain any substantial results in this case, since the condition $h(x,\xi) \to 0$ as $x_n^{-1} + |\xi| \to \infty$ does not hold.

Melrose has found a (local) imbedding of operators of type III with $s^0 = 0$ into the usual algebra of pseudodifferential operators with symbols satisfying the estimates (6.1) with the metric (6.11) and with kernels concentrated on the set $x_n \geq 0$, $y_n \geq 0$, $|x_n - y_n| \leq x_n/2$ (cf. Hörmander 1985, Sect. 18.4; actually a special class of pseudodifferential operators with double symbols was studied). The calculus of such pseudodifferential operators, called *totally characteristic*, is constructed following the outline of the calculus of the pseudodifferential operators $A \in L_{1,0}^m$. One of the principal differences from the calculus of standard pseudodifferential operators is that the kernel of the operator A can have singularities, even when A is an operator of order $-\infty$ in the sense of this calculus. Thus there is no way of studying operators "up to operators of order $-\infty$." A detailed description of the nature of the singularities of kernels and a study of wave fronts of solutions form one of the main purposes of the Melrose calculus.

The algebras of pseudodifferential operators constructed using the Mellin transform on the variable y_n have been applied in several other papers studying the singularities of solutions of equations of type III, cf. Rempel 1987; Rothschild, Stein 1976, and the bibliography found there.

Operators with coefficients having singularities at isolated points or on a submanifold, which are close to operators of type III, have served as models in constructing an algebra of pseudodifferential operators with symbols having singularities (cf. Plamenevskij 1986 and the bibliography found there). Such an algebra, like the usual algebra of pseudodifferential operators, is constructed using the Fourier transform, but the singularities of the symbols lead to an essential alteration in the properties of the algebra in comparison with the usual algebra. Plamenevskij (1986) discusses several variants of the symbolic calculus and establishes necessary and sufficient conditions for a pseudodifferential operator to be a Fredholm operator: the algebra constructed is studied from the point of view of the theory of noncommutative C^*-algebras.

6.4. Algebras of Pseudodifferential Operators Connected with Operators of Type IV. The survey of Glushko and Savchenko (1985) studies the weighted Fourier transform that maps functions on $\mathbb{R}_+$ into functions on $\mathbb{R}$. Using this transform (and the Fourier transform along directions tangent to the boundary) a class of pseudodifferential operators on $\mathbb{R}^n_+$ is defined that exhibits nonpower degeneracy on the boundary. These pseudodifferential operators are used to study a class of operators that degenerate strongly into elliptic operators on the boundary (the derivatives along directions tangent to the boundary are not degenerate). The class of differential operators studied by Glushko and Savchenko (1985) is found among the classes of Sect. 3.3.

The idea of Sect. 3 makes it possible to construct the parametrix of a boundary value problem for a hypoelliptic pseudodifferential operator that satisfies the transmission condition and degenerates strongly on the boundary of the domain into an elliptic pseudodifferential operator of lower order (Levendorskij 1988). For such operators one can construct an algebra analogous to the algebra of boundary value problems for elliptic pseudodifferential operators.

It is of undoubted interest to construct algebras containing operators for boundary value problems for other classes of operators of type IV with smooth coefficients and their parametrices. One such algebra was constructed by Melrose and Mendoza (1983). This algebra includes parametrices to certain boundary value problems for operators of type IV with $k_{jl} = l + j$. The conditions obtained in Sect. 2.4 for an operator to be a Fredholm operator show that to describe such operators requires three types of symbols: the usual symbol, an operator-valued symbol $A_{\mathrm{IV}}(y', \eta')$, and an auxiliary scalar symbol $M_0(y', \lambda)$. In the algebra of Mendoza and Melrose these same three symbols are used, only the last of them is used in a slightly altered form. Melrose and

Mendoza (1983) also study the singularities of the kernels of pseudodifferential operators.

6.5. Algebras of Pseudodifferential Operators and Integral Transforms Connected with the Singular Integral Operators of Bessel. Replacing the Fourier transform on x_n by the Bessel transform in the construction of standard pseudodifferential operators on $\mathbb{R}^n$, Kipriyanov (1971, 1973, cf. also the bibliography there) constructed an algebra of pseudodifferential operators on $\mathbb{R}^n_+$ having the same properties as the algebra of pseudodifferential operators on $\mathbb{R}^n$ and studied the Fredholm property for operators of this algebra in a scale of weighted spaces constructed by him. A typical example of an operator from Kipriyanov's algebra is the operator

$$A = \sum_{|\alpha'|+2\alpha_n \leq 2m} a_\alpha(x)\partial_{x'}^{\alpha'} B_\nu^{\alpha_n}, \tag{6.12}$$

where $B_\nu = \partial_n^2 + (2\nu+1)x_n^{-1}\partial_n$.

Katrakhov (1980) has applied transformations of Poisson and Sonin types (P_ν and S_ν) instead of the direct and inverse Bessel transform. The operator P_ν is defined on $C_0^\infty(\bar{\mathbb{R}}^+)$ and satisfies the equality $B_\nu P_\nu v = P_\nu \partial_n v$; the operator S_ν is the inverse of P_ν. It can be established using these transforms that operators of the form (6.12) with constants a_α are similar to operators of the form

$$\sum a_\alpha \partial_{x'}^{\alpha'} \partial_n^{2\alpha_n}$$

on $\mathbb{R}^n_+$ (rather than $\mathbb{R}^n$, as in the approach of Kipriyanov). This makes it possible to pose general weighted boundary conditions on the characteristic surface, and the number of such conditions is m, in contrast to the formulation of the boundary problems in Sect. 2.4.

References to Chapter 1

An extensive list of the papers on degenerate elliptic operators that were published up to 1984 can be found in the survey of Glushko and Savchenko (1985); the monograph of Triebel (1978) contains a bibliography of papers up to 1977. On second-order operators see the survey of Paneah (1981). Various classes of weighted spaces connected with degenerate operators are studied in the monographs of Olejnik and Radkevich (1971) and Triebel (1978), in the survey of Glushko and Savchenko (1985), and in the papers of Grushin (1970; 1971; 1972).

The elementary classes of type I are studied in Chapter VI of the monograph of Triebel (1978); model class of types II and III are studied in the papers of Fursikov (1971bc); various class of type IV are studied in Chapter VII of the book of Triebel (1978) and in the papers of Grushin (1970; 1971; 1972). Some classes of type IV were studied by Glushko and Savchenko (1985) in the case of nonpower degeneracy; Levendorskij (1988c) has studied larger classes of such operators. The general outline of Sect. 3 was proposed by Levendorskij (1988a).

Rojtberg and Sheftel' (1980) have given an application of a special regularization of elliptic equations that degenerate strongly to equations of lower order on the boundary.

The theory of pseudodifferential operators with symbols of class $S^m_{\rho,\delta}$ is expounded in the books of Shubin (1978), Hörmander (1985), Taylor (1981), and Trèves (1980); more general calculi are discussed in the papers of Beals (1975; 1981), Beals and Fefferman (1974), Hörmander (1979), and in Sect. 18 of the book of Hörmander (1985). The calculus of Sect. 6.2 is a weak variant of the calculus of Levendorskij (1988a).

Various classes of hypoelliptic operators have been studied by Hörmander (1979), Taylor (1981), Trèves (1980), Beals (1975; 1981), and Beals and Fefferman (1974); inequalities of Gårding type have been studied by Kolesnikova (1984), Beals and Fefferman (1974), Hörmander (1985), Taylor (1981), and Trèves (1980).

A very thorough analysis of the different methods of studying hypoelliptic and degenerate operators was carried out by Fefferman (1983).

For the asymptotics of the spectrum of degenerate and hypoelliptic operators see the surveys of Levendorskij (1988b) and Fefferman (1983).

Chapter 2
Degenerate Elliptic Boundary Value Problems

By B. Paneah

§1. Introduction

1.1. Contents of the Chapter. In this chapter we study degenerate elliptic (noncoercive) boundary value problems, i.e., boundary value problems for elliptic differential equations in a domain for which the coercivity conditions (the Shapiro-Lopatinskij conditions, complementarity conditions) are violated on certain subsets α of the boundary of the domain. The nature of the modifications the operator of the boundary value problem undergoes as a consequence turns out to be very unstable and is connected not so much with the "size" of α (although size does play a role) as with the behavior of the symbol of a certain pseudodifferential operator that is canonically associated with the boundary value problem (the Calderón operator) in a neighborhood of the set α. We shall trace these modifications in detail mostly through the example of boundary value problems for a scalar second-order elliptic differential operator in a bounded domain with a smooth boundary. In the more general situation (for example, operators on bundles and operators of order higher than 2) the available results are significantly less advanced and will be mentioned discussed only in general terms.

1.2. Basic Definitions and Notation. Throughout this chapter we shall use the letter Ω to denote a bounded domain in the Euclidean space $\mathbb{R}^{n+1}$, $n \geq 1$. We assume that the boundary $\partial\Omega = M$ of the domain is a C^∞-hypersurface, and that Ω lies on one side of M. Let $\mathcal{T}$ denote a sufficiently small open tubular neighborhood of the boundary manifold M in $\mathbb{R}^{n+1}$ and let $(t, x) = (t, x_1, x_2, \ldots, x_n)$ be *normal coordinates* in $\mathcal{T}$. The latter means that to each point $p \in \mathcal{T}$ there is assigned a pair (t, x) in which t is the geodesic distance from p to M, taken to be positive for $p \in \Omega$ and negative for $p \notin \Omega$, and x are the local coordinates of the projection $\pi(p)$ of the point p on M. We shall use the notation $\partial_t = \partial/\partial_t$, $\partial_j = \partial/\partial x_j$, $1 \leq j \leq n$, and $\partial_x = (\partial_1, \partial_2, \ldots, \partial_n)$. We shall consider a linear elliptic differential operator $\mathcal{L}$ with smooth real coefficients in the closure $\overline{\Omega}$ of the domain Ω and write it inside $\mathcal{T}$ as

$$\mathcal{L}u = (l^0 \partial_t^2 + 2l^{0j} \partial_t \partial_j + l^{ij} \partial_i \partial_j) u + l^k \partial_k u + lu,$$

where $1 \leq i, j \leq n$, $0 \leq k \leq n$, $\partial_0 = \partial_t$, and here (and throughout the following) summation over repeated indices is understood.

It is assumed that for all $(t, x) \in \mathcal{T}$ and $(\tau, \xi) = (\tau, \xi_1, \xi_2, \ldots, \xi_n) \in \mathbb{R}^{n+1}$ we have

$$l^0(t, x)\tau^2 + 2l^{0j}(t, x)\tau\xi_j + l^{ij}(t, x)\xi_i\xi_j > 0, \quad \text{if } \tau^2 + |\xi|^2 = 1,$$

and $l \leq 0$ everywhere in $\overline{\Omega}$.

We are studying the boundary value problem

$$\mathcal{L}u = F \text{ in } \Omega, \quad \mathcal{A}u = f \text{ on } M, \tag{1.1}$$

in which the operator $\mathcal{A}$ has the form

$$\mathcal{A} = \partial_i a^{ij} \partial_j + a^{0j} \partial_j + a\partial_t + a_0 \tag{1.2}$$

in the neighborhood $\mathcal{T}$. Here $a^{ij}(x)$ are the coordinates of a contravariant symmetric C^∞-tensor on M of type $(2, 0)$ and $a^{0j}(x)$, $a(x)$, and $a_0(x)$ are C^∞-functions on M. This class of problems includes, on the one hand, the well-known oblique derivative problem (cf. Egorov and Shubin 1987, Sect. 2.18) if $a^{ij} \equiv 0$ on M and, on the other hand, a problem connected with the description of diffusion processes in a closed domain in the context of probability theory when the matrix $\|a^{ij}(x)\|$ and the coefficient $a(x)$ are non-negative on M. In particular the very popular operator $\mathcal{A} = \sum_{j=0}^{n} X_j^2 + X_0 + a_0$, where $X_0, X_1, \ldots, X_n$ are vector fields on M, falls into this class. Hörmander (1967) attracted interest to this problem, and he was the first to clarify the role played by the Lie algebra (over $\mathbb{R}$) of all the commutators of the fields $X_0, X_1, \ldots, X_n$ in questions of regularity of solutions of the equation $\mathcal{A}u = f$ (and hence also of solutions of the problem (1.1)).

Before specializing our considerations, we reduce the problem (1.1)–(1.2) in an equivalent manner to a pseudodifferential equation on the boundary manifold M (cf. Egorov, Shubin 1988, Sect. 2.7), which enables us, in particular to give a very brief description of the condition for nonellipticity of this problem. In what follows we shall denote by $\mathfrak{B}F$ the solution of the problem

$$\mathcal{L}u = F \text{ in } \Omega, \quad u = 0 \text{ on } M$$

and by $\mathfrak{D}f$ the solution of the problem

$$\mathcal{L}u = 0 \text{ in } \Omega, \quad u = f \text{ on } M.$$

Both operators are well-defined, because of the condition $\mathcal{L}1 = l \leq 0$. We also introduce the notation P^c for the operator that is the formal adjoint of the pseudodifferential operator P. The equation of interest to us on M has the form

$$Av \equiv \partial_i a^{ij}(x)\partial_j v + a^j(x)\partial_j v + a(x)\Lambda(x, \partial_x)v + a^0(x)v = \Phi, \tag{1.3}$$

where $a^j(x) = a^{0j}(x) + al^{0j}(0, x)$, $\Lambda(x, \partial_x)$ is a first-order pseudodifferential operator on M for which the real part of the principal symbol is equal to

$$\Lambda^0(x, \xi) = -(1/l^0(0, x))(l^0 l^{ij}(0, x)\xi_i\xi_j - (l^{0k}\xi_k)^2)^{\frac{1}{2}}, \tag{1.4}$$

and Φ is a function that depends on f and F. It is important to note that

$$\Lambda^0(x, \xi) < 0 \text{ for all } (x, \xi) \in T_0^*(M). \tag{1.5}$$

The operator $A = A(x, \partial_x)$ is called the *projection* or the *Calderón operator* for the boundary value problem (1.1)–(1.2). The connection between the solutions of the problems (1.1)–(1.2) and (1.3) is realized (in the case, for example, when the operator A is invertible) by the relation

$$u = \mathfrak{B}F + \mathfrak{D}A^{-1}(f - a\partial_t \mathfrak{B}F\big|_M). \tag{1.6}$$

The condition for the boundary value problem (1.1)–(1.2) to be elliptic (the Shapiro-Lopatinskij condition, cf. Egorov, Shubin 1987) is equivalent to the condition that the operator $A(x, \partial_x)$ be elliptic, which in turn is equivalent to the condition that the principal symbol $A^0(x, \xi)$ of the operator A (Egorov, Shubin 1988, Sect. 1.3) does not vanish on $T_0^*(M)$. Thus if $a^{ij}(x) \not\equiv 0$ on M, then $A^0(x, \xi) = a^{ij}(x)\xi_i\xi_j$, and the condition for ellipticity of this problem has the form

$$a^{ij}(x)\xi_i\xi_j \neq 0, \quad (x, \xi) \in T_0^*(M). \tag{1.7}$$

But if $a^{ij}(x) \equiv 0$ on M, the condition for ellipticity $A^0(x, \xi) \equiv -a(x)\Lambda^0(x, \xi) + ia^j(x)\xi_j \neq 0$ for all $(x, \xi) \in T_0^*(M)$ is always satisfied for $n = 1$, and for $n \geq 2$ it reduces to the inequality

$$a(x) \neq 0 \text{ on } M. \tag{1.8}$$

In what follows, regardless of the form of $A^0(x, \xi)$, we assume that $n \geq 2$ and use the notation

$$\alpha = \{x \in M \big| a(x) = 0\}.$$

§2. A Second-Order Boundary Operator $\mathcal{A}$ with the Form $a^{ij}\xi_i\xi_j$ of Fixed Sign

In the present section we assume that

$$a(x) \leq 0, \quad a^{ij}(x)\xi_i\xi_j \leq 0 \text{ when } (x,\xi) \in T_0^*(M). \tag{2.1}$$

We recall that when conditions (1.7) hold, i.e., in the case when the problem (1.1)–(1.2) is elliptic, its solvability properties are described, for example, in Egorov, Shubin 1987, Theorem 3.5. In this situation

(i) the operator $\mathfrak{A} : u \mapsto (\mathcal{L}u, \mathcal{A}u\big|_M)$ is a *Fredholm operator*: $H^{s+2}(\Omega) \to H^s(\Omega) \oplus H^{s-\frac{1}{2}}(M)$, i.e., its range of values is closed and the kernel and cokernel are finite-dimensional;
(ii) the index $\operatorname{ind}\mathfrak{A}$ of the operator is 0;
(iii) the kernel ($\ker\mathfrak{A}$) and the cokernel ($\operatorname{coker}\mathfrak{A}$) belong to the spaces $H^\infty(\Omega)$ and $H^\infty(\Omega) \oplus H^\infty(M)$ respectively;
(iv) for any $s > \frac{1}{2}$ the *a priori estimate*

$$\|u,\Omega\|_{s+2} \leq c\big(\|\mathcal{L}u,\Omega\|_s + \|\mathcal{A}u, M\|_{s-\frac{1}{2}} + \|u,\Omega\|\big),$$

in which c is a constant depending on s but not on u, holds for the solutions u of the problem (1.1)–(1.2).

Now suppose that for certain values $x \in M$ the quadratic form $a^{ij}(x)\xi_i\xi_j$ vanishes on the vectors $\xi = \xi(x) \neq 0$. Then parts (i) and (iv) demonstrably fail for the corresponding problem (1.1)–(1.2). Nevertheless under certain restrictions of a geometric nature on the operator $\mathcal{A}$ many important properties of elliptic problems, such as the Fredholm property (but in different spaces), vanishing of the index of the problem, and the imbedding $\ker\mathfrak{A} \subset H^\infty(\Omega)$, carry over to the more general situation. In what follows we shall attempt to trace the connection between the deviations of the formal properties of the operator $\mathcal{A}$ from ellipticity and the failure of the qualitative picture of solvability of the problem (1.1)–(1.2) in comparison with the elliptic situation.

We associate with the operator $\mathcal{A}$ a set of vector fields on M

$$X_0 = a^j\partial_j, \quad X_i = a^{ij}\partial_j, \quad 1 \leq i,j \leq n. \tag{2.2}$$

We denote by $\mathcal{L} = \mathcal{L}(X_0, X_1, \ldots, X_n)$ the *Lie algebra over* $\mathbb{R}$ with respect to the multiplication $[X,Y] = XY - YX$ generated by the fields $X_0, X_1, \ldots, X_n$. In other words $\mathcal{L}$ is the real vector space generated by the successive commutators of the fields X_k, $k = 0, 1, \ldots, n$. At each point $x \in M$ an element of $\mathcal{L}$ is a first-order differential operator with constant coefficients in $\mathbb{R}^n$ and the set of such operators (for a given point x) forms a real vector space. The dimension of this space $r_{\mathcal{L}}(x)$ is called the *rank* of the algebra $\mathcal{L}$ at the point x. We now introduce another concept connected with the algebra $\mathcal{L}$. We shall say that the algebra $\mathcal{L}$ is *transversal* to the submanifold $\nu \subset M$ at the point x

if among the vector fields belonging to $\mathcal{L}$ there is one that does not belong to the tangent space $T_x(\nu) \subset T_x(M)$ at the point x. This, of course, is equivalent to the condition that at least one of the fields

$$X_{j_1}, [X_{j_1}, X_{j_2}], \dots, [X_{j_1}, [X_{j_2}, \dots, [X_{j_{k-1}}, X_{j_k}] \dots]], \quad j_s = 0, 1, \dots, n,$$

does not belong to $T_x(\nu)$. It can be shown (Paneah 1985a) that the concepts of the rank of the algebra $\mathcal{L}$ and its transversality to a submanifold $\nu \subset M$ are independent of the local coordinates on M. Both concepts play important roles in the following theorem, which describes the solvability conditions of a large class of nonelliptic boundary value problems (1.1)–(1.2) and (2.1).

Theorem 2.1 (Paneah 1985a). *Suppose that the rank $r_{\mathcal{L}}(x) = n$ at each point $x \in \alpha$ except for perhaps a submanifold $\mu \subset M$ of codimension 1. Let $\mu = \mu^1 \supset \mu^2 \supset \cdots \supset \mu^k \supset \mu^{k+1} = \varnothing$ be a system of nested compact closed orientable submanifolds of decreasing dimensions, and suppose that at each point $x \in \mu^j \setminus \mu^{j+1}$, $1 \le j \le k$, contained in α the algebra $\mathcal{L}$ is transversal to μ^j. Then*

(i) *for each $s > 0$ the problem has a solution $u \in H^{s+\frac{1}{2}}(\Omega)$, for the functions $F \in H^{s-\frac{1}{2}}(\Omega)$, $f \in H^s(M)$ satisfying the N orthogonality conditions*

$$(F, \mathfrak{D}^c(raw_j))_\Omega + (f, w_j)_M = 0, \quad j = 1, 2, \dots, N, \tag{2.3}$$

and only for such functions. Here $r > 0$ is a smooth function on M depending only on $\mathcal{L}$ and M, and the $w_j \in H^\infty(M)$ form a basis in the solution space of the pseudodifferential equation

$$A^c w = 0 \ \text{ on } M.$$

Here $\mathfrak{D}^c$ is an operator of the form $\mathfrak{D}$ (cf. Sect. 1.2), but related to the equation $\mathcal{L}^c u = 0$, and $(\cdot,\cdot)_G$ denotes the inner product in $L_2(G)$.

(ii) *When $F = 0$ and $f = 0$, the problem has exactly N linearly independent solutions, all belonging to the space $H^\infty(\Omega)$.*

(iii) *For all values $\lambda > \lambda(s) > 0$ the problem (1.1)–(1.2) with $\mathcal{A}_\lambda = \mathcal{A} + \lambda$ in place of $\mathcal{A}$ has a unique solution $u \in H^{s+\frac{1}{2}}(\Omega)$ for any $(F, f) \in H^{s-\frac{1}{2}}(\Omega) \oplus H^s(M)$ and the inverse operator R_λ of the problem is compact in the corresponding spaces.*

(iv) *If $u \in H^{s+\frac{1}{2}}(\Omega)$ is the solution of the problem under consideration in which $(F, f) \in H^{s-\frac{1}{2}+\delta}(\Omega) \oplus H^{s+\delta}(M)$, then actually $u \in H^{s+\frac{1}{2}+\delta}(M)$ for $\delta > 0$. In particular $F \in H^\infty(\Omega)$, $f \in H^\infty(M) \Rightarrow u \in H^\infty(M)$.*

(v) *The estimate*

$$\|u, \Omega\|_{s+\frac{1}{2}} \le c(s)\big(\|\mathcal{L}u, \Omega\|_{s-\frac{1}{2}} + \|\mathcal{A}u, M\|_s + \|u, \Omega\|_s\big) \tag{2.4}$$

holds for solutions u of the problem (1.1)–(1.2).

An earlier result can be found in the paper of Taira (1982), which also contains further references. We note that the main thrust of this paper is the application of the solvability theory of the problem (1.1)–(1.2) to the description of diffusion processes in closed domains in the context of probability theory. In this connection the boundary operator $\mathcal{A}$ in Taira's paper has the very special form

$$\mathcal{A} = a^{ij}\partial_i\partial_j + a^{0j}\partial_j + a\partial_t + a_0 - \delta\mathcal{L},$$

where $a_0 \geq 0$, $a^{ij}\xi_i\xi_j \leq 0$, $a \leq 0$, and in addition $\delta \leq 0$ and $a(x) + \delta(x) < 0$ everywhere on M. In the terminology used in probability theory the terms $a^{ij}(x)\partial_i\partial_j u + a^{0j}(x)\partial_j u$ correspond to the diffusion of a particle along the boundary, $a(x)\partial_t u$ to the reflection of a particle from the boundary, and $a_0(x)$ to the absorption of a particle; finally, $\delta(x)\mathcal{L}$ corresponds to viscosity on the boundary.

The proof of Theorem 2.1 is quite typical in the circle of noncoercive problems under consideration and can be reduced completely to the study of the corresponding Calderón operator A. We sketch an outline of the proof. If $u \in H^{s+\frac{1}{2}}(\Omega)$ is a solution of the problem (1.1)–(1.2) in which $(F, f) \in H^{s-\frac{1}{2}}(\Omega) \oplus H^s(M)$, then by Theorem 3.5 of Egorov, Shubin 1987 we have $\mathfrak{B}F \in H^{s+\frac{3}{2}}(\Omega)$ and since $(\mathcal{A} - a\partial_t)\mathfrak{B}f \equiv 0$ on M, using the usual embedding theorems we have $\mathcal{A}\mathfrak{B}F\big|_M = a\partial_t\mathfrak{B}F\big|_M \in H^s(M)$. Therefore the function $w = u - \mathfrak{B}F$ is a solution of the problem (1.1)–(1.2) with $F = 0$, $f = f - a\partial_t\mathfrak{B}F\big|_M$ and its boundary value $w_M = w\big|_M$ belongs to the space $H^s(M)$ and satisfies the equation

$$Aw_M = f - a\partial_t\mathfrak{B}F\big|_M. \tag{2.5}$$

Under the restrictions imposed on the operator $\mathcal{A}$ in the hypothesis of the theorem we can deduce the a priori estimates

$$\|u\|_s \leq \sigma\|Au\|_s + c(\sigma, s)\|u\|_{s-1}, \quad \|u\|_s \leq c(s)\|A_\lambda u\|_s \tag{2.6}$$

for all functions $u \in H^\infty(M)$, in which the numbers $s \in \mathbb{R}^1$ and $\sigma > 0$ are arbitrary and $\lambda = \lambda(s) > 0$ is sufficiently large. This is the most difficult part of the proof technically. It is here that we make full use of all the postulated properties of the operator $\mathcal{A}$. In particular inequality (1.5) makes it possible to deal with the pseudodifferential part A. This is achieved, as often happens in such situations, using *Hörmander's inequality*, which is known as "the sharp Gårding inequality," and consists of the following. If $P(x, \partial_x)$ is a first-order pseudodifferential operator on M and its principal symbol $P_0(x, \xi)$ is nonnegative for $(x, \xi) \in T_0^*(M)$, then for all functions $u \in C^\infty(M)$ we have the estimate

$$\operatorname{Re}(Pu, u) + c\|u\|^2 \geq 0, \tag{2.7}$$

in which c is a constant independent of u and $(\cdot, \cdot)$ and $\|\cdot\|$ are the inner product and norm in the space $L_2(M)$. We shall exhibit a typical example of such reasoning in an elementary situation. Consider the quadratic form $\operatorname{Re}(Au, u)$

on functions $u \in C_0^\infty(\omega)$ that are concentrated on some coordinate neighborhood $\omega \subset M$. Integration by parts, taking account of the zero boundary value for u leads immediately to the estimate

$$-(a^{ij}\partial_j u, \partial_j u) - \operatorname{Re}(a\Lambda u, u) + c\|u\|^2 \leq -\operatorname{Re}(Au, u) \tag{2.8}$$

with some constant c. If we now use the inequality

$$-a^{ij}(x)\xi_i\xi_j \geq c\sum_i (a^{ij}\xi_j)^2, \quad c > 0 \text{ constant}, \tag{2.9}$$

which follows from the classical inequality

$$|\operatorname{grad}_\xi f(x,\xi)|^2 \leq cf(x,\xi), \quad x \in \omega, \quad |\xi| = 1, \tag{2.9'}$$

for nonnegative functions $f(x,\xi)$, along with the estimate (2.7) for $P = a(x)\Lambda(x,\partial_x)$ and take account of condition (1.5), (2.1), we obtain immediately from (2.8) an elementary variant of the energy estimate

$$\sum_{j\geq 1} \|X_j u\|^2 \leq -c(\operatorname{Re}(Au, u) + \|u\|^2), \quad c > 0, \tag{2.10}$$

where $u \in C_0^\infty(\omega)$. And if $a(x) < 0$ in ω, then by the usual Gårding inequality

$$\operatorname{Re}(Pu, u) \geq c\|u\|_{\frac{1}{2}}^2 - c_1\|u\|^2, \quad u \in C_0^\infty(\omega);$$

and by the same considerations as above

$$\|u\|_{\frac{1}{2}}^2 \leq -c\operatorname{Re}(Au, u) + c_1\|u\|^2.$$

Since the formally conjugate operator A^c has the same structure as A and satisfies conditions connected with the rank and transversality of the algebra $\mathcal{L}$ if and only if A does, the operator A^c also satisfies estimates (2.6). By the second of these estimates, using the Hahn-Banach theorem, we find that the equation $A_\lambda u = f$ has a unique solution in the space $H^s(M)$ for any function $f \in H^s(M)$ when $\lambda = \lambda(s)$ is sufficiently large. The first of the estimates enables us to conclude that the inverse operator $R_\lambda = A_\lambda^{-1}$ is compact in $H^s(M)$. Using the Riesz-Fredholm theory we deduce from this in the usual manner that

$$\dim\ker A = \dim\ker A^c < \infty$$

and that the condition for solvability of Eq. (2.5) is the equalities

$$(f - a\partial_t \mathfrak{B}F, w_j)_M = 0, \quad j = 1, 2, \ldots, \dim\ker A, \tag{2.11}$$

where the set of w_j forms a basis of the space $\ker A^c$.

We note that in the course of this discussion we have studied the solvability of the (pseudo)differential equation $Au = f$ on a manifold without boundary under the hypotheses of Theorem 2.1. To transform the condition (2.11) into the form (2.3) we use *Green's formula*

$$(\mathcal{L}u, v)_\Omega - (u, \mathcal{L}^c v)_\Omega = (\partial_{\mathbf{n}} u, hv)_M - (hu, \partial_{\mathbf{n}} v)_M + (ru, v)_M,$$

in which $\mathbf{n}$ is the interior conormal of the boundary M, and $h > 0$ and r are smooth functions depending only on $\mathcal{L}$ and M. Let $\partial_{\mathbf{n}} = q\partial_t + c_p(x)\partial_p$ on M, and $q > 0$. We introduce the functions

$$\Phi_j = -\mathfrak{D}^c(a(qh)^{-1}w_j), \quad 1 \leq j \leq \dim\ker A.$$

Substituting $u = \mathfrak{B}F$ and $v = \Phi_j$ in Green's formula and using the fact that $\partial_{\mathbf{n}} u = q\partial_t u$ on M, we obtain

$$-(a\partial_t \mathfrak{B}F, w_j)_M = (F, \Phi_j)_\Omega,$$

and this makes it possible to rewrite (2.11) in the form (2.3)

In the proof of the second assertion of part (ii) and also in part (iv), by known theorems on the regularity of solutions of the Dirichlet problem for the operator $\mathcal{L}$, it suffices to verify that all the solutions of the equation $Aw = 0$ belong to $H^\infty(M)$. But each such solution w satisfies the equation $w - \lambda R_\lambda w = 0$, and by what has been proved the operator R_λ is compact in both spaces $H^s(M)$ and $H^{s+1}(M)$ for $\lambda > \lambda(s)$, and its values in the two spaces coincide on the elements $w \in H^{s+1}(M)$. By properties of compact operators it follows from this that

$$H^s(M) \cap \ker(I - \lambda R_\lambda) = H^{s+1}(M) \cap \ker(I - \lambda R_\lambda),$$

and this means that w belongs to all of the spaces $H^s(M)$, i.e., $w \in H^\infty(M)$.

Once the preceding parts of the theorem have been proved the a priori estimate (v) turns out to be a simple corollary of the following fact from the theory of linear operators. If E, F, and G are Banach spaces, E is compactly imbedded in G, and $T : E \to F$ is a closed operator, then the a priori estimate

$$\|u, E\| \leq c\big(\|Tu, F\| + \|u, G\|\big), \quad u \in E,$$

is equivalent to the condition that the range of values of the operator T is closed in F and its kernel and co-kernel are finite-dimensional. To prove (v) in our case it suffices to set $T = \mathfrak{A}$, $E = H^{s+\frac{1}{2}}(\Omega)$, $F = H^{s-\frac{1}{2}}(\Omega) \oplus H^s(M)$, $G = H^{s-1}(\Omega)$.

Comparing the results of Theorem 2.1 with the results relating to the solvability of the elliptic problem (1.1)–(1.2), (1.7), we discover that eliminating the ellipticity of the problem by weakening the condition (1.7) while the restrictions connected with the rank and transversality of the algebra $\mathcal{L}$ hold leads only to a worsening of the regularity of the solutions. If $F = 0$, then in the elliptic case the condition $f \in H^s(M)$ implies that $u \in H^{s+\frac{5}{2}}(\Omega)$, and in the nonelliptic case it implies that $u \in H^{s+\frac{1}{2}}(\Omega)$. A decrease in the smoothness of the solution by 2 is said to occur. The problem remains a Fredholm problem (although in different spaces), as do the properties of vanishing of the index, infinite differentiability of the elements of the kernel and cokernel, the possibility of increasing the smoothness (as part (iv) of the theorem is expressed), and compactness of the inverse operator (when it exists). The following theorem clarifies the influence of the transversality condition on this parallelism of properties.

Theorem 2.2 (Paneah 1985a). *If the transversality condition of the algebra $\mathcal{L}$ is violated on a curve $l \subset \alpha$, then the inverse operator of the problem cannot be compact.*

This is an example of a worsening of the properties of the operator of the problem (1.1)–(1.2) in comparison with the result of Theorem 2.1. The following theorem is concerned with the improvement of these properties, expressed in an increase in the smoothness of the solution. We first agree on some notation connected with the algebra $\mathcal{L}(X_0, X_1, \dots, X_n)$. Let $I = (i_1, i_2, \dots, i_k)$ be a multi-index with components $i_m = 0, 1, \dots, n$, and let $|I| = \lambda_1 + \lambda_2 + \dots + \lambda_k$, where $\lambda_m = 1$ if $i_m \neq 0$ and $\lambda_m = 2$ if $i_m = 0$. To each multi-index I we assign a vector field

$$X_I = [X_{i_1}, [X_{i_2}, \dots, [X_{i_{k-1}}, X_{i_k}] \dots]]].$$

Theorem 2.3 (Paneah 1985a, Hörmander 1985). *If the rank $r_{\mathcal{L}}(x)$ equals n at all points $x \in \alpha$, then on the left-hand side of the estimate (2.4) the norm $\|\cdot\|_{s+\frac{1}{2}}$ can be replaced by the norm $\|\cdot\|_{s+\frac{1}{2}+\varkappa}$, where $\varkappa \geq 2^{1-m}$ and the integer m has the property that the tangent spaces $T_x(M)$, $x \in \alpha$, are generated by the vector fields X_I with $|I| \leq m$.*

This result follows immediately from the estimate

$$\|u\|_{s+\varkappa} \leq c(s)\big(\|Au\|_s + \|u\|_s\big), \quad u \in C^\infty(M), \tag{2.12}$$

which simultaneously plays a key role in the problem of hypoellipticity for the operator A. (In particular, if

$$A = \sum_{j=1}^{n} X_j^2 + X_0 + a_0(x), \tag{2.13}$$

where, as before, $X_0, X_1, \dots, X_n$ are vector fields on M and $a_0(x)$ is a scalar-valued function, then inequality (2.12) by itself guarantees that A is hypoelliptic (Olejnik, Radkevich 1971; Trèves 1980)). Inequality (2.12) in turn is closely connected with the possibility of estimating the norms of the commutators of the X_I in terms of the norm of the operator A. The following theorem is an example of such an estimate for operators of general form (1.3) involving no additional hypotheses on the rank r and the transversality of $\mathcal{L}$.

Theorem 2.4 (Paneah 1985a; Olejnik, Radkevich 1971; Hörmander 1985). *If $|I| = m$, then for all functions $u \in C^\infty(M)$ and any $s \in \mathbb{R}^1$ we have the estimate*

$$\|X_I u\|_{s-1+2^{1-m}} \leq c(s,m)\big(\|Au\|_s + \|u\|_s\big). \tag{2.14}$$

To exhibit the mechanism of the application of the rank condition $r_{\mathcal{L}}(x) = n$, $x \in M$, explicitly we shall prove the following proposition.

Theorem 2.5 (Olejnik, Radkevich 1971; Hörmander 1985). *If the rank $r_{\mathcal{L}}(x)$ of the algebra $\mathcal{L}(X_0, X_1, \ldots, X_n)$ is n at each point of some domain $\omega \subset M$ and the tangent spaces $T_x(M)$, $x \in \omega$, are generated by the vector fields X_I for which $|I| \leq m$, then for all functions $u \in C_0^\infty(\omega)$ with fixed compact support K and any $s \in \mathbb{R}^1$ we have the estimate*

$$\|u\|_{s+2^{1-m}} \leq c(s,m)\big(\|Au\|_s + \|u\|_s\big).$$

Indeed, by the condition $r_{\mathcal{L}}(x) = n$ for $x \in \omega$ the pseudodifferential operator Q with symbol $Q(x,\xi) = \sum_{|I|\leq m} |X_I(x,\xi)| + 1$ is elliptic. Hence

$$\|u\|_{s+1} \leq c(s)\Big(\|Qu\|_s + \|u\|_s\Big), \quad u \in C_0^\infty(\omega).$$

Replacing s by $s - 1 + 2^{1-m}$ here and using the obvious inequality

$$\|Qu\|_s \leq c\Big(\sum_{|I|\leq m} \|X_I u\|_s + \|u\|_s\Big), \quad u \in C_0^\infty(\omega),$$

we arrive at the required estimate (2.12) using Theorem 2.4.

We note further that under the assumption that the vector fields $X_I(x)$ with $|I| \leq m$ generate the tangent spaces $T_x(M)$ for $x \in$M estimate (2.14) is proved for the operators $A = \sum_{j=1}^{n} X_j^2$ with optimal exponent $\frac{1}{m}$ instead of 2^{1-m} (Fefferman, Phong 1981; Rothschild, Stein 1976). The methods applied in these papers are very special and far removed from the elementary techniques that suffice to obtain inequality (2.14). It is also worth pointing out that the question of estimate (2.14) with $\frac{1}{m}$ instead of 2^{1-m} (and hence also the question of estimate (2.12) with $\varkappa = \frac{1}{m}$) for operators A (even purely differential operators) of general form remains open at present. In spaces with Hölder norms estimates of the form (2.12) and (2.14) with $\frac{1}{m}$ for operators of the form (2.13) were obtained in a paper of Hörmander (1967), which gave a new stimulus to studies connected with operators of type A.

We now consider several examples that illustrate what has been said. We shall assume that Ω is the half-space $x_4 > 0$ in $\mathbb{R}^4$ and $M = \{x \in \mathbb{R}^4 \,|\, x_4 = 0\}$.

1) Let $A(x, \partial_x) = x_3^2\partial_1^2 + x_1^2\partial_2^2 + x_2^2\partial_3^2 + \partial_2 + (x_1x_2x_3)^2\Lambda$. In this case $X_0 = \partial_2$, $X_1 = x_3^2\partial_1$, $X_2 = x_1^2\partial_2$, $X_3 = x_2^2\partial_3$, and $\alpha = \bigcup_j \{x \in M \,|\, x_j = 0\}$. We shall show that in this case $r_{\mathcal{L}}(x) = 3$ everywhere on α (even on M). To do this it suffices to present iterated commutators constructed from the fields X_0, X_1, X_2, and X_3 and equal to ∂_1 and ∂_3. It is easy to see that up to constant multiples

$$\partial_3 = [X_0,[X_3,X_0]], \quad \partial_1 = [X_0,[X_0,[X_3,[X_1,[X_0,[X_3,X_0]]]]]].$$

In this case $m = 11$, so that $\varkappa \geq \frac{1}{2^{10}}$ and, by Theorem 2.5,

$$\|u\|_{s+\frac{1}{2^{10}}} \leq c(s)\big(\|Au\|_s + \|u\|_s\big), \quad u \in C^\infty(M).$$

2) Let $A = x_3^2\partial_1^2 + e^{-\frac{1}{x_1^2}}\partial_2^2 + \partial_3 + (x_1x_2x_3)^2\Lambda$. It is obvious that the algebra $\mathcal{L}$ formed by the fields $X_0 = \partial_3$, $X_1 = x_3^2\partial_1$, $X_2 = e^{-\frac{1}{x_1^2}}\partial_2$ contains no operator equal to ∂_2 on the manifold $\mu = \{x|\, x_1 = 0\}$, so that the rank $r_{\mathcal{L}}(x) = 2$ for $x \in \mu$. However, since $\partial_1 \notin T_x(\mu)$ for any $x \in \mu$ and $[X_0,[X_1,X_0]] = 2\partial_1$, the algebra $\mathcal{L}$ is transversal to the submanifold μ in this case and Theorem 2.1 applies.

3) Let $A = \partial_1^2 + (x_2^2 + x_3^2)\partial_2^2 + e^{-\frac{1}{x_2^2}}\partial_3^2 + \partial_1 + x_3\partial_2 + (x_1x_2x_3)^2\Lambda$. Then $X_0 = \partial_1 + x_3\partial_2$, $X_1 = \partial_1$, $X_2 = (x_2^2 + x_3^2)\partial_2$, $X_3 = e^{-\frac{1}{x_2^2}}\partial_3$, $\alpha = \{x|\, x_1 = 0\}$. The following facts are obvious: outside the set $\mu = \{x|\, x_2 = 0\}$ the rank $r_{\mathcal{L}}(x) = 3$; at all points of μ except the line $l\{t,0,0\}$, $t \in \mathbb{R}^1$, the field X_2 is transversal to μ; on the line l all the fields X_j belong to the tangent space $T(l)$. In this case (although nothing is known about the solvability of the corresponding problem) the inverse operator cannot be compact (Theorem 2.2).

Returning to the discussion of the solvability properties of the problem (1.1)–(1.2), we pause to discuss an important result that is characteristic for the classical Dirichlet and Neumann problems. The subject is the maximum principle, which, among other things, guarantees the uniqueness of the solution of the problem (1.1)–(1.2) In an elementary situation it is given in Egorov, Shubin 1987, Sect. 2.4. It turns out that this principle remains valid in the general case of problem (1.1)–(1.2) under certain conditions on A. Since the index of the problem is zero, it follows that the problem is solvable for any right-hand sides F and f from the corresponding spaces. The following result holds.

Theorem 2.6 (Olejnik, Radkevich 1971; Paneah 1985a; Taira 1982). *Let α and μ be as in Theorem 2.1. Suppose also that $a + a_0 \not\equiv 0$, $a_0 \geq 0$ on M, $a_0 > 0$ on $\alpha \cap \mu$, and the rank of the algebra $\mathcal{L}(X_1, X_2, \ldots, X_n)$ is n on $\alpha \setminus \mu$. If $\mathcal{L}u \geq 0$ in Ω and $Au \leq 0$ on M, then either $u \leq 0$ or $u \equiv$ const on Ω.*

We give a brief proof, which provides a different characterization of the condition $r_{\mathcal{L}}(x) = n$.

Assume that $u \not\equiv$ const in $\overline{\Omega}$ and that the function u has a positive maximum equal to γ and attained at the point p. We introduce the set $N = \{q \in M|\, u(q) = \gamma\}$. Since $\mathcal{L}1 \leq 0$, the Hopf theorem (Miranda 1955) excludes the possibility $p \in \Omega$. It follows from this that the set N is nonempty. We shall show that it is open in M. By hypothesis at each point $q \in M$

$$a^{ij}(q)\partial_i\partial_j u + a^i(q)\partial_i u + a(q)\partial_t u + a_0(q)u \leq 0. \tag{2.15}$$

We shall verify that if $q \in N$, then all the terms in this sum are nonnegative. Indeed, the inequality $a^{ij}(q)\partial_i\partial_j u \geq 0$ follows from the fact that both matrices $\|a^{ij}(q)\|$ and $\|\partial_i\partial_j u(q)\|$ are nonpositive; the inequality $a(q)\partial_t u(q) \geq 0$ holds by Giraud's theorem (Miranda 1955) and the condition $a \leq 0$. The nonnegativity of $a_0(q)u(q)$ follows from the inequality $a_0 \geq 0$; and $a^i(q)\partial_i u(q) = 0$, since u attains an extremum at the point q. Thus in fact at each point $q \in N$ we have

$$a^{ij}(q)\partial_i\partial_j u(q) = 0, \quad a(q)\partial_t u(q) = 0, \quad a_0(q)u(q) = 0. \tag{2.16}$$

The second of these equalities, together with Giraud's theorem makes it possible to conclude that $N \subset \alpha$, and the condition $a_0 > 0$ on $\mu \cap \alpha$ and the third relation in (2.16) excludes the possibility that N contains points of $\mu \cap \alpha$. Consequently $N \subset \alpha \setminus \mu$, and hence by hypothesis $r_{\mathcal{L}}(q) = n$ for any $q \in N$. This last equality carries over by continuity to some neighborhood U of the point q, which can be assumed so small that the inequality $\partial_t u < 0$ holds in it, so that $a\partial_t u \leq 0$ also. In this situation we have

$$a^{ij}\partial_i\partial_j u + a^j\partial_j u + a_0 u \leq 0$$

in U. By a theorem of Rashevskij (1939) any two points of U can be joined by a chain of trajectories of the vector fields X_1, X_2,..., X_n, and on each such chain, as proved by Amano (1979) we have $u \equiv \gamma$. Thus $u \equiv \gamma$ in the neighborhood U, and so the set N is open in M. Since it is obviously also closed, we find that $N = M$. Thus $u \equiv \gamma$ on M, which by Giraud's theorem, is possible only when $a \equiv 0$ on M; but then by hypothesis $a_0(q) > 0$ at some point $q \in M$, and we have arrived at a contradiction with (2.15), since $a^{ij}\partial_i\partial_j\gamma \equiv 0$ on M.

To conclude this section we note that the hypothesis of Theorem 2.1 can be significantly weakened by allowing the inequality $r_{\mathcal{L}}(x) \neq n$ on a set of more general nature than a manifold. All the assertions of the theorem also remain in effect in the case when M has a system of submanifolds $\mu_{i_1}, \mu_{i_1 i_2}, \ldots, \mu_{i_1 i_2 \ldots i_k}$ of codimensions $1, 2, \ldots, k$ respectively and for any integer values $1 \leq i_s \leq m_s$, $s \leq k \leq n$

$$\mu_{i_1} \supset \mu_{i_1 i_2} \supset \cdots \supset \mu_{i_1 i_2 \ldots i_k} \supset \mu_{i_1 i_2 \ldots i_k i_{k+1}} = \varnothing.$$

The algebra $\mathcal{L}(X_0, X_1, \ldots, X_n)$ is required to be such that at each point $x \in \alpha \setminus \left(\bigcup_{i_1=1}^{m_1} \mu_{i_1}\right)$ the rank $r_{\mathcal{L}}(x) = n$, while at the points of the sets $(\mu_{i_1 i_2 \ldots i_m} \setminus \mu_{i_1 i_2 \ldots i_m i_{m+1}}) \cap \alpha$ the algebra is transversal to $\mu_{i_1 i_2 \ldots i_m}$ (Paneah 1985a). It is clear from this description that the submanifolds in question can be piecewise smooth, intersecting, and the like.

§3. The Boundary Operator $\mathcal{A}$ of Second Order. The Form $a^{ij}\xi_i\xi_j$ Not of Fixed Sign

In this section the form $a^{ij}\xi_i\xi_j$ and the coefficient $a(x)$ that occur in the operator $\mathcal{A}$ can be nonnegative on one subset M^+ of the boundary M and nonpositive on another subset M^-. In this case the solvability properties of the problem (1.1)–(1.2) differ sharply from those of § 2: at least one of the quantities $\dim\ker\mathfrak{A}$ and $\dim\operatorname{coker}\mathfrak{A}$ becomes infinite. To obtain a well-posed formulation, consequently, the problem must be modified. It is of course desirable to do this in such a way that the original conditions (1.1)–(1.2) are violated as little as possible. We now pass to precise statements.

In a tubular neighborhood $\mathcal{T}$ of the boundary M we consider an operator $\mathcal{A}$ of the form

$$\mathcal{A}u = \partial_i \rho a^{ij}\partial_j u + a^{0j}\partial_j u + \rho a \partial_t u + a_0 u, \tag{3.1}$$

where $\rho \in C^\infty(M)$ and the coefficients a, a_0, a^{0j}, and a^{ij} are defined in § 2. We assume that M contains open subsets M^+ and M^-, as well as submanifolds μ_1 and μ_2 of codimension 1 such that

1) $M = M^+ \cup M^- \cup \mu_1 \cup \mu_2$; $M^+ \cap M^- = \varnothing$; $\partial M^+ = \partial M^- = \mu_1 \cup \mu_2$.

2) $\rho^2 + |d\rho|^2 \neq 0$; $\rho^{-1}(0) = \mu_1 \cup \mu_2$; $\rho > 0$ on M^+, $\rho < 0$ on M^-.

3) $a^{ij}(x)\xi_i\xi_j \geq 0$, $a(x) \geq 0$, $(x,\xi) \in T_0^*(M)$.

4) The vector field $X_0 = a^{0j}\partial_j$ is tranversal to $\mu_1 \cup \mu_2$ and oriented in the direction of M^- on μ_1 and in the direction of M^+ on μ_2.

We also require certain rank and transversality conditions to hold.

5) At all points $x \in \alpha$ except possibly a submanifold μ of codimension 1 the rank of the algebra $\mathcal{L}(X_0, X_1, \ldots, X_n)$ equals n, where, as above, $\alpha = \{x \in M \,|\, a(x) = 0\}$.

6) At the points $x \in \mu \cap \alpha$ the algebra $\mathcal{L}$ is transversal to μ. We note that in contrast to § 2 we have $X_j = \rho a^{ij}(x)\partial_j$ here. It can be shown that if $\mu_1 = \varnothing$, the kernel of problem (1.1)–(1.2) is infinite-dimensional; and if $\mu_2 = \varnothing$, the cokernel of the problem is infinite-dimensional. As will be clear from what follows, the following modification of problem (1.1)–(1.2), in which a function v on μ_2 is sought in addition to the function u, does not suffer from these defects:

$$\mathcal{L}u = F \text{ in } \Omega, \quad \mathcal{A}u = f + v \otimes \delta(\mu_2) \text{ on } \mu_1. \tag{3.2}$$

(Here $\delta(\mu_2)$ is the surface measure on μ_2.) That the problem is well-posed in this form—a fact established by Theorem 2.1—means that the infinite dimensionality of the cokernel of the problem (1.1)–(1.2) is neutralized under conditions 1)–6) by the violation of the boundary condition $\mathcal{A}u = f$ on the submanifold μ_2 and the infinite-dimensional kernel that arises can be parametrized by the values of its elements on the submanifold μ_1. This approach was suggested by the results of Egorov and Kondrat'ev (1969), Maz'ya and Paneah (1974), and Malyutov (1969), in which the analogous situation arose in connection with the oblique derivative problem (cf. § 4). It is natural to expect that the solution of problem (3.2) has a singularity on μ_{12}. To detect the singularity it is convenient to introduce the weighted Sobolev-Slobodetskij spaces on M and in Ω. Let $\mathcal{T}_\Theta \subset \mathcal{T}$ be a tubular neighborhood of the submanifold μ_{12} in $\mathbb{R}^{n+1}$ whose projection $\Theta = \pi(\mathcal{T}_\Theta)$ coincides with a tubular neighborhood of μ_{12} in M. If $x = (x_1, x') = (x_1, x_2, \dots, x_n)$ are normal coordinates in Θ, then coordinates (t, x_1, x') are thereby defined in $\mathcal{T}_\Theta$, and in these coordinates $M = \{p \in \overline{\Omega} \,|\, t = 0\}$, $\mu_{12} = \{p \in \overline{\Omega} \,|\, t = 0, x_1 = 0\}$. Let $\varphi + \psi = 1$ be a smooth partition of unity on M in which $\varphi \in C_0^\infty(\Theta)$ and $\varphi \equiv 1$ near μ_{12}. The space $\mathcal{H}_s(M)$ is defined as the completion of $C^\infty(M)$ in the norm

$$\langle u \rangle_s^2 = \sum_{k=0}^{s} \sum_{|\alpha|=0}^{k} \|x_1^{k-|\alpha|} \partial_1^{k-|\alpha|} \partial_{x'}^{\alpha} \varphi u\|^2 + \|\psi u\|_s^2,$$

where $\partial_{x'}^\alpha = \partial_2^{\alpha_2} \partial_3^{\alpha_3} \dots \partial_n^{\alpha_n}$ and $|\alpha| = \alpha_2 + \alpha_3 + \dots + \alpha_n$. Let $\Phi + \Psi \equiv 1$ be a smooth partition of unity in $\overline{\Omega}$, where $\Phi \in C_0^\infty(\mathcal{T}_\Theta \cap \overline{\Omega})$ and $\Phi \equiv 1$ in some neighborhood of μ_{12}. For any $m \in \mathbb{R}^1$ the space $H_{s,m}(\Omega)$ is defined as the completion of $C^\infty(\overline{\Omega})$ in the norm

$$\ll u \gg_{s,m} = \sum_{k=0}^{s} \Big(\|t^k \Phi u\|_{k+m} + \|x_1^k \Phi u\|_{k+m} + \sum_{|\alpha|=0}^{k} \|\partial_{x'}^{\alpha} \Phi u\|_m \Big) + \|\Psi u\|_{s+m}.$$

We set $\mathbf{H}_s(\Omega) = H_{s,0}(\Omega) \oplus \mathcal{H}_s(M) \oplus H^s(\mu_1)$. Under conditions 1)–6) the following theorem holds.

Theorem 3.1 (Paneah 1987). *Let $s \geq 0$ and $(F, f, g) \in \mathbf{H}_s(\Omega)$. Then the following assertions hold.*

(i) *The problem (3.2) has a solution $u \in H_{s,\frac{1}{2}}(\Omega)$ if and only if the function (F, f, g) satisfies the orthogonality conditions*

$$(F, \Phi_j)_\Omega + (f, \varphi_j)_M + (g, \psi_j)_{\mu_1} = 0, \quad j = 1, 2, \dots, J,$$

with functions $\Phi_j \in L_2(\Omega)$, $\varphi_j \in L_2(M)$, $\psi_j \in L_2(\mu_1)$.

(ii) *The subpace $\mathcal{N}$ of solutions of the problem in $H_{s,\frac{1}{2}}(\Omega)$ is finite dimensional when $F = 0$, $f = 0$, and $g = 0$, and is independent of s, so that $\mathcal{N} \subset H_{\infty,\frac{1}{2}}(\Omega) = \bigcap_s H_{s,\frac{1}{2}}(\Omega)$. The dimension of this subspace is J.*

(iii) *If $\dim \mathcal{N} = 0$, then the inverse operator of the problem (3.2) is compact: $\mathbf{H}_s(\Omega) \to H_{s,\frac{1}{2}}(\Omega)$.*

(iv) *The following* a priori estimate *holds:*

$$\ll u \gg_{s,\frac{1}{2}}^{\Omega} \leq c\big(\ll \mathcal{L}u \gg_{s,0}^{\Omega} + \langle \mathcal{A}u \rangle_s^M + \|u\|_s^{\mu_1} + \|u\|_{-1}^{\Omega} \big).$$

Thus the essential properties of elliptic boundary value problems are preserved in this situation, although in a slightly modified form. The choice of the spaces $H_{s,r}(\Omega)$ and $\mathcal{H}_s(M)$ is determined by the following mutual property that they possess.

Lemma 3.1 (Paneah 1988). *If $v \in \mathcal{H}_s(M)$, then $u = \mathfrak{D}v \in H_{s,\frac{1}{2}}(\Omega)$ and $\ll u \gg^{\Omega}_{s,\frac{1}{2}} \leq c(s)\langle v\rangle^M_s$. If $F \in H_{s,0}(\Omega)$, then $u = \mathfrak{B}F \in H_{s,2}(\Omega)$ and $\ll u \gg^{\Omega}_{s,2}$ $+\langle \mathcal{A}u\rangle^M_s \leq c(s) \ll F \gg^{\Omega}_{s,0}$.*

To prove Theorem 3.1 we reduce problem (3.2) in an equivalent manner to the following problem on M:

$$Au = f + v \otimes \delta(\mu_2) \text{ on } M, \quad u = g \text{ on } \mu_1, \tag{3.3}$$

where A is a Calderón operator. The lower a priori estimates in $L_2(M)$ of both the operator A itself and the perturbed operator $A_\lambda = A + \lambda\rho$ for $\lambda > 0$ play a decisive role in the study of this problem by making it possible to invoke the standard reasoning connected with the Riesz-Fredholm theory. To obtain the needed estimates we construct a special quadratic form of the operator A_λ that gives an upper estimate for the norm $\|\cdot\|^2$. All the constructions have a local character. In particular the quadratic form just mentioned has quite different forms in the tubular neighborhoods Θ_1 and Θ_2 of the submanifolds μ_1 and μ_2. We give the derivation of the estimate of the operator A_λ in the space $L_2(\Theta_1)$ in order to demonstrate the typical difficulties that arise in working with pseudodifferential operators that are neither positive- nor negative-definite and the way these difficulties are overcome. Let $x = (x_1, x')$ be normal coordinates in Θ_1 with $x_1 > 0$ in M^+, $x_1 < 0$ in M^-, and $|x_1| < \theta$ for all points $x \in \Theta_1$. We introduce the function

$$\varkappa = -x_1 + 2\theta \operatorname{sgn} x_1$$

and we show that for sufficiently small values of θ the estimate

$$\|u\|^2 + \sum_{j\geq 1} \|x_1^{-\frac{1}{2}} X_j u\|^2 + \lambda\theta\|x^{\frac{1}{2}}u\|^2 \leq c\big(\operatorname{Re}(A_\lambda u, \varkappa_1 u) + \theta\|u, \mu_1\|^2\big) \tag{3.4}$$

holds for any function $u \in H^1(\Theta_1) \cap H^2(\Theta_1 \setminus \mu_1)$ that vanishes along the boundary $\partial\Theta_1$.

Besides standard integration by parts on the "differential terms" of the operator A_λ the proof requires a lower bound on the quadratic form of a pseudodifferential operator that is typical for noncoercive problems. The typical example of such an estimate is inequality (2.7). In this case it is a question of obtaining the estimate

$$-\operatorname{Re}\left((2\theta|x_1| - x_1^2)a(x)\Lambda^0 u, u\right) + c\theta\|u\|^2 \geq 0, \quad u \in C_0^\infty(\Theta_1). \tag{3.5}$$

The difficulty is that although the symbol $P(x,\xi) = -(2\theta|x_1| - x_1^2)a(x)\Lambda^0$ of the pseudodifferential operator on the left-hand side is nonnegative, the application of inequality (2.7) is prevented by the fact that the function $P(x, \xi)$ is not smooth. (Nevertheless if $a(x) \equiv 0$ on μ_1, then $\operatorname{grad} a = 0$ on μ_1 and $P(x, \xi)$ has the necessary regularity anyway.) Thus estimate (3.5) requires a special proof, a sketch of which is given below.

It suffices to limit ourselves to the case $-\Lambda^0(x, \xi) = |\xi|$ and functions $u(x) \in C_0^\infty(\omega)$, where ω is a domain in $\mathbb{R}^n$, while $|x_1| < \theta/2$ in ω. Consider a bounded domain Q in the half-space $\mathbb{R}_+^{n+1} = \{(t, x)|t > 0\}$ whose smooth boundary contains ω, while $|x_1| < \theta$ everywhere on Q. We set $\rho = \theta(x_1^2 + t^2)^{\frac{1}{2}} - x_1^2$, and let $\mathcal{U}$ be a harmonic function in Q equal to u on ω and to zero on $\partial Q \setminus \omega$ (cf. Maz'ya 1972). As noted in § 1, we have $\partial_t \mathcal{U}\big|_\omega = -|\partial_x| u$, where $|\partial_x|$ is the pseudodifferential operator with the symbol $|\xi|$. Therefore integration by parts in the equality $(\Delta \mathcal{U}, \rho \mathcal{U})_Q = 0$ gives

$$(\operatorname{grad} \mathcal{U}, \operatorname{grad}(\rho \mathcal{U}))_Q = (\rho u, |\partial_x| u)_\omega = \operatorname{Re}((\theta|x_1| - x_1^2)u, |\partial_x| u)_\omega.$$

On the other hand, taking account of the explicit form of the functions $\partial\rho/\partial x_1$ and $\partial\rho/\partial t$ and again integrating by parts, we find that

$$(\operatorname{grad} \mathcal{U}, \operatorname{grad} \rho \mathcal{U})_Q = (\rho \operatorname{grad} \mathcal{U}, \operatorname{grad} \mathcal{U})_Q + \|\mathcal{U}\|_Q^2 - \theta\|\mathcal{U}(x_1^2 + t^2)^{-\frac{1}{4}}\|_Q^2.$$

Hence $-\operatorname{Re}((\theta|x_1| - x_1^2)u, |\partial_x| u)_\omega \geq c\theta\|\mathcal{U}(x_1^2 + t^2)^{-\frac{1}{4}}\|_Q^2$ and the required estimate now follows from the inequality $\|\mathcal{U}(x_1^2 + t^2)^{-\frac{1}{4}}\|_Q \leq c\|t^{\frac{1}{2}} \operatorname{grad} \mathcal{U}\|_Q$ (cf. Maz'ya 1972) and the obvious identity $2\|t^{\frac{1}{2}} \operatorname{grad} \mathcal{U}\|_Q^2 = \|u\|_\omega^2$.

In the neighborhood $\Theta_2 \supset \mu_2$ the necessary energy estimate of the operator A_λ is obtained in a completely different way. We mention only that in this case also there are difficulties in handling the pseudodifferential component of A, since the function $\varkappa$ must also be chosen nonsmooth, namely $\varkappa = |x_1|^{2\delta} \operatorname{sgn} x_1$, where δ $(0 < \delta < \frac{1}{2})$ is a sufficiently small number.

Up to this point we have been discussing a priori estimates for the operator A. However, by themselves such estimates can guarantee only the regularity of a solution or its uniqueness, or, for example, the compactness of the inverse operator. To prove that the problem (3.3) is solvable it is necessary either to apply information on the adjoint problem, as in § 2, or, if the adjoint problem is difficult to study for technical reasons, construct a solution u_ε of the regularized equation containing a small parameter ε. In this case the availability of a priori estimates for the regularized operator $A_{\lambda\varepsilon}$ that are uniform on ε makes it possible to prove the convergence of the functions u_ε to a solution of the original equation as $\varepsilon \to 0$. It is in this way that the proof of Theorem 3.1 is constructed; moreover the regularized operator $A_{\lambda\varepsilon}$ has a discontinuous principal part and is equal to $A_\lambda + \varepsilon(\operatorname{sgn} \rho)G$, where G is an elliptic operator on M.

Returning to the statement of Theorem 3.1, we note that, as follows from the a priori estimate and the definition of the norm $\ll \cdot \gg_{s,r}$, the solution u has a singularity near μ_{12} in the general case. Examples show that in the class of operators A of this type the results connected with the regularity of the solutions of the problem (3.2) are sharp: even when $F=0$, $f\in C^\infty(M)$, and $g\in C^\infty(\mu_1)$ the solution need not belong to the space $H^1(\Omega)$. However, when the operator $\mathcal{A}$ has a certain structure near μ_{12}, all solutions are rather smooth functions. This is reflected in the following two theorems (cf. Paneah 1987).

Theorem 3.2. *Let* $\mu_{12}=\mu_1$ *and* $a^1(x)>(m-\frac{1}{2})a^{11}(x)$ *at all points* $x\in\mu_1$, *where* $m>0$ *is an integer. Then the following assertions hold for any integer value of* s, $1\le s\le m$.

(i) *The problem* (3.2) *has a solution* $u\in H^{s+\frac{1}{2}}(\Omega)$ *for those functions and only those functions* $(F,f,g)\in\mathbf{H}^s=H^{s+\frac{1}{2}}(\Omega)\oplus H^s(M)\oplus H^s(\mu_1)$ *that are orthogonal (in the metric of* L_2*) to a finite-dimensional subspace of* $\mathbf{H}^s$ *of the same dimension as the kernel.*

(ii) *If the kernel of the problem is empty, then the inverse operator is compact.*

(iii) *If* $u\in H^{s+\frac{1}{2}}(\Omega)$ *and* $(F,f,g)\in\mathbf{H}^t$ *for integer values* $s<t\le m$, *then* $u\in H^{t+\frac{1}{2}}(\Omega)$. *In particular, if* $a^{11}\equiv 0$ *on* μ_1, *then properties* (i)–(iii) *hold for any integer* $s\ge 0$; *in this case all the solutions of the problem* (3.2) *belong to* $H^\infty(\Omega)$ *if* $(F,f,g)\in\mathbf{H}^\infty$.

(iv) *The following estimate holds for solutions of the problem:*

$$\|u,\Omega\|_{s+\frac{1}{2}}\le c(s)\big(\|\mathcal{L}u,\Omega\|_{s+\frac{1}{2}}+\|\mathcal{A}u,M\|_s+\|u,\mu_1\|_s+\|u,\Omega\|_{-1}\big).$$

It is worth mentioning here that although the coefficients a^{ij} and a^j depend on the choice of local coordinates x' on μ_1, the quantities $a^{11}\big|_{\mu_1}$ and $a^1\big|_{\mu_1}$ are uniquely determined in any *normal* coordinate system (x_1,x') in Θ_1.

Theorem 3.3. *If* $s\ge 1$, *then each solution* (u,v) *of the problem* (3.2) *has the property that* $u\big|_M\in H^1(M\setminus\mu_2)$ *and* v *coincides with the jump of the function* $a^1u\big|_M$ *on* μ_2. *If in addition* $a=O(x_1^{[s/2]})$ *in* Θ_2, *then* $u\big|_M\in H^{[s/2]}(M\setminus\mu_2)$ *under the condition* $f\in H^s(M\setminus\mu_2)$.

In the proof of these theorems the idea of the correct presentation of the principal part of the operator (operator A in our case), which arises frequently in nonelliptic problems, shows through clearly. This idea should be taken as a guide in choosing the corresponding quadratic form.

§4. Boundary Operator $\mathcal{A}$ of First Order. The Oblique Derivative Problem

4.1. Statement of the Problem. We now pass to one of the most popular objects of the theory of partial differential equations – the oblique derivative problem. Due to the apparent simplicity and geometric intituitiveness of the problem, it has become a sort of testing ground for the latest methods of the theory (pseudodifferential operators, Fourier integral operators, and others). It produced the first example of a boundary value problem for an elliptic equation with infinite index (Bitsadze 1964, Borrelli 1966), moreover against the background of the beautiful achievements of the 1960's in the theory of elliptic boundary value problems. The Calderón operator of the oblique derivative problem turned out to be the first example of a subelliptic operator (Hörmander 1966).

As already pointed out, the oblique derivative problem is the special case of the problem (1.1)–(1.2) obtained by setting $a^{ij} \equiv 0$. Therefore the results of §§ 2–3 extend to it. The restrictions on the operator $\mathcal{A}$ under which these results are obtained referred exclusively to the behavior of the vector fields $X_0, X_1, \ldots, X_n$ on the set $\alpha = \{x \in M \,|\, a(x) = 0\}$. In the case of the oblique derivative problem $X_1 = X_2 = \cdots = X_n \equiv 0$ on M; therefore conditions 1)–6) of § 3 reduce to the condition that α is a submanifold in M, $\operatorname{codim} \alpha = 1$, and the field X_0 is transversal to α. The connected components of α are μ_1 and μ_2, and also the submanifold μ_0 on which the function a vanishes and on a neighborhood of which it is of constant sign. It is under these assumptions that the degenerate oblique derivative problem was studied in the original papers devoted to its solvability (Egorov, Kondrat'ev 1969; Maz'ya, Paneah 1974; Malyutov 1969; Hörmander 1966; Kato 1982; Soga 1975). Therefore the results of these papers are contained in the theorems of §§2 and 3. In what follows we consider a more general case of degeneracy of ellipticity than degeneracy on a manifold.

In the study of the oblique derivative problem it is natural and convenient from the technical point of view to give the boundary condition in invariant form. For this purpose we consider a nonsingular vector field $\mathbf{l} \in C^\infty(M, T_M(\mathbb{R}^{n+1}))$, and along with it the boundary value problem

$$Lu = F \text{ in } \Omega, \quad \partial u/\partial \mathbf{l} + hu = f \text{ on } M, \tag{4.1}$$

where h is a function in $C^\infty(M)$. We represent $\mathbf{l}$ in the form

$$\mathbf{l} = \tau + a\nu,$$

where $\tau \in C^\infty(M, T(M))$ is a vector field on M, ν is a unit vector field of interior normals to M, and a is a function in $C^\infty(M)$. We denote by (z, x) the normal coordinates in a tubular neighborhood $\mathcal{T}$ of the boundary M (in contrast to the coordinates (t, x) in §§2 and 3). Then the boundary condition in (4.1) can be written in the form

$$a^j(x)\partial_j u + a(x)\partial_z u + h(x)u = f \text{ on } M,$$

and the principal symbol of the corresponding Calderón operator A turns out to be (cf. (1.5))

$$A^0(x,\xi) = ia^j(x)\xi_j + a(x)\Lambda^0(x,\xi).$$

For $n = 1$ this function, which equals $ia^1(x)\xi_1 - a(x)|\xi_1|$, cannot vanish if $|\xi_1| = 1$, since the field $\mathbf{l}$ is nondegenerate. Therefore the two-dimensional oblique derivative problem is always elliptic. For $n \geq 2$, as already noted above, a necessary and sufficient condition for the problem (4.1) to be elliptic is that the inequality $a \neq 0$ hold on M, or, what is the same, that $\alpha = \varnothing$. Thus we can assume in what follows that the set α of points where the field $\mathbf{1}$ is tangent to the boundary manifold M is nonempty. It will cause no loss of generality to assume that the field τ is also nonsingular (provided, of course, the field $\mathbf{1}$ is; the case of a degenerate field $\mathbf{1}$ is handled in Theorem 4.6). We make the convention in what follows that $\gamma_p(t) : \mathbb{R}^1 \to M$ denotes a *maximal trajectory of the field* τ passing through the point $p \in M$. In other words

$$\partial_t \gamma_p(t) = \tau \circ \gamma_p(t), \quad \gamma_p(0) = p, \quad -\infty < t < \infty.$$

Let M^+, M^-, μ_1, μ_2 be as in § 3 and suppose the following conditions hold.

1) $M = M^+ \cup M^- \cup \mu_1 \cup \mu_2$; $M^+ \cap M^- = \varnothing$; $\partial M^+ \cap \partial M^- = \mu_1 \cup \mu_2$.

2) $a \geq 0$ on M^+, $a \leq 0$ on M^-.

3) The field τ is transversal to $\mu_1 \cup \mu_2$ and oriented toward M^- on μ_1 and toward M^+ on μ_2. These conditions coincide with the corresponding conditions of § 3. The final conditions 4)–5) are specific to the problem under consideration (and hold automatically if α is a submanifold in M).

4) (cf. Melin and Sjöstrand 1976). The set α does not contain any complete semitrajectories $\gamma_p(t)$, $p \in \alpha$.

In connection with the compactness of the inverse operator of the problem (4.1) (and only in this connection) we need the following condition.

5) (Paneah 1981). The set α does not contain any intervals of trajectories $\gamma_p(t)$, $p \in \alpha$.

As in § 3, if $\mu_1 \cup \mu_2 = \mu_{12} \neq \varnothing$, we study the modified problem

$$\mathcal{L}u = F \text{ on } \Omega, \quad \partial u/\partial \mathbf{1} + hu - v \otimes \delta(\mu_2) = f \text{ on } M, \quad u = g \text{ on } \mu_1 \quad (4.2)$$

instead of (4.1).

Thus no structural restrictions are imposed on a set of "degenerate ellipticity." It may contain quite arbitrary subsets of dimension $\leq n$, subject only to condition 4) (or 5)). As in § 3, the condition that the field $\mathbf{1}$ be transversal to certain submanifolds μ_1 and μ_2 of codimension 1 contained in α remains, but the location of these submanifolds on M is in general not dictated a priori by the field $\mathbf{1}$, but may be chosen to a certain extent arbitrarily. Condition 4) means that in moving along the trajectory $\gamma_p(t)$ from a point $p \in \alpha$ we leave the set α both as $t \to +\infty$ and as $t \to -\infty$.

4.2. Solvability of the Problem in the Spaces H^s. We begin by describing the spaces $\mathcal{H}_k(M, \mu_2)$ in which it is convenient to state the results relating to the problem (4.2). Let $\psi_1 + \psi_2 \equiv 1$ be a C^∞-partition of unity on M for which $\operatorname{supp} \psi_j \subset \Theta_j$, $\psi_j \equiv 1$ in some neighborhood of μ_j, $j = 1, 2$. (We recall that Θ_1 and Θ_2 are disjoint tubular neighborhoods of the submanifolds μ_1 and μ_2

respectively in M.) The space $\mathcal{H}_k(M, \mu_2)$ is defined as the subset of functions u in $L_2(M)$ for which the following norm is finite:

$$\langle u \rangle_k^2 = \|\psi_1 u\|_k^2 + \sum_{p=0}^{k} \sum_{|\alpha|=0}^{p} \|(s\partial_s)^{k-|\alpha|}\partial_y^\alpha \psi_2 u\|^2 + \|\psi u\|_k^2,$$

where $(s, y) = (s, y_1, y_2, \ldots, y_{n-1})$ are normal coordinates in Θ_2. Analogously, but with the roles of μ_1 and μ_2 reversed, we define the space $\mathcal{H}_k(M, \mu_1)$. We set $\mathcal{H}_\infty(M, \mu_2) = \bigcap_k \mathcal{H}_k(M, \mu_2)$. In contrast to the spaces $\mathcal{H}_k(M)$ of § 3, the elements $u \in \mathcal{H}_k(M, \mu_2)$ have no singularities near μ_1. Introducing the corresponding modification in the description of the norm $\ll \cdot \gg_{s,r}$, we can define the space $H_{s,r}(\Omega, \mu_2)$, whose elements have singularities only on μ_2. For the norm in this space we retain the previous notation $\ll \cdot \gg_{s,r}$.

The following theorem generalizes information on the solvability of the modified oblique derivative problem contained in the papers of Egorov and Kondrat'ev (1969), Maz'ya and Paneah (1974), Malyutov (1969), Paneah (1981, 1988), Soga (1975), and Taira (1979).

Theorem 4.1. (Paneah 1981, 1988). *Suppose that conditions* 1)–4) *hold. Then the following assertions are true.*

(i) *For any integer value $s \geq 0$ the problem* (4.2) *has a solution $(u, v) \in H_{s,\frac{1}{2}}(\Omega, \mu_2) \oplus H^s(\mu_2)$ for those functions and only those functions $(F, f, g) \in H_{s,0}(\Omega, \mu_2) \oplus \mathcal{H}_s(M, \mu_2) \oplus H^s(\mu_1)$ that satisfy the N orthogonality conditions*

$$(F, \mathfrak{D}^c(raw_j))_\Omega + (f, w_j)_M + (g, \{w_j\}_{\mu_1})_{\mu_1} = 0, \quad 1 \leq j \leq N. \quad (4.3)$$

Here $r < 0$ is a smooth function depending only on $\mathcal{L}$ and M, $\{w_j\}_{\mu_1}$ is the jump of the function $w_j \in H^1(\Theta_1 \setminus \mu_1)$ on μ_1 and the functions $w_j \in \mathcal{H}_\infty(M, \mu_1) \cap H^\infty(M \setminus \mu_1)$ form a basis of the solution space of the problem

$$A^c w + \{w\}_{\mu_1} \otimes \delta(\mu_1) = 0 \text{ on } M, \quad w = 0 \text{ on } \mu_2.$$

(ii) *If $F = 0$, $f = 0$, and $g = 0$, then problem* (4.2) *has exactly N linearly independent solutions, all belonging to the space $H_{\infty,\frac{1}{2}}(\Omega, \mu_2)) \oplus H^\infty(\mu_2)$.*

(iii) *The estimate*

$$\ll u \gg_{s,\frac{1}{2}} + \|v, \mu_2\|_s \leq c\big(\ll F \gg_{s,0} + \langle f \rangle_s + \|g, \mu_1\|_s + \|u\| + \|v, \mu_2\|_{-1}\big)$$

holds for solutions of problem (4.2) *with the constant c depending only on s.*

(iv) *If $(u, v) \in H_{s,\frac{1}{2}}(\Omega, \mu_2) \oplus H^s(\mu_2)$ is a solution of problem* (4.2) *in which $(F, f, g) \in H_{t,\frac{1}{2}}(\Omega, \mu_2) \oplus \mathcal{H}_t(M, \mu_2) \oplus H^t(\mu_1)$, and $t > s$, then $u \in H_{t,\frac{1}{2}}(\Omega, \mu_2)$ and $v \in H^t(\mu_2)$. In particular when $(F, f, g) \in$*

$H_{\infty,\frac{1}{2}}(\Omega,\mu_2)\oplus\mathcal{H}_\infty(M,\mu_2)\oplus H^\infty(\mu_2)$, *any solution* (u,v) *belongs to the space* $H_{\infty,\frac{1}{2}}(\Omega,\mu_2)\oplus H^\infty(\mu_2)$.

(v) *If* $s\geq 1$, *then* $u\big|_M\in H^1(M\setminus\mu_2)$ *and* $v=\{u\}_{\mu_2}$.

Thus the problem (4.2) is a Fredholm problem and has index zero, while its kernel and cokernel consist of C^∞-functions on $\overline{\Omega}\setminus\mu_2$ and the boundary values of its solutions on M have a finite jump at μ_2. Any solution u of this problem satisfies the conditions of problem (4.1) everywhere in $\overline{\Omega}$ except on the submanifold μ_2. We note further that when $\mu_{12}=\varnothing$, the elliptic degeneracy on a set $\alpha\subset M$ connected with $\mathbf{1}$ only by condition 4) does not invalidate the assertions of the theorem (with the natural modifications) for the original problem (4.1). In this case the solutions loses only smoothness and only on the set α compared with the nondegenerate oblique derivative problem.

The proof of Theorem 4.1, as in § 3, is based on a detailed study of the Calderón operator A corresponding to the boundary operator $\partial/\partial\mathbf{1}+h$. In obtaining the corresponding estimates the correct choice of the partition of unity on M to localize the problem plays a central role, and both the form of the neighborhoods covering M and the properties of the functions subordinate to this covering are essential here. While in the elliptic case any sufficiently small spheres covering M could be used as the neighborhoods, and when there was degeneracy on a submanifold α, it was necessary to distinguish a small tubular neighborhood $\Theta\supset\alpha$ (cf., for example, Maz'ya, Paneah 1974), in the present situation with a massive set α it turns out to be convenient to localize the problem along trajectories $\gamma_p(t)$, $p\in\alpha$. In this situation the neighborhoods are formed by cylinders of the trajectories of the field τ, whose bases are situated in $\mathbb{C}\alpha$. The existence of such cylinders, a finite number of which cover α, is guaranteed by condition 4). The extent of each cylinder Γ (i.e., the time t_Γ required to traverse Γ along its axis $\gamma(t)$) can of course be arbitrary, so that the term "localization" is being used very loosely here. In particular (and this corresponds to the very general conditions 1)–4)) Γ can intersect the components of the submanifold μ_{12} an arbitrary number of times, but no additional localization is carried out near μ_{12}. The convenience of such a choice of neighborhoods is that in "cylindrical" coordinates (t,x) in Γ (t is time, x are local coordinates on the base Γ^+ that is transversal to the field τ and situated in $\mathbb{C}\alpha$) the operator $\partial/\partial\tau$ becomes $\partial/\partial t$, and the operator A becomes $\partial_t+a(t,x)\Lambda+h$. The union of such cylinders Γ, supplemented by the set $E=M\setminus\alpha$ (on which the operator A is now elliptic), forms the desired covering of M. To show the possibilities connected with such a localization, we deduce an estimate of the operator A in $L_2(\Gamma)$, which is the heart of the entire proof. Let $\Gamma=\{(t,x)\big|\,0\leq t\leq 1,\,|x|<\delta\}$, and let $t=\xi_j(x)$, $1\leq j\leq s$, $t=\eta_k(x)$, $1\leq k\leq p$, be the equations of the components of μ_1 and μ_2 respectively that intersect Γ. It is obvious that the functions ξ_j and η_k form an increasing sequence and alternate with one another, so that if, for example, $\xi_1=\min\{\xi_j,\eta_k\}$, then the inequalities $0<\xi_1(x)<\eta_1(x)<\xi_2(x)<\cdots<1$

hold. We set $\rho_1 = \prod_j (t - \xi_j(x))$, $\rho_2 = \prod_k (t - \eta_k(x))$. Then the function $\varkappa = \rho_2/\rho_1$, like $a(t,x)$, changes sign only when $t = \xi_j(x)$ and $t = \eta_k(x)$, so that $\operatorname{sgn}(a\varkappa) = \text{const}$ in Γ. It can also be shown that $\operatorname{sgn}(\partial\varkappa/\partial t) = \text{const}$ in Γ (which is not completely obvious). Therefore for some normalization of the function $\varkappa$ we have

$$a\varkappa \leq 0, \quad \partial\varkappa/\partial t < -c\rho_1^2 \text{ in } \Gamma, \tag{4.4}$$

where $c > 0$ is a constant. Consider the quadratic form $J = \operatorname{Re}(Au, \varkappa u) = \operatorname{Re}(\partial_t u, \varkappa u) + \operatorname{Re}(a\varkappa \Lambda u, u)$ on the functions $u \in \overset{\circ}{H}{}^1(\Gamma \setminus \mu_{12})$. By the second of inequalities (4.4) integration by parts yields the estimate $\operatorname{Re}(\partial_t u, \varkappa u) \geq c\|u/\rho_1\|^2$. As usual, difficulties arise in working out the pseudodifferential part of the operator, i.e., the second term in J. Since the smooth function $a\varkappa$ is nonpositive, $(a\varkappa)^{\frac{1}{2}} \in \text{Lip}$. Therefore the commutator $P = [\Lambda, (a\varkappa)^{\frac{1}{2}}]$ is bounded in $L_2(\Gamma)$ (Maz'ya 1972, Calderón 1965). Applying Gårding's inequality ($\operatorname{Re}(\Lambda w, w) \geq c\|w\|_{\frac{1}{2}}^2 - c_1\|w\|^2$) to the operator $(-\Lambda)$, we find

$$\begin{aligned}&\operatorname{Re}(a\varkappa\Lambda u, u) = -\operatorname{Re}(\Lambda(-a\varkappa)^{\frac{1}{2}}u, (-a\varkappa)^{\frac{1}{2}}u) + \operatorname{Re}(Pu, (-a\varkappa)^{\frac{1}{2}}u) \geq \\ &- c\|(a\varkappa)^{\frac{1}{2}}u\|^2 - c_1\|u\| \cdot \|(a\varkappa)^{\frac{1}{2}}u\|^2 \geq (-c/\lambda)\|u/\rho_1\|^2 - (c_1 + \lambda/2)\|(a\varkappa)^{\frac{1}{2}}u\|^2.\end{aligned}$$

Together with the estimate for the form $(\partial_t u, \varkappa u)$ this leads to the required estimate of the perturbed operator $A_\lambda = A + \lambda a$ for large values of λ:

$$\operatorname{Re}(A_\lambda u, \varkappa u) \geq c\|u/\rho_1\|^2 + \lambda\|(a\varkappa)^{\frac{1}{2}}u\|^2, \quad u \in \overset{\circ}{H}{}^1(\Gamma \setminus \mu_{12}).$$

As noted in § 3, when local estimates are pieced together the norm $\|u, M\|$ is estimated in terms of the sum of the norms $\|Au, M\|$ and $\|T_\varphi u, M\|$, where $T_\varphi = [A, \varphi]$ is the commutator of the operator A with an element of the partition of unity that admits the rough estimate $\|T_\varphi u\| \leq c(\varphi)\|u\|$, which does not allow us to obtain the inequality $\|u\| \leq c(\|Au\| + \|u\|_{-1})$. It is also not possible to handle T_φ by choosing a large parameter λ, as in § 3. However, one peculiarity of Γ, namely the fact that its base belongs to $\complement\alpha$, makes it possible to choose the corresponding functions φ_Γ so that the essential part of the function $T_\varphi u$ is concentrated outside α, where the operator A is elliptic. This makes it possible to neutralize the increase in the norms $\|T_\varphi u\|$ by using the elliptic estimate $\|T_\varphi u\|_1 \leq c(\varphi)(\|Au\| + \|u\|)$ and the interpolation inequality $\|T_\varphi u\| \leq \sigma\|T_\varphi u\|_1 + c(\sigma)\|u\|_{-1}$.

Another approach to the nonelliptic oblique derivative problem in the case when α is a submanifold of codimension 1 in M and $\mu_2 \neq \varnothing$ was broached by Eskin (1971a). The boundary condition on M in (4.2) is replaced by the condition

$$\partial u/\partial \mathbf{1} + hu + G(\rho \otimes \delta(\mu_2)) = f, \tag{4.5}$$

where G is a pseudodifferential operator on M, $\operatorname{ord} G < -\frac{1}{2}$, and ρ is a new unknown function on μ_2. The term $G(\rho \otimes \delta)$ is called a *coboundary term* or an operator *of potential type.* In the simplest situation ($\mathcal{L}$ is the Laplacian and τ is the field of geodesic normals to $\mu_2 = \alpha$) the Calderón operator of the problem is $A = \partial_t + a(\partial_t^2 + |\partial_x|^2)^{\frac{1}{2}}$ and can be represented up to small terms in the form $e\circ(\partial_t - a|\partial_x|)$, where e is an elliptic pseudodifferential operator on M and $\operatorname{ord} e = 0$. Using the fact that (in normal coordinates (t, x) in Θ_2)

$$G(t, x; \tau, \xi) = G(t, x; a|\xi|, \xi) + (\tau - a|\xi|)G_0,$$

where $|G_0| < c(|\tau| + |\xi|)^{-\frac{3}{2}}$ and also the equality $G(t, x; a|\partial_x|, \partial_x)(\rho \otimes \delta(t)) = G(0, x; 0, \partial_x)\rho(x) \otimes \delta(t)$ based on the identity $a(0, x) \equiv 0$, equation (4.5) can be brought into the form (up to unimportant terms)

$$e\circ(\partial_t - a|\partial_x|)(u + G_0(\rho(x) \times \delta(t))) + \mathcal{E}(x, \partial_x)\rho(x) \otimes \delta(t) = \Phi, \qquad (4.6)$$

where $\mathcal{E}(x, \xi) = G(0, x; 0, \xi)$. If we require that $\mathcal{E}(x, \xi) \neq 0$, $(x, \xi) \in T_0^*(\mu_2)$ – the analog of the Lopatinskij-Shapiro condition – then since the operator $\partial_t - a|\partial_x|$ is a differential operator on t it is not difficult to construct the right and left regularizers for the operator of problem (4.6), which leads to the theorem that the problem (4.1), (4.5) is a Fredholm problem, although there is no information on its index. We note also that the function $G(\rho \otimes \delta)$ is not concentrated on μ_2 (in contrast to $\rho(x) \otimes \delta(\mu_2)$). Therefore for the solutions u of the modified problem (4.1), (4.5) the original condition $\partial u/\partial \mathbf{1} + hu = f$ is violated, again in contrast to problem (4.2), on an open subset in M.

The methods just discussed for studying the oblique derivative problem are based on the theory of pseudodifferential operators, in contrast to the original papers of Egorov and Kondrat'ev (1969) and Malyutov (1969). Egorov and Kondrat'ev (cf. also Egorov 1984) carried out their investigation in the context of the classical theory of differential equations and relied on the possibility of extending the field $\mathbf{1}$ to a neighborhood of the boundary of M and on sharp estimates of the solutions of the Dirichlet problem. Malyutov (1969) made partial use of probabilistic methods. We note that for $\mu_2 = \varnothing$ Melin and Sjöstrand (1976) have constructed a parametrix of the problem (4.2) under the assumption that no trajectory $\gamma_p(t)$, $p \in \alpha$, is entirely contained in the set α.

Continuing the discussion of problems (4.1) and (4.2), we pause to discuss some more delicate questions such as the uniqueness of the solution, the validity of the boundary condition in (4.1) on all of M, the compactness of the inverse operator, and the maximum principle. In what follows we assume that $h \geq 0$ on M^- and $h \leq 0$ on M^+.

Theorem 4.2 (maximum principle, cf. Maz'ya, Paneah 1974; Paneah 1981; Taira 1979; Winzell 1977, 1979a). *If u is a solution of the problem* (4.2), *$\mathcal{L}u \geq 0$ in Ω, $\partial u/\partial \mathbf{1} + hu \geq 0$ on M^+, $\partial u/\partial \mathbf{1} + hu \leq 0$ on M^-, and $u \leq 0$ on μ_1, then either $u \leq 0$ in Ω or $u \equiv$* const.

In the case when $\mu_2 = \varnothing$, the proof does not go beyond the scope of classical methods. Let $u \not\equiv$ const in Ω and $\max u = u(p) > 0$. Then the classical theorems of Hopf and Giraud (cf. Miranda 1955), as well as the condition $u \le 0$ on μ_1 exclude the possibility that $p \in \operatorname{Int} \Omega$, $p \in M \setminus \alpha$, $p \in \mu_1$. It remains to prove that $p \notin \alpha \setminus \mu_1$. Suppose the contrary and for definiteness assume $p \in M^+$. Consider a trajectory $\gamma_p(t)$, and let t_+ be the infimum of the values of $t \ge 0$ for which $\gamma_p(t) \in \complement\alpha$. Set $p_+ = \gamma(t_+)$. The existence of the point p_+ is guaranteed by condition 4). Setting $v_\gamma(t) = v \circ \gamma(t)$, we can rewrite the boundary condition in (4.2) in the form

$$\partial_t u_\gamma = f_\gamma - h_\gamma u_\gamma - (a\partial_\nu u)_\gamma. \tag{4.7}$$

If $p_+ = p$, then by Giraud's theorem, the inequality $\partial_\nu u < 0$ holds at the point p_+, hence on a neighborhood of it in M as well. Together with the hypothesis of the theorem, this means that $\partial_t u_\gamma \ge 0$ in a sufficiently small neighborhood of the point $t = t_+$. On the other hand there exists a sequence of points $t_k > t_+$ converging to t_+ for which $a(t_k) > 0$, as a consequence of which $\partial_t u_\gamma(t_k) > 0$. But this is impossible, since $u_\gamma(t_+) = \max u$. If the points p_+ and p do not coincide, the interval $[p, p_+]$ of the trajectory $\gamma_p(t)$ is entirely contained in α. Hence $a(t) \equiv 0$ for $0 \le t \le t_+$ and, by (4.7), $\partial_t u_\gamma + h_\gamma u_\gamma \ge 0$ on this closed interval. It follows immediately from this that $\partial_t u_\gamma \ge 0$ for $0 \le t \le t_+$, and hence $u(t) \equiv u(p) = \max u > 0$ for the same values of t. In particular, this means, by virtue of the hypothesis of the theorem, that $\mu_1 \cap [p, p_+] = \varnothing$, $p_+ \in \operatorname{Int} M^+$, and $u(p_+) = \max u > 0$. To obtain a contradiction it now suffices to repeat the reasoning given in the case $p_+ = p$. The theorem is now proved.

Theorem 4.3 (cf. Maz'ya, Paneah 1974; Paneah 1981). *Suppose one of the following three conditions holds: μ_1 is nonempty; $\mathcal{L}1 \not\equiv 0$; $h \not\equiv 0$. Then the problem* (4.2) *has a unique solution in the space $H_{s,\frac{1}{2}}(\Omega, \mu_2) \oplus H^s(\mu_2)$ for any functions (F, f, g) belonging to $H_{s,0}(\Omega, \mu_2) \oplus \mathcal{H}_s(M, \mu_2) \oplus H^s(\mu_2)$ and the inverse operator of the problem is continuous. This operator is completely continuous (compact) if and only if the field* **1** *satisfies condition* 5). *When $\mu_1 = \varnothing$, $\mathcal{L}1 \equiv 0$, and $h \equiv 0$, the problem* (4.2) *has no solution for $F = 0$, $f = 0$ except $u \equiv$* const.

The unique solvability of the problem (4.2) follows from the maximum principle and the vanishing of the index of the problem. The continuity of the inverse operator is no longer amenable to a discussion in general terms, much less its compactness, since these properties are based on rather delicate estimates. The final assertion also follows from Theorem 4.2.

As follows from Hörmander's theorem, if $\mu_2 \neq \varnothing$, then problem (4.1) is unsolvable even locally in a neighborhood of μ_2, although, of course, there exist solutions for a rather "thin" set of functions F and f. The following theorem contains a description of the functions F, f, and g that guarantee the solvability of the problem (4.2) with $v = 0$.

Theorem 4.4 (Maz'ya, Paneah 1974; Paneah 1981). *Let $s \ge 1$ and $F \in H_{s,0}(\Omega, \mu_2)$, $f \in \mathcal{H}_s(M, \mu_2)$, $g \in H^s(\mu_1)$. Then the problem* (4.2) *has a solution $u \in H^{\frac{3}{2}}(\Omega)$ if and only if*

$$(F, \mathfrak{D}^c(raw_\chi))_\Omega + (f, w_\chi)_M + (g, \{w_\chi\}_{\mu_1})_{\mu_1} = 0$$

for any solutions $w_\chi \in \mathcal{H}_\infty(M, \mu_1)$ *of the problem*

$$A^c w + \{w\}_{\mu_1} \otimes \delta(\mu_1) = 0 \ \ on \ \ M, \ \ w = \chi \ \ on \ \ \mu_2$$

with arbitrary $\chi \in C^\infty(\mu_2)$. *In particular, for* $\mu_1 = \varnothing$ *the original problem* (4.1) *has a solution* $u \in H^{\frac{3}{2}}(\Omega)$ *if and only if* $(F, \mathfrak{D}^c(raw_\chi))_\Omega + (f, w_\chi)_M = 0$ *for all* w_χ, *satisfying the conditions* $A^c w_\chi = 0$, $w_\chi = \chi$ *on* μ_2.

In all these situations the field τ has been required to be transversal to a submanifold that separates the sets M^+ and M^-. The only reasonably general study of the oblique derivative problem without such a restriction has been carried out by Maz'ya (1972). Consider a system $M = M_0 \supset M_1 \supset \cdots \supset M_k$ of compact C^∞-submanifolds without boundary, and let the field $\mathbf{1}$ be tangent to each manifold M_j at the points of M_{j+1} and only at those points, but not tangent to M_k. It is then obvious that $M_1 \setminus M_2 \supset \mu_1 \cup \mu_2$, where, as before, the field τ is oriented toward M^- (resp. M^+) on μ_1 (resp. μ_2). We consider the problem

$$\mathcal{L}u = 0 \ \text{in} \ \Omega, \quad \partial u/\partial \mathbf{1} + hu = f \ \text{on} \ M \setminus \mu_2, \quad u = 0 \ \text{on} \ \mu_1. \tag{4.8}$$

A solution of this problem is a function $u \in H^{\frac{1}{2}}(\Omega)$ such that $a^{\frac{1}{2}} u\big|_M \in H^{\frac{1}{2}}(M)$ and $\rho^{-1} u\big|_M \in L_2^{\text{loc}}(M \setminus \mu_2)$, where $\rho(x) = \operatorname{dist}(x, \mu_1)$; the equality $(u, \mathcal{L}^c v)_\Omega + (u, Rv)_M = (a^{-1} A\mathbf{n} \cdot \mathbf{n} v, \varphi)_M$, in which R is a fixed first-order differential operator on M and $\mathbf{n}$ is the interior normal to M, holds for all $v \in C^2(\overline{\Omega})$ that vanish in a neighborhood of μ_2 and are such that $a^{-1} v \in C^2(M)$.

Theorem 4.5. *If* $\mathcal{L}1 \leq -c$ *for a sufficiently large positive constant c, then for any function* $\varphi \in L_2(M)$ *there exists a unique solution of the problem* (4.8).

A model encompassed by this theorem is provided by the domain $\Omega = \{x \in \mathbb{R}^n \,|\, x_n > 0\}$ and the field $\mathbf{1} = \partial_1 + \sum_{j \geq 2} x_{j-1} \partial_j$. In this case $M^\pm = \{x \,|\, x_n = 0, x_{n-1} \gtrless 0\}$, $\mu_{1(2)} = \{x \,|\, x_n = x_{n-1} = 0, x_{n-2} \gtrless 0\}$, $M_j = \{x \,|\, x_n = x_{n-1} = \cdots = x_{n-j-1} = 0\}$, $j = 0, 1, \ldots, n-2$.

Up to now we have assumed that $\mathbf{1} \neq \mathbf{0}$ everywhere on M. Taking account of the fact that the Dirichlet problem can be interpreted as the case $\mathbf{1} \equiv 0$ on M, it is natural to consider intermediate situations also. The following theorem gives a typical example.

Theorem 4.6 (Taira 1976). *Suppose the boundary condition in problem* (4.1) *has the form* $a\partial u/\partial \mathbf{1} + hu = f$, *where* $a \geq 0$. *If* $h > 0$ *on the set* α, *then problem* (4.1) *has a unique solution in* $H^s(\Omega)$ *for any* $F \in H^{s-1}(\Omega)$, $f \in H^{s-\frac{1}{2}}(M)$, $s \geq 2$.

4.3. Solvability of the Problem in the Spaces C^λ. Another natural class of spaces for studying the oblique derivative problem, besides H^s, is provided by the Hölder spaces C^λ with norm $|\cdot|_\lambda$. The majority of results on this topic are discussed in the papers of Winzell (1977, 1979a, 1981), in which problem (4.2) is also studied under assumptions 1)–4). Suppose at first that $\mu_1 = \varnothing$, $\mu_2 = \varnothing$, and $a \geq 0$ on M. Then the following result holds.

Theorem 4.7 (Winzell 1981). *For any functions $F \in C^\lambda(\overline{\Omega})$ and $f \in C^{1+\lambda}(M)$ satisfying a finite number of linear conditions, the problem* (4.2) *has a solution $u \in C^{2+\lambda}(\overline{\Omega})$. The index of this problem is zero, and if $\mathcal{L}1 \leq 0$ and $\mathcal{L}1 \not\equiv 0$, the solution is unique.*

In the general case $\mu_{12} \neq \varnothing$ the following theorem holds.

Theorem 4.8 (Winzell 1977, 1979a, 1981). *If the functions $F \in C^\lambda(\overline{\Omega})$, $f \in C^{1+\lambda}(M)$, $h \in C^{2+\lambda}(\mu_2)$ satisfy a finite number of linear conditions, then problem* (4.2) *has a finite number of linearly independent solutions $u \in C^{2+\lambda}_{\text{loc}}(\overline{\Omega} \setminus \mu_2) \cap L_\infty(\Omega)$.*

4.4. Sharp Regularity Theorems for Solutions of the Oblique Derivative Problem. We have seen in § 1 that the condition $r_{\mathfrak{L}}(x) = n$ for all $x \in \alpha$ has as a consequence an increase in the smoothness of the solution of the problem (1.1)–(1.2) compared with the general result that holds in the case when $r_{\mathfrak{L}}(x) < n$, and this extra smoothness depends on the maximal order $|I|$ of the commutators in $\mathfrak{L}$ that generate the whole algebra $\mathfrak{L}$. An analogous effect of improving the regularity of solutions holds in the case of the oblique derivative problem. In this case it is connected with the order of vanishing of the function a in a neighborhood of the set α, which will be studied in the present section. Before turning to precise statements, we note that the traditional terminology treats the facts just mentioned not as a gain in smoothness (compared with the weaker result), but as a loss of smoothness (compared with the elliptic case). Thus Theorem 2.5 guarantees a loss of smoothness in the amount $2 - 2^{1-m}$ (since $\operatorname{ord} \mathcal{A} = 2$) rather than a gain of 2^{1-m}. We shall not depart from tradition in this section.

The first result in this direction was obtained by Hörmander (1966), who described all the pseudodifferential operators of order m in a domain $\mathcal{U}$ for which the estimate

$$\|u\|_{m-\frac{1}{2}} \leq c(K)\big(\|Au\| + \|u\|_{m-1}\big)$$

holds for any $u \in C_0^\infty(K)$, K being a compact subset of $\mathcal{U}$. We also note that the Calderón operator of the oblique derivative problem has this property when the coefficient $a = a(t,x)$ admits the estimate $|a(t,x)| > c|t|$ in a neighborhood $\Theta_2 \supset \mu_2$. Subsequently (Egorov 1969, 1984; Hörmander 1985) a complete description was obtained for the operators of order m that admit an optimal estimate

$$\|u\|_{m-\delta} \leq c(K)\big(\|Au\| + \|u\|_{m-1}\big), \quad u \in C_0^\infty(K), \tag{4.9}$$

when $\delta = k/(k+1)$, $k > 0$ being an integer. In the same papers it was proved that in the case of a smooth symbol $A(x,\xi)$ there can be no loss of smoothness except $k/(k+1)$. Such operators have come to be called *subelliptic.* The same Calderón operator for which the function $a(t,x)$ admits the estimate $|a(t,x)| \geq c|t|^k$ in Θ_2 with odd k was given as an example of an operator with a loss of $k/(k+1)$. It is worth mentioning that the estimate (4.9) in the neighborhood Θ_2 plays only a limited role in the context of the oblique derivative problem, since the solutions of this problem, as a rule, are discontinuous on μ_2 (cf. Theorem 4.1). We shall see below that in this case also estimates are possible that register the exact loss in smoothness, but apply to solutions u that are discontinuous on μ_2. Furthermore, in the case when $\mu_1 \neq \varnothing$, the kernel of the oblique derivative problem is infinite-dimensional, and no estimate of the form (4.9) can hold. However, if on the right-hand side of (4.9) we invoke the norm of the function $u\big|_{\mu_1}$ in the corresponding space, we can obtain the precise loss of smoothness in the neighborhood Θ_1 also. Now suppose that the zeros of the function a are located on the submanifold $\mu = \{(t,x)\big|\, t = 0\}$ and the field τ is transversal to μ. It can happen that on the trajectories $\gamma_p(t)$, $p \in \mu$, the function $a(t,x)$ tends to zero at different rates as $t \to +0$ and $t \to -0$. It is natural to expect that this property will be reflected in the regularity properties of the solution u. Finally, the rate at which $a(t,x)$ tends to zero as $t \to 0$ need not be measured on the power function scale: what degree of smoothness is possessed by the solution of the equation $\partial_t u + e^{-1/t^2}\Lambda u = f$ for $f \in L_2(\Theta)$? We give below some precise statements of the problem and we formulate more or less definitive answers to the questions just raised as Theorems 4.9–4.11. All of these theorems were discussed in detail in the papers of Maz'ya and Paneah (1970, 1974), but seem not to have been noticed by later authors (Alimov 1981, 1987; Winzell 1979b).

For the sake of brevity we limit ourselves to estimates for the pseudodifferential operator $A = \partial_t + a(t,x)\Lambda$ in a tubular neighborhood Θ of a submanifold $\mu \subset M$ of codimension 1, where (t,x) are normal coordinates in Θ. As we have seen, the Calderón operator of the oblique derivative problem can be brought into this form in a cylindrical neighborhood Γ. It is assumed that $a(0,x) \equiv 0$ on μ. We now define a class of functions of t that qualify the degeneracy of a on μ. We shall say that $\gamma(t) \in \mathfrak{D}_+$ if $\gamma(t) \in C([0,\infty])$, $\gamma(t)$ is nondecreasing, and $\gamma(t) > 0$ for $t > 0$. We set

$$\Gamma(t) = \int_0^t \gamma(s)\,ds$$

and let

$$\varkappa(z) = \frac{1}{\Gamma^{-1}\left(\frac{1}{z}\right)}, \quad 0 < z < \infty,$$

where Γ^{-1} is the inverse of the Γ-function. We shall say that $\gamma(t) \in \mathfrak{D}$ if

$$\gamma(t) \in C(\mathbb{R}^1), \quad \gamma_+(t) \in \mathfrak{D}_+, \gamma_-(-t) \in \mathfrak{D}_+,$$

where $\gamma_\pm(t)$ are the restrictions of the function $\gamma(t)$ to the half-lines $t \geq 0$ and $t \leq 0$. We denote by $\varkappa_+(z)$ and $\varkappa_-(z)$ the functions determined by $\gamma_+(t)$ and $\gamma_-(t)$. We now introduce the class $\mathcal{R}$ of functions $\nu(z)$ satisfying the conditions

$$\nu(z) \in \operatorname{Lip}(0,\infty); \;\; \nu(z) \nearrow \infty \text{ as } z \to \infty; \;\; 0 \leq z\partial\nu/\partial z \leq \nu(z).$$

The regularity theorems below are stated in terms of the functions $\nu \in \mathcal{R}$. It is easily verified that the functions $\varkappa(z)$ and $\Gamma^{-1}(z)$ introduced above belong to the class $\mathcal{R}$, as do $\sqrt{\varkappa(z)}$, $z/\varkappa(z)$, $\sqrt{\varkappa_+(z)+\varkappa_-(z)}$, and $K(z) = \min\{\varkappa_+(z), \varkappa_-(z)\}$. For example, if $\gamma_1(t) = |t|^k/(k+1)$, then $\varkappa_1(z) = z^{\frac{1}{(k+1)}}$, and if $\gamma_2(t) = t^{-2}\exp(-|t|^{-1})$, then $\varkappa_2(z) = \ln z$.

For any function $\nu \in \mathcal{R}$ we define the space $H^\nu(\mathbb{R}^n_x)$ in which the norm has the form $|w|_\nu = \|\nu(\xi)\hat{w}(\xi)\|$, where $\hat{w}(\xi)$ is the Fourier transform of the function $w(x)$. Using a partition of unity on μ we now introduce the space $H^\nu(\mu)$ in the standard way. Finally, we set

$$\|u, \Theta^+\|^2_\nu = \int_0^\theta |u(t,\cdot)|^2_\nu \, dt, \quad \|u, \Theta^-\|^2_\nu = \int_{-\theta}^0 |u(t,\cdot)|^2_\nu \, dt,$$

where $\Theta^+ = \{(t,x) \in \Theta \,|\, 0 < t < \theta\}$, $\Theta^- = \{(t,x) \in \Theta \,|\, -\theta < t < 0\}$.

Theorem 4.9. *Assume that the inequality*

$$|a(t,x)| \geq \gamma(t), \quad (t,x) \in \Theta,$$

holds for some function $\gamma(t) \in \mathfrak{D}$. Then for all $u \in C_0^\infty(\Theta)$ the following assertions are true.

1) *If $ta(t,x) \geq 0$ in Θ, then*

$$\|u,\Theta^+\|_{\varkappa_+} + \|u,\Theta^-\|_{\varkappa_-} + |u(0,\cdot)|_{\sqrt{\varkappa_+ + \varkappa_- + 1}} \leq c\big(\|Au, M\| + \|u, M\|\big).$$

2) *If $ta(t,x) \leq 0$ in Θ, then*

$$\|u,\Theta^+\|_{\varkappa_+} + \|u,\Theta^-\|_{\varkappa_-} \leq c\big(\|Au, M\| + |u(0,\cdot)|_{\sqrt{\varkappa_+ + \varkappa_- + 1}} + \|u, M\|\big).$$

3) *If $a(t,x) \geq 0$ in Θ, then*

$$\|u,\Theta^+\|_{\varkappa_+} + \|u,\Theta^-\|_{K} + |u(0,\cdot)|_{\sqrt{\varkappa_+}} \leq c\big(\|Au, M\| + \|u, M\|\big).$$

If $a(t,x) \leq 0$ in Θ, then $+$ and $-$ must be interchanged in the subscripts and superscripts in the last inequality.

These results are sharp in the class of operators we are considering. To be specific, the following result holds.

Theorem 4.10. *In the case of operators*

$$A = \partial_t + \gamma(t)\Delta, \quad \gamma(t) \in \mathfrak{D},$$

estimates 1)–3) *are optimal: if they hold with some function* ν *instead of* $\varkappa$, *then*

$$\nu(z) \le c\varkappa(z), \quad z > 0.$$

Strictly speaking, the last two theorems were really about regularity with respect to the "tangential" variables x. It is not difficult, however, to deduce results from the estimates obtained that characterize global regularity (i.e., in (t, x)). The following result is an example.

Theorem 4.11. *If* $\gamma(t) \in \mathfrak{D}$ *and* $ta(t,x) \ge 0$ *in* Θ, *then for all functions* $u \in C_0^\infty(\Theta)$ *the estimate*

$$\|u, H^K(\Theta)\| \le c\big(\|Au, M\| + \|u, M\|\big) \tag{4.10}$$

holds, in which the norm on the left-hand side is defined as the L_2*-norm of the function*

$$\hat{u}(\tau,\xi)K\big(\sqrt{\tau^2 + |\xi|^2}\big).$$

Similar assertions hold also in the cases $ta \le 0$, $a \ge 0$, $a \le 0$.

This theorem, among other things, admits the following interpretation. Let μ be an arbitrary submanifold in M with $\operatorname{codim}\mu = 1$, and let (t, x) be normal coordinates in a tubular neighborhood $\Theta \supset \mu$. Suppose for definiteness that $a(t,x) \ge 0$ in Θ and $a(t,x) = 0 \Leftrightarrow t = 0$. Theorem 4.1 then guarantees the estimate

$$\|u\| \le c\big(\|Au, M\| + \|u\|_{-1}\big), \quad u \in C_0^\infty(M),$$

for the corresponding operator A, signifying a "complete" loss of smoothness. But for each function $a(t,x)$ of this form there exists a (nonunique) minorant $\gamma(t) \in \mathfrak{D}$, i.e., a function such that $\gamma(t) \le a(t,x)$ for all $(t,x) \in \Theta$. By Theorem 4.11, therefore, it follows from the relation $Au \in L_2(M)$ that the function u has a certain degree of regularity determined by the space $H^K(\Theta)$ to which it belongs.

In Theorems 4.9–4.11 we have dealt with the case of degeneracy of the ellipticity of the operator A when the function a tends to zero at an arbitrary rate, but on a smooth submanifold. We now give details of a more complicated geometric picture of the distribution of zeros of a. We recall (cf. Sect. 4.2) that $\gamma_p(t)$ denotes the maximal trajectory of the field τ passing through the point $p \in M$ and $a_\gamma(t) = a \circ \gamma(t)$.

Theorem 4.12. *Suppose that the function* $a_{\gamma_p}(t)$ *never changes sign from "$-$" to "$+$" at any point* $p \in M$ *and* $\partial_t^s a_{\gamma_p}(t) \ne 0$ *when* $s = s(p)$ *at the points with* $a_{\gamma_p}(t) = 0$. *If* $\sup s(p) = k < \infty$, *then estimate* (4.11) *holds.*

This result is merely an illustration, using the oblique derivative problem, of a general theorem that contains a criterion for subellipticity of an arbitrary pseudodifferential operator on M (cf. Egorov 1984 and also Hörmander 1985, Vol. IV). A simple proof of Theorem 4.12 in the case of a polynomial $a(t,x)$ was given by Ehskin (1971b).

We now consider several examples.

1) Let $\gamma(t) = |t|^k$, $k \geq 0$, k not necessarily an integer. As we have seen, in this case $\varkappa_\pm(z) = z^{1/(k+1)}$ (up to a constant factor), and so $K(z) = z^{1/(k+1)}$. By Theorem 4.11 we have the estimate

$$\|u\|_{1/(k+1)} \leq c(\|Au\| + \|u\|), \quad u \in C_0^\infty(M). \tag{4.11}$$

For integer values of k, as noted, this result was obtained by Egorov (1969). In the general case estimate (4.11) proves the fact already mentioned that when the function $a(t,x)$ has only finite smoothness the loss of smoothness of the solution (equal to $k/(k+1)$ in the present case) can be any number δ with $0 \leq \delta < 1$.

2) Let $\gamma(t) = t^{-2}\exp(-|t|^{-1})$. In this case $K(z) = \ln z$, which, by Theorem 4.11, leads to the estimate

$$\|\hat{u}(\tau,\xi)\ln(|\tau| + |\xi| + 1)\| \leq c(\|Au\| + \|u\|), \quad u \in C_0^\infty(\Theta).$$

According to Theorem 4.10 this estimate is optimal, and, in particular, no estimate of the form (4.11) is possible. Allowing the usual abuse of language in such situations, one could say that, although $u(t,x)$ has no derivatives of fractional order, there does exist a derivative of "logarithmic" order. We note that in this case $a(t,x) = \gamma(t)$ belongs to C^∞.

3) Let $\gamma(t) = t^\beta$ for $t > 0$ and $\gamma(t) = |t|^\delta$ for $t < 0$; $\beta, \delta \geq 0$. We set $K^0 = \max(\beta, \delta)$. Then

$$\|u, \Theta^+\|_{1/(\beta+1)} + \|u, \Theta^-\|_{1/(\delta+1)} \leq c(\|Au\| + \|u\|)$$

and in addition

$$\|u\|_{1/(K^0+1)} \leq c(\|Au\| + \|u\|).$$

4) Let $a(t,x) = t^k + \sum b_j(x)t^j$ be a nonnegative polynomial in (t,x) of degree k. Then for all $u \in C_0^\infty(\Theta)$, according to Theorem 4.12,

$$\|u\|_{1/k} \leq c(\|Au\| + \|u\|).$$

We now pause to discuss two-sided estimates of the operator A, or, as they are called by analogy with the elliptic case, *coercive estimates.* In the case of an arbitrary operator $P(x,\partial_x) = \sum P_j(x,\partial_x)$ such estimates have the form

$$c^{-1}\sum \|P_j(x,\partial_x)u\| \leq \|P(x,\partial_x)u\| \leq c\sum \|P_j(x,\partial_x)u\|$$

and play an important role in questions of the regularity and existence of solutions of differential equations. In particular the estimates

$$\|P_1(x,\partial_x)u\| + \|P_2(x,\partial_x)u\| \leq c(\|P(x,\partial_x)u\| + \|u\|)$$

are of interest, where $P_1(x,\xi) = \operatorname{Re} P(x,\xi)$, $P_2(x,\xi) = \operatorname{Im} P(x,\xi)$. Such estimates show, among other things, the equivalence of the graphs of the operators

$P(x, \partial_x)$ and $P^c(x, \partial_x)$. The following theorem contains necessary and sufficient conditions that guarantee such an estimate for the operator A of the oblique derivative problem.

We shall say that the function $\gamma(t) \in \mathfrak{D}_+$ belongs to $\mathfrak{D}'_+$ if $t\gamma(t) \leq c\Gamma(t)$ for some constant $c > 0$. Suppose that the function $a(t, x)$ admits the two-sided estimate

$$c^{-1}\gamma(t) \leq |a(t,x)| \leq c\gamma(t), \quad (t,x) \in \Theta,$$

in which $\gamma(t) \in \mathfrak{D}$. We then have the following result.

Theorem 4.13 (Maz'ya, Paneah 1970a).

(i) *If $ta(t,x) \geq 0$ in Θ, then*

$$\|\partial_\tau u\| + \|a\Lambda u\| \leq c\big(\|Au\| + \|u\|\big), \quad u \in C_0^\infty(\Theta). \tag{4.12}$$

(ii) *Let $ta(t,x) \leq 0$. Then the conditions $\gamma_+(t) \in \mathfrak{D}'_+$ and $\gamma_-(-t) \in \mathfrak{D}'_+$ are necessary and sufficient for the validity of the estimate*

$$\|\partial_\tau u\| + \|a\Lambda u\| \leq c\big(\|Au\| + |u(0,\cdot)|_{\sqrt{\varkappa_+ + \varkappa_- + 1}} + \|u\|\big)$$

for all $u \in C_0^\infty(\Theta)$.

(iii) *Let $a(t,x) \geq 0$. Then estimate* (4.12) *holds for all $u \in C_0^\infty(\Theta)$ if and only if*

$$\gamma_-(-t) \in \mathfrak{D}'_+ \quad \text{and } \gamma_+(t) \leq c\gamma_-(-t).$$

If $a(t,x) \leq 0$, the roles of $\gamma_+(t)$ and $\gamma_-(-t)$ must be reversed in these conditions.

Regularity questions for solutions of the oblique derivative problem have been studied not only in the scale H^s (and the scale H^ν adjacent to it, with $\nu \in \mathcal{R}$) but also in the spaces L_p, $1 < p \leq \infty$ (Maz'ya 1972), W_{p^s} (Kolesnikova 1984), the spaces H^φ of variable order of smoothness (Egorov, Nguyen Minh Chieng 1984), and the spaces C^λ (Alimov 1981, 1987; Winzell 1979b). The following theorem is representative of the results obtained.

Theorem 4.14 (Alimov 1981, Winzell 1979b, cf. Theorem 4.11). *Suppose that $|a(t,x)| \geq c|t|^k$ for $(t,x) \in \Theta$ and an integer value of $k \geq 0$.*

If $a \geq 0$ (resp. $a \leq 0$) or $ta(t,x) \geq 0$ in Θ, then each solution $u \in C^{1+\lambda}(\overline{\Omega}) \cap C^2(\Omega)$ of the problem (4.2) *belongs to the space $C^{1+\lambda+1/(k+1)}(\overline{\Omega}) \cap C^{2+\lambda}(\overline{\Omega} \setminus \mu_2)$ and the following estimate holds:*

$$|u|^{\Omega}_{1+\lambda+1/(k+1)} \leq c\big(|F|^{\Omega}_{\lambda} + |f|^{M}_{1+\lambda}\big).$$

The result given by Alimov (1987) generalizes to a more general case than the case of power degeneracy of the function $a(t,x)$. This generalization can be achieved by introducing the spaces $C^\mu_\lambda(\Omega)$ analogous to the spaces H^λ_μ in which the norm is defined for noninteger values $\lambda > 1$ as follows:

$$|u|^{\Omega}_{C^\lambda_\mu} = |u|^{\Omega}_{\lambda} + \min |\partial^\alpha u(x+h) - 2\partial^\alpha u(x) + \partial^\alpha u(x+h)|/|h|^{\lambda - |\lambda|}\nu(|h|),$$

where $|\alpha| = [\lambda]$ and $x \pm h \in \Omega$.

§5. Boundary Operator $\mathcal{A}$ of Arbitrary Order

The results of this section are mainly due to Eskin (1971a) and are discussed in the work cited (cf. also Vishik, Grushin 1969a; Maz'ya, Paneah 1970ab). To simplify the discussion we assume, as before, that the operator $\mathcal{L}$ is of order 2, although everything that will be said below also holds for Douglis-Nirenberg elliptic systems. Thus we return to the problem (1.1), in which the boundary operator $\mathcal{A}$ can be written in the form

$$\mathcal{A} = \sum_{|\alpha|=0}^{m} a_\alpha(t,x)\partial_x^\alpha \partial_t^{m-|\alpha|} + \mathcal{A}'(\partial_t, \partial_x), \quad \operatorname{ord}\mathcal{A}' < m$$

in a normal coordinate system (t, x) in a tubular neighborhood $\mathcal{T}$ of the boundary manifold M. The principal part of the Calderón operator corresponding to problem (1) in this case is

$$A = \sum_{|\alpha|=0}^{m} i^{m-|\alpha|} a_\alpha(0,x)\partial_x^\alpha \Lambda_0^{m-|\alpha|}(x, \partial_x),$$

where Λ^0 is defined by formula (1.4). Problem (1.1) is now an elliptic problem if and only if the inequality $A(x,\xi) \neq 0$ holds on $T_0^*(M)$, where $\xi = (\tau, \lambda) = (\tau, \lambda_2, \dots, \lambda_n)$ is a covector at the point (x_1, x'). Suppose that this inequality fails to hold on the submanifold $\alpha = \{x = (x_1, x') \big| \, x_1 = 0\}$ and only on that submanifold, i.e., $A(x,\xi) \neq 0$ when $x_1 \neq 0$ and $\xi \neq 0$, and

$$(\forall\, x' \in \alpha)\, (\exists\, \xi = \xi(x') \neq 0 :\ A(0, x', \xi) = 0).$$

We require that

$$A(0, x_0'; \xi_0) = 0 \Rightarrow \partial_\tau A(0, x_0', \xi_0) \neq 0.$$

Under these conditions, as proved by Eskin (1971a) the symbol $A(x,\xi)$ admits the following factorization on $T_0^*(\Theta)$:

$$A(x,\xi) = \prod_{k=1}^{p_+} (\tau - \varkappa_k(x,\lambda)) \prod_{k=1}^{p_-} (\tau - \zeta_k(x,\lambda)) B(x,\xi), \tag{5.1}$$

in which $\varkappa_k$ and ζ_k are complex-valued functions homogeneous of degree 1 in λ and infinitely differentiable in some neighborhood U of the point $(0, x_0'; \xi_0)$, and B is an arbitrarily smooth function in U that is homogeneous in (τ, λ). Here the functions $\operatorname{Im}\varkappa_k$ and $\operatorname{Im}\zeta_k$ have no zeros for $x_1 \neq 0$, $\lambda \neq 0$, and

$$x_1 \operatorname{Im}\varkappa_k(x,\lambda) \geq 0, \quad x_1 \operatorname{Im}\zeta_k(x,\lambda) \leq 0$$

for all $x \in \Theta$, $|\lambda| = 1$. As for the function B, in a neighborhood of its zero

$(0, x'_0; \xi_0)$ it admits a factorization $B(x,\xi) = (\tau - \rho(x,\lambda))B_1(x,\xi)$ analogous to (5.1), in which $\operatorname{Im}\rho(x,\lambda) \neq 0$ when $x_1 \neq 0$, $|\lambda| = 1$, but $\operatorname{Im}\rho(x,\lambda)$ is of constant sign in a neighborhood of the submanifold α. As in § 3, and also in the oblique derivative problem, the kernel of the problem (1.1) is infinite-dimensional when $p_+ > 0$ and the cokernel is infinite-dimensional when $p_- > 0$. In this connection when $p_+ + p_- > 0$ we consider the modified problem

$$\begin{gathered}\mathcal{L}u = F \text{ in } \Omega,\\ \mathcal{A}u + \sum_{j=1}^{p_-} G_j(\rho_j(x') \otimes \delta(\mu_2)) = f \text{ on } M,\\ \mathcal{F}_j u = g_j \text{ on } \mu_1, \quad j = 1,\dots,p_+,\end{gathered} \tag{5.2}$$

in which G_j are pseudodifferential operators on M of orders $\gamma_j < -s - \frac{1}{2}$, $\mathcal{F}_j$ are differential operators on M of orders $\beta_j < s - \frac{1}{2}$, s is any nonnegative number, ρ_j are new unknown functions on $\mu_2 = \{x \,|\, \operatorname{Im}\varkappa_k = 0\}$, and $\mu_1 = \{x \,|\, \operatorname{Im}\zeta_k = 0\}$. The condition for ellipticity of this problem is that for all $(x',\lambda) \in T_0^*(\mu_1 \cup \mu_2)$

$$\begin{gathered}\det \|G_j(0,x',\varkappa_i(x',\lambda),\lambda)\|_{i,j=1}^{p_-} \neq 0,\\ \det \|\mathcal{F}_j(0,x';\zeta_i(x',\lambda),\lambda)\|_{i,j=1}^{p_+} \neq 0.\end{gathered} \tag{5.3}$$

Theorem 5.1. *To the problem* (5.2) *we assign the closed unbounded operator*

$$\mathfrak{A} : (u,\rho_1,\rho_2,\dots,\rho_{p_-}) \mapsto \Big(\mathcal{L}u; \mathcal{A}u + \sum_{j=1}^{p_-} G_j(\rho_j \otimes \delta(\mu_2))\big|_M; B_1 u\big|_{\mu_1}, \dots, B_{p_+} u\big|_{\mu_1}\Big)$$

mapping the space $\mathcal{H}^s_{(1)} = H^s(\Omega) \oplus \sum_{j=1}^{p_-} \oplus H^{s+\gamma_j-m+1}(\mu_2)$ *into the space* $\mathcal{H}^s_{(2)} = H^{s-1}(\Omega) \oplus H_s^{-m+\frac{1}{2}}(M) \oplus \sum_{j=1}^{p_+} \oplus H^{s-\beta_j}(\mu_1)$, *where* $\max\{m, \beta_j + \frac{1}{2}\} + \frac{1}{2} < s < -\gamma_j + m$. *If the symbols* $A(x,\xi)$, $G_j(x,\xi)$, *and* $\mathcal{F}_k(x,\xi)$ *satisfy the conditions* (5.1) *and* (5.3), *then the operator* $\mathfrak{A}$ *is a Fredholm operator:* $\mathcal{H}^s_{(1)} \to \mathcal{H}^s_{(2)}$. *If in a neighborhood* Θ *all the functions* $\operatorname{Im}\varkappa$, $\operatorname{Im}\zeta$, *and* $\operatorname{Im}\rho$ *are bounded below by the quantity* $c|t|^k$, $c = \text{const}$, *then in* $\mathcal{H}^s_{(1)}$ *one can replace* $s + \gamma_j + 1$ *by* $s + \gamma_j + 1 - (2(k+1))^{-1}$ *and* $\mathcal{H}^s_{(2)}$ *by* $H^{s-1+k/(k+1)}(\Omega) \oplus H^{s-m+\frac{1}{2}+k/(k+1)}(M) \oplus \Sigma \oplus H^{s-\beta_j-\frac{1}{2(k+1)}}(\mu_1)$.

The proof of this theorem, after the problem (5.2) has been reduced to the problem

$$Aw + \sum G_j(\rho_j \otimes \delta(\mu_2)) = \varphi \text{ on } M, \quad B_k w = g_k \text{ on } \mu_1, \tag{5.4}$$

is based on the construction of right and left regularizers of the operator corresponding to the problem. To do this we introduce a finite partition of unity $\{\varphi_j, \mathcal{U}_j\}$ in $T_0^*(M)$ such that in each neighborhood $\mathcal{U}_j$ either $A(x,\xi) \neq 0$ or the surface $A(x,\xi) = 0$ is given by one of the equations $\tau - \varkappa_k = 0$, $\tau - \zeta_j = 0$, $\tau - \rho_e = 0$ (by virtue of (5.1)). If the surface $A(x,\xi) = 0$ does not intersect $\mathcal{U}_j$, then (for example) the right regularizer $\mathcal{R}_j$ corresponding to $\mathcal{U}_j$ has the form $1/A_0(x, \partial_x)$. If $\mathcal{U}_j$ intersects the surface $\tau - \rho_e = 0$, then as $\mathcal{R}_j\varphi$ we choose a solution of the problem

$$\partial_1 u - \rho(x', \partial_{x'})u = \varphi \text{ in } \Theta, \quad u(\theta, x') = 0 \text{ (resp. } u(-\theta, x') = 0), \tag{5.5}$$

if $\operatorname{Im} \rho \geq 0$ (resp. $\operatorname{Im} \rho \leq 0$) on $[-\theta, \theta] \times T_0^*(\mu_0)$. If the intersection is with the surface $\tau - \varkappa_j = 0$, the role of the right regularizer $\mathcal{R}_j(f, g)$ is played by the solution of the problem

$$\partial_1 u - \varkappa_j(x, \partial_{x'})u = f \text{ in } \Theta, \quad u(0, x') = g, \tag{5.6}$$

and if it is with the surface $\tau - \zeta_j = 0$, the right regularizer is the solution of the problem

$$\partial_1 u - \zeta_j(x, \partial_{x'})u = f + G(\rho(x') \otimes \delta(x_1)), \quad u(\pm\theta, x') = 0. \tag{5.7}$$

The existence of solutions of the problem (5.5)–(5.7) and their properties can be obtained by the same methods that were described above in connection with the oblique derivative problem.

For solutions of the problem (5.4) precise estimates have been obtained (cf. Maz'ya, Paneah 1970b) in the case of qualified vanishing of the functions $\varkappa_k$, ζ_j, and ρ_l analogous to the estimates of Sect. 4.4. To gain an appreciation of the nature of these results we consider the model problem

$$Au \equiv \sum_{k=0}^{m} a_k(x, \partial_{x'})\partial_1^k u = f \text{ in } \Theta^+ = \{x \,|\, 0 \leq x_1 \leq \theta\}. \tag{5.8}$$

Here $a_k(x, \lambda)$ is a homogeneous symbol in $\lambda \in \mathbb{R}^{n-1}$ of degree k of a pseudodifferential operator on $\alpha = \{x \in \Theta^+ \,|\, x_1 = 0\}$ and $a_m \neq 0$. Assume that the roots $\tau = \varkappa_j(x, \lambda)$ of the polynomial $A = \sum a_k(x, \lambda)\tau^k$ are simple and that $\operatorname{Im} \varkappa_j \geq 0$ when $1 \leq j \leq l$ and $\operatorname{Im} \varkappa_j \leq 0$ when $l+1 \leq j \leq m$ in $\Theta^+ \times S^{(n-1)}$. Suppose also that for $1 \leq j \leq m$

$$|\operatorname{Im} \varkappa_j(x, \lambda)| \geq \gamma(x_1)|\lambda|, \quad \text{where } \gamma \in \mathfrak{D}_+.$$

We introduce the polynomials $A_j(x, \tau, \lambda) = A(x, \tau, \lambda)/(\tau - \varkappa_j(x, \lambda))$. It follows easily from Theorem 4.9 that the precise estimates

$$\|A_j u\|_{\mu+1} \leq c(\|Au\| + |A_j u(0, \cdot)|_{\sqrt{\mu+1}} + \|u\|_{m-1}), \quad 1 \leq j \leq l, \tag{5.9}$$

$$\|A_j u\|_{\mu+1} + |A_j u(0, \cdot)|_{\sqrt{\mu+1}} \leq c(\|Au\| + \|u\|_{m-1}), \quad l+1 \leq j \leq m, \tag{5.10}$$

hold for all $u \in C^\infty(\bar{\Theta}^+)$ that vanish near $x_1 = \theta$. The first of these inequalities shows that the question of the unique solvability of problem (5.8) is connected with additional conditions on the boundary values $u(0, x')$. We shall solve problem (5.8) under the conditions $B_j(x', \partial_1, \partial_{x'})u\big|_{x_1=0} = g_j$, where $B_j = \sum_{k=0}^{m_j} b_{jk}(x', \lambda)\tau^k$ are polynomials in τ and λ and $\mathrm{ord}_\lambda b_{jk} = m_j - k$, $m_j < m$. The analogue of the ellipticity condition (5.3) here is the requirement that the polynomials $A_1(0, x'; \tau, \lambda), \ldots, A_l(0, x'; \tau, \lambda)$ be linearly expressible in terms of the polynomials $B_1, B_2, \ldots, B_l, A_{l+1}, \ldots, A_m$ when $|\lambda| = 1$ (compare with the condition for ellipticity Ell$_2$ in Egorov, Shubin 1987, p. 113). It then follows immediately from (5.10) that for $1 \le j \le l$

$$|A_j u(0,\cdot)|_{\sqrt{\mu+1}} \le c\Big(\sum_1^l |B_j u(0,\cdot)|_{m-1-m_j,\sqrt{\mu+1}} + \|Au\| + \|u\|_{m-1}\Big),$$

and this, by (5.9), implies the analogous estimate with the norm $\sum_1^m \|A_j u\|_{\mu+1}$ on the left. (The symbol $|u|_{s,\nu}$ is understood as the norm $|(1 + |\partial_{x'}|)^s u|_\nu$). Since the roots $\varkappa_j$ are distinct, the polynomials $A_1, A_2, \ldots, A_m$ form a basis of the space of polynomials of degree at most m when $|\lambda| = 1$. Therefore $\sum_{k=0}^{m-1} \|\partial_1^k u\|_{m-1-k,\mu+1} \le c \sum_{j=1}^m \|A_j u\|_{\mu+1}$, from which it follows that

$$\sum_{k=0}^{m-1} \|\partial_1^k u\|_{m-1-k,\mu+1} \le c\Big(\sum_1^l |B_j u(0,\cdot)|_{m-1-m_j,\sqrt{\mu+1}} + \|Au\| + \|u\|_{m-1}\Big). \tag{5.11}$$

When $\gamma(x_1) = |x_1|^k$, this estimate assumes the form

$$\|u\|_{m-1+1/(k+1)} \le c\Big(\|Au\| + \sum_j |B_j u(0,\cdot)|_{m-m_j-1+1/2(k+1)} + \|u\|\Big).$$

But if $\gamma(x_1) \in \mathfrak{D}'_+$, then for the same functions $u(x)$ as above two-sided estimates of type (4.12) have been obtained. As an example (compare (5.11)) we offer the estimate

$$\sum_{k=0}^{m-1} \|\gamma(x_1)\partial_1^k u\|_{m-k} \le c\Big(\|Au\| + \sum_{j=1}^l |B_j u(0,\cdot)|_{m-1-mj,\sqrt{\mu+1}} + \|u\|\Big).$$

References to Chapter 2

After the first examples of degenerate elliptic boundary value problems with infinite index (Bitsadze 1964; Borrelli 1966) Hörmander (1966) studied a large class of such problems, including the degenerate oblique derivative problem. The reduction of a boundary value problem to a pseudodifferential equation on the boundary in conjunction with the technique of pseudodifferential operators first used by Hörmander (1966), promoted rapid progress in the theory of solvability of boundary value problems (cf. also Hörmander 1985, Vol. III).

In studying a variety of boundary value problems of the type considered in § 2 Taira (1982) applied the machinery of vector fields on a manifold in the spirit of Olejnik and Radkevich (1971) and Hörmander (1967). The concept of transversality of a Lie algebra $\mathfrak{L}$ to a submanifold was introduced by Paneah (1985a).

A voluminous literature has been devoted to the problem of solvability of the oblique derivative problem (Bitsadze 1964; Vishik, Grushin 1970b; Egorov 1984; Egorov, Kondrat'ev 1969; Egorov, Nguyen Minh Chieng 1984; Maz'ya 1972; Maz'ya, Paneah 1974; Malyutov 1969; Paneah 1981, 1988; Eskin 1971a; Borrelli 1966; Hörmander 1966; Kato 1982; Melin, Sjöstrand 1976; Soga 1975; Taira 1976; Winzell 1977, 1979ab, 1981). In the case of degeneracy on a submanifold the first well-posed problems with infinite-dimensional kernel or cokernel were exhibited by Egorov and Kondrat'ev (1969) and Malyutov (1969), and theorems on unique solvability of the modified problem were obtained by Maz'ya and Paneah (1974). For the case of degeneracy on an arbitrary subset of the boundary condition 4) of § 4 plays an important role; this condition seems to have been introduced by Melin and Sjöstrand (1976). General results on the solvability of the modified problem in this case were discussed by Paneah (1981, 1988) for the spaces H^s and by Winzell (1977, 1979a, 1981) for the spaces C^λ.

References*

Alimov, Sh. A. (1981): On a problem with oblique derivative. Differ. Uravn. **17**(10), 1738–1751. English transl.: Differ. Equations **17**, 1073–1083 (1982), Zbl. 488.35018.

Alimov, Sh. A. (1987): On the smoothness of solutions of the degenerate oblique derivative problem. Differ. Uravn. **23**(1), 10–22. English transl.: Differ. Equations **23**, 1–10 (1987), Zbl. 651.35025.

Amano, K. (1979): Maximum principle for degenerate elliptic-parabolic operators. Indiana Univ. Math. J. **28**(3) 545–557, Zbl. 423.35023.

Bagirov, L. A. (1987): On a class of elliptic equations of higher order that are degenerate on part of the boundary. Sib. Mat. Zh. **28**(6), 12–25. English transl.: Sib. Math. J. **28**, 872–884 (1987), Zbl. 656.35036.

Beals, R. (1975): A general calculus of pseudodifferential operators. Duke Math. J. **42**(1), 1–42, Zbl. 343.35078.

* For the convenience of the reader references to reviews in *Zentralblatt für Mathematik* (Zbl.), compiled using the MATH database, have, as far as possible, been included in this bibliography.

Beals, R. (1981): Weighted distribution spaces and pseudodifferential operators. J. Anal. Math. **39**, 131–187, Zbl. 474.35089.

Beals, R., Fefferman, C. (1974): Spatially unhomogeneous pseudodifferential operators. I, II. Commun. Pure Appl. Math. **27**(1), 1–24, Zbl. 279.35071; (2), 161–205, Zbl. 283.35071.

Bitsadze, A. V. (1964): The oblique derivative problem with polynomial coefficients. Dokl. Akad. Nauk SSSR **157**(6), 1273–1275. English transl.: Sov. Math., Dokl. **5**, 1102–1104 (1964), Zbl. 196,417.

Borrelli, R. (1966): The singular second-order oblique derivative problem. J. Math. Mech. **16**, 51–82, Zbl. 143.146.

Boutet de Monvel, L., Grigis, A., Helffer, B. (1976): Parametrixes d'opérateurs pseudodifférentiels à caractéristiques multiples. Astérisque, 34–35, 93–121, Zbl. 344.32009.

Calderón, A. (1965): Commutators of singular integral operators. Proc. Natl. Acad. Sci. USA **53** (5), 1092–1099, Zbl. 151,169.

Egorov, Yu. V. (1969): On subelliptic pseudodifferential operators. Dokl. Akad. Nauk SSSR **188**(1), 20–22. English transl.: Sov. Math., Dokl. **10**, 1056–1059 (1970), Zbl. 195,146.

Egorov, Yu. V. (1984): Linear Differential Equations of Principal Type. Nauka, Moscow. English transl.: Contemp. Sov. Math., New York 1986, Zbl. 574.35001.

Egorov, Yu. V., Kondrat'ev, V. V. (1969): On the oblique derivative problem. Mat. Sb., Nov. Ser. **78**(1), 148–176. English transl.: Math. USSR, Sb. **7**, 139–169 (1969), Zbl. 165.122.

Egorov, Yu. V., Nguyen, Minh Chieng (1984): The oblique derivative problem in Sobolev spaces of variable order. Differ. Uravn. **20**(12), 2163–2164 (Russian), Zbl. 599.35047.

Egorov, Yu. V., Shubin, M. A. (1987): Linear partial differential equations. Foundations of the classical theory. Itogi Nauki Tekh., Ser. Sovrem. Probl. Mat., Fundam. Napravleniya **30**. English transl.: Encycl. Math. Sci. **30**, Springer-Verlag, Berlin - Heidelberg - New York, 1992, Zbl. 657.35002.

Egorov, Yu. V., Shubin, M. A. (1988): Linear partial differential equations. Elements of the modern theory. Itogi Nauki Tekh., Ser. Sovrem. Probl. Mat., Fundam. Napravleniya **31**, 5–125. English transl.: Encycl. Math. Sci. **31**, Springer-Verlag, Berlin - Heidelberg - New York (in preparation), Zbl. 686.35005.

Eskin, G. I. (1970): Degenerate elliptic pseudodifferential equations of principal type. Mat. Sb., Nov. Ser. **82**(4) 585–628. English transl.: Math. USSR, Sb. **11**, 539–582 (1972), Zbl. 203,414.

Eskin, G. I. (1971): Elliptic pseudodifferential operators with first-order degeneracy in the space variables. Tr. Mosk. Mat. O-va **25**, 83–118. English transl.: Trans. Mosc. Math. Soc. **25**, 91–130 (1973), Zbl. 233.35041.

Fefferman, C. L. (1983): The uncertainty principle. Bull. Am. Math. Soc., New Ser. **9**, 129–206, Zbl. 526.35080.

Fefferman, C. L., Phong, D. (1981): The uncertainty principle and sharp Gårding inequalities. Commun. Pure Appl. Math. **34**, 285–311, Zbl. 458.35099.

Fursikov, A. V. (1971a): Boundary value problems for some classes of degenerate elliptic operators. Dokl. Akad. Nauk SSSR **197**(3), 535–538. English transl.: Sov. Math., Dokl. **12**, 506–510 (1971), Zbl. 227.35041.

Fursikov, A. V. (1971b): On degenerate Euler elliptic operators in a bounded domain. Vestn. Mosk. Univ., Ser. 1, No. 1, 36–43. English transl.: Mosc. Univ. Math. Bull. **26**, No. 112, 28–34 (1972), Zbl. 238.35033.

Fursikov, A. V. (1971c): On the global smoothness of solutions of a class of degenerate elliptic equations. Usp. Mat. Nauk **26**(5), 227–228 (Russian), Zbl. 219.35042.

Glushko, V. P., Savchenko, Yu. B. (1985): Degenerate elliptic equations of high order: spaces, operators, boundary value problems. Itogi Nauki Tekh., Ser. Mat. Anal. **23**, 125–218. English transl.: J. Sov. Math. **39**, No. 6, 3088–3148 (1987), Zbl. 642.35029.

Grushin, V. V. (1970): On a class of hypoelliptic operators. Mat. Sb., Nov. Ser. **83**(3), 456–473. English transl.: Math. USSR, Sb. **12**, 458–476 (1971), Zbl. 211,405.

Grushin, V. V. (1971): On a class of elliptic pseudodifferential operators that are degenerate on a submanifold. Mat. Sb., Nov. Ser. **84**(2), 163–195. English transl.: Math. USSR, Sb. **13**, 155–185 (1971), Zbl. 215,492.

Grushin, V. V. (1972): Hypoelliptic differential equations and pseudodifferential operators with operator-valued symbols. Mat. Sb., Nov. Ser. **88**(4), 504–521. English transl.: Math. USSR, Sb. **17**, 497–514 (1973), Zbl. 243.35020.

Helffer, B. (1976): Invariants associés à une classe d'opérateurs pseudodifférentiels et application à l'hypoellipticité. Ann. Inst. Fourier **26**, No. 2, 55–70, Zbl. 301.35026.

Hörmander, L. (1966): Pseudo-differential operators and non-elliptic boundary value problems. Ann. Math., II. Ser. **83**, 129–209, Zbl. 132,74.

Hörmander, L. (1967): Hypoelliptic second order differential equations. Acta Math. **119**, 147–171, Zbl. 156,107.

Hörmander, L. (1979): The Weyl calculus of pseudodifferential operators. Commun. Pure Appl. Math. **32**, 355–443, Zbl. 388.47032.

Hörmander, L. (1985): The Analysis of Linear Partial Differential Operators. Vols. III, IV. Springer-Verlag, Berlin - Heidelberg - New York, Zbl. 601.35001, Zbl. 612.35001..

Kato, Y. (1982): On non-elliptic boundary problems. Nagoya Math. J. **86**, 1–38, Zbl. 502.35037.

Katrakhov, V. V. (1980): General boundary value problems for a class of singular and degenerate elliptic equations. Mat. Sb., Nov. Ser. **112**(3), 354–379. English transl.: Math. USSR, Sb. **40**, 325–347 (1981), Zbl. 446.35048.

Kipriyanov, I. A. (1971): On a class of singular elliptic operators. I. Differ. Uravn. **7**, 2066–2077. English transl.: Differ. Equations **7**, 1560–1568 (1974), Zbl. 241.35033.

Kipriyanov, I. A. (1973): On a class of singular elliptic operators. II. Sib. Mat. Zh. **14**, 560–568. English transl.: Sib. Math. J. **14**, 388–394 (1974), Zbl. 319.35040.

Kolesnikova, E. A. (1984): Subelliptic estimates for the oblique derivative problem in L_p. Vestn. Mosk. Univ., Ser. 1, 1984, No. 4, 87–88. English transl.: Mosc. Univ. Math. Bull. **39**, No. 4, 94–97 (1984), Zbl. 576.35035.

Korenyako, V. V. (1985): Gårding's inequality. Dokl. Akad. Nauk SSSR **283**(4), 806–810. English transl.: Sov. Math., Dokl. **32**, 203–207 (1985), Zbl. 596.35130.

Levendorskij, S. Z. (1982): The asymptotic distribution of eigenvalues. Izv. Akad. Nauk SSSR, Ser. Mat. **46**(4), 810–852. English transl.: Math. USSR, Izv. **21**, 119–160 (1983), Zbl. 529.35060.

Levendorskij, S. Z. (1988a): On the symbols of degenerate elliptic differential and hypoelliptic pseudodifferential operators. Izv. Akad. Nauk SSSR, Ser. Mat. **52**(4), 541–588. English transl.: Math. USSR, Izv. **32**, No. 3, 543–561 (1989), Zbl. 672.35075.

Levendorskij, S. Z. (1988b): Nonclassical spectral asymptotics. Usp. Mat. Nauk **43**(1), 123–157. English transl.: Russ. Math. Surv. **43**, 149–192 (1988), Zbl. 671.35064.

Levendorskij, S. Z. (1988c): Coercivity of degenerate elliptic operators. Dokl. Akad. Nauk SSSR **298**, 29–32. English transl.: Sov. Math., Dokl. **37**(1), 21–24 (1988), Zbl. 722.47039.

Lizorkin, P. K. (1985): On the theory of degenerate elliptic equations. Tr. Mat. Inst. Steklova **172**, 235–251. English transl.: Proc. Steklov Inst. Math. **172**, 257–274 (1987), Zbl. 591.35024.

Lizorkin, P. K., Nikol'skij S. M. (1981): Coercive properties of elliptic equations with degeneracy. Variational methods. Tr. Mat. Inst. Steklova **157**, 90–118. English transl.: Proc. Steklov Inst. Math. **157**, 95–125 (1983), Zbl. 475.35050.

Malyutov, M. B. (1969): On the Poincaré boundary value problem. Tr. Mosk. Mat. O.-va **20**, 173–204. English transl.: Trans. Mosc. Math. Soc. **20**, 173–204 (1971), Zbl. 181, 379.

Maz'ya, V. G. (1972): On the degenerate oblique derivative problem. Mat. Sb., Nov. Ser. **87**(3), 417–454. English transl.: Math. USSR, Sb. **16**, 429–469 (1972), Zbl. 262.35024.

Maz'ya, V. G., Paneyakh, B. P. (= B. Paneah) (1970a): Coercive estimates and regularity of solutions of degenerate elliptic pseudodifferential equations. Funkts. Anal. Prilozh. **4**(4), 41–56. English transl.: Funct. Anal. Appl. **4** (1971), 299–311, Zbl. 227.35046.

Maz'ya, V. G., Paneyakh, B. P. (=B. Paneah) (1970b): Degenerate elliptic pseudodifferential operators with simple complex characteristics. Usp. Mat. Nauk **25**(1), 193–194 (Russian), Zbl. 134,451.

Maz'ya, V. G., Paneyakh, B. P. (= B. Paneah) (1974): Degenerate elliptic pseudodifferential operators and the oblique derivative problem. Tr. Mosk. Mat. O-va **31**, 237–295. English transl.: Trans. Mosc. Math. Soc. **31**, 247–305 (1974), Zbl. 331.35056.

Melin, A., Sjöstrand, J. (1977): Opérateurs intégraux de Fourier et le problème de la dérivée oblique. Semin. Goulaouic-Schwartz 1976/77, Exp. No. VI, Zbl. 361.35063.

Melrose, R., Mendoza, G. (1983): Elliptic operators of totally characteristic type. Preprint MSRI 047-83, Berkeley.

Miranda, C. (1955): Equazioni alle derivate parziali di tipo ellittico. Springer-Verlag, Berlin - Heidelberg - New York, Zbl. 65,85.

Nikol'skij, S. M. (1977): Approximation of Functions of Several Variables and Imbedding Theorems. 2nd ed. Nauka, Moscow. English transl.: Springer-Verlag, Berlin - Heidelberg - New York, 1975, Zbl. 496.46020, Zbl. 185, 379.

Olejnik, O. A., Radkevich, E. V. (1971): Second-order equations with nonnegative characteristic form. Itogi Nauki Tekh., Ser. Mat. Anal. English transl.: Am. Math. Soc. (1973), Zbl. 217,415.

Paneyakh, B. P. (= B. Paneah) (1981): On the theory of solvability of the oblique derivative problem. Mat. Sb., Nov. Ser. **114**, 226–268. English transl.: Math. USSR Sb. **42**, 197–235 (1982), Zbl. 457.35028. .

Paneyakh, B. P. (= B. Paneah) (1985a): Some boundary value problems for elliptic equations and the Lie algebras connected with them. Mat. Sb., Nov. Ser. **126**, 215–246. English transl.: Math. USSR Sb. **54**, 207–237 (1986), Zbl. 589.35035.

Paneyakh, B. P. (= B. Paneah) (1985b): On the solvability of a class of noncoercive boundary-value problems. Usp. Mat. Nauk **40**, 209–210. English transl.: Russ. Math. Surv. **40**, 231–232 (1985). Zbl. 588.35011.

Paneyakh, B. P. (= B. Paneah) (1987): Some boundary value problems for elliptic equations and the Lie algebras connected with them. II. Mat. Sb., Nov. Ser. **133**, 508–538. English transl.: Math. USSR, Sb. **61**, 495–527 (1988), Zbl. 658.35028.

Paneyakh, B. P. (= B. Paneah) (1988): On the solvability of some noncoercive boundary value problems for elliptic equations. Usp. Mat. Nauk **43**, 209–210. English transl.: Russ. Math. Surv. **43**, 250–251 (1988), Zbl. 677.35026.

Plamenevskij, B. A. (1986): Algebras of Pseudodifferential Operators. Nauka, Moscow. English transl.: Kluwer, Dordrecht 1989, Zbl. 615.47038.

Rashevskij, P. K. (1938): On the connectability of any two points of a completely nonholonomic space by an admissible line. Uch. Zap. Pedagog. Inst. (Moscow), Ser. Fiz.-Mat., **2**, 83–94 (Russian).

Rempel, S. (1987): Regularity for a class of degenerate operators. Prepr. Akad. Wiss. DDR Karl-Weierstrass-Inst. Math., No. 6, 1–34. Appeared in: Commun. Partial Differ. Equations **14**, 833–865 (1989), Zbl. 719.35011.

Rojtberg, Ya. A., Sheftel', Z. G. (1980): On general elliptic problems with strong degeneracy. Dok. Akad. Nauk SSSR **254**(6), 1334–1342. English transl.: Sov. Math., Dokl. **22**, 602–608 (1980), Zbl. 465.35037.

Rothschild, L., Stein, E. (1977): Hypoelliptic differential operators and nilpotent groups. Acta Math. **137**, 247–320, Zbl. 346.35030.

Schulze, B.-W. (1987): Mellin expansions of pseudodifferential operators and conormal asymptotics of solutions. Lect. Notes Math. **1256**, 378–401, Zbl. 637.58088.

Shubin, M. A. (1978): Pseudodifferential Operators and Spectral Theory. Nauka, Moscow. English transl.: Springer-Verlag, Berlin - Heidelberg - New York 1987, Zbl. 451.47064.

Soga, H. (1975): Boundary value problems with oblique derivative. Publ. Res. Inst. Math. Scil, , Kyoto Univ. **10**, 619–668, Zbl. 309.35030.

Taira, K. (1976): On some degenerate oblique derivative problems. J. Fac. Sci. Univ. Tokyo, Sec. IA **23**, No. 2, 259–287, Zbl. 359.35020.

Taira, K. (1979): Sur le problème de la dérivée oblique. II. Ark. Mat. **17**, 177–191, Zbl. 431.35030.

Taira, K. (1982): Semigroups and boundary value problems. Duke Math. J. **49**, No. 2, 287–320, Zbl. 504.47042.

Taylor, M. E. (1981): Pseudodifferential Operators. Princeton Univ. Press, Zbl. 453.47026.

Trèves, F. (1980): Introduction to Pseudodifferential and Fourier Integral Operators, 1, 2. Plenum Press, New York, Zbl. 453.47027.

Triebel, H. (1978): Interpolation Theory. Function Spaces. Differential Operators. Deutscher Verlag der Wissenschaften, Berlin, Zbl. 387.46033.

Vishik, M. I., Grushin, V. V. (1969a): On a class of degenerate elliptic equations of higher orders. Mat. Sb., Nov. Ser. **79**(1), 3–36. English transl.: Math. USSR, Sb. **8**, 1–32 (1969), Zbl. 177,143.

Vishik, M. I., Grushin, V. V. (1969b): Boundary value problems for elliptic equations that are degenerate on the boundary of the domain. Mat. Sb., Nov. Ser. **80**(4), 455–491. English transl.: Math. USSR, Sb. **9**, 423–454 (1969), Zbl. 186,171.

Vishik, M. I., Grushin, V. V. (1970a): Degenerate elliptic differential and pseudodifferential operators. Usp. Mat. Nauk **25**(4), 29–56. English transl.: Russ. Math. Surv. **25**, No. 4, 21–50 (1970), Zbl. 209,409.

Vishik, M. I., Grushin, V. V. (1970b): Elliptic boundary value problems that are degenerate on a submanifold of the boundary. Dokl. Akad. Nauk SSSR **190**(2), 255–258. English transl.: Sov. Math., Dokl. **11**, 60–64 (1970), Zbl. 238.35032.

Winzell, B. (1977): The oblique derivative problem. I. Math. Ann. **229**, 267–278, Zbl. 362.35025.

Winzell, B. (1979a): The oblique derivative problem. II. Ark. Mat. **17**, No. 1, 107–122, Zbl. 414.35025.

Winzell, B. (1979b): The sub-ellipticity with best exponent for the oblique derivative problem. Report LiTH-MAT-R-79-11, 1–6.

Winzell, B. (1981): A boundary value problem with an oblique derivative. Commun. Partial Differ. Equations **6**, No. 3, 305–328, Zbl. 473.35037.

Yanushauskas, A. I. (1985): The Oblique Derivative Problem of Potential Theory. Nauka, Novosibirsk. English transl.: Contemp. Sov. Math., New York 1990, Zbl. 581.35002.

III. Parabolic Equations

S. D. Ejdel'man

Translated from the Russian
by M. Capinski

Contents

Preface

Parabolic equations are the subject of practically limitless number of investigations. With years their stream does not decrease, covering new substantial mathematical objects and a continuously growing number of quite various applications.

Incompatibility between the length of the article and the amount of important information on the theory of parabolic equations required making difficult decisions on the choice of the material included. Regrettably, many results deserving attention had to be left out. Being aware of all that, the author nevertheless did not consider possible to exclude from the text the analysis of simple examples of illustrative and substantial character, assuming that they can help the reader in better understanding of definitions and results which cannot be quite so simply formulated. The core of the article is the theory of the Cauchy and boundary-value problems for linear parabolic equations and systems though the article on qualitative properties of solutions of parabolic equations would not be less interesting. We have also touched upon them when possible but in fact this is a subject for another article of not lesser length.

The author shared his longstanding interest in parabolic equations with S.D.Ivasishen, N.V.Zhitarashu, M.I.Matijchuk, V.D.Repnikov, and F.O.Porper, and many aspects of the theory were discussed with them during the joint creative collaboration.

Section 2 of Chapter 3 was written jointly with N.V.Zhitarashu. I am obliged to A.N.Kochubej for some interesting information. V.A.Solonnikov's important papers and continuous encouragement greatly stimulated my efforts. I am deeply grateful to them all.

Chapter 1
Systems. Problems. Spaces

§1. Parabolic Systems

1.1. Introduction. *Parabolic equations* appeared in the study of the phenomenon of heat conduction and diffusion by means of mathematical tools. One of the simplest but most important examples of parabolic equations, the one in fact studied first, is the heat equation. It was first derived by Fourier in his classical treatise 'Théorie analytique de la chaleur' which came out in 1822 though the main results were obtained considerably earlier. The derivation of the heat equations is contained, as a rule, in all textbooks on mathematical physics and it is also present in the article of Egorov and Shubin (1988, p.11-12).

Here we briefly dwell on some concrete examples taken from various domains and at least in some way illustrating a variety of applications of parabolic equations.

We note that the *heat equation*

$$\frac{\partial u}{\partial t} = a^2 \left[\frac{\partial^2 u}{\partial x^2} + \frac{\partial^2 u}{\partial y^2} + \frac{\partial^2 u}{\partial z^2} \right] = a^2 \Delta u \tag{1.1}$$

arises also in the theory of nuclear chain reactions (see (Sneddon 1951)).

In the chain reaction the nuclei of uranium split into parts and as a result the neutrons are freed. The speed of freed neutrons is great but when they pass through a moderator (for example a graphite one) it decreases rapidly and after that for some time the neutrons diffund through the continuum and are next absorbed. The absorption of neutrons in majority of cases leads to the partition of nuclei that have absorbed a neutron. As a result, the neutrons of the next generation are freed, etc. When we investigate the process of moderation of the neutrons an equation for the age of neutrons appears and it has the form (1.1), where $u(t, x, y, z)$ represents the number of neutrons at the point (x, y, z) that have achieved the age t, the number related to the unit of volume and the unit of time. After having passed the moderator the neutrons cease loosing energy and their motion can be described by the diffusion equation (1.1) with respect to the density u of the neutrons.

And now we present some simple problems of the modern theory of signals where in a natural way interesting parabolic equations arise.

Example 1.1. We consider electric RC and RL circuits, Fig. 1. Such circuits play the roles of filters in various radiotechnic devices. We are concerned with the voltage λ on the condensator, where to the RC circuit the voltage F is applied (Fig. 1a) or the voltage λ on the resistor of the RL circuit under the voltage f (Fig. 1b).

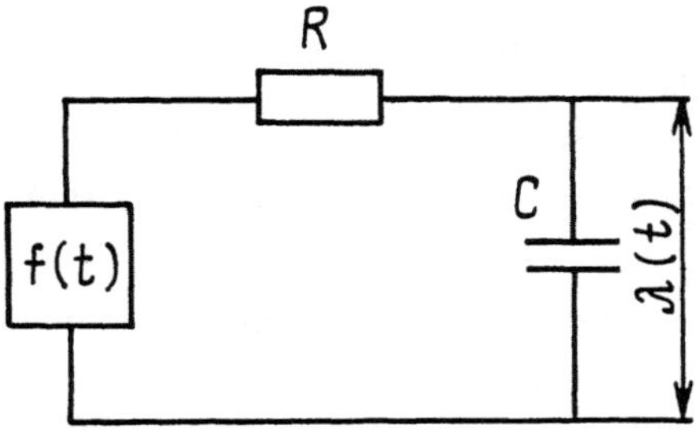

Fig. 1a

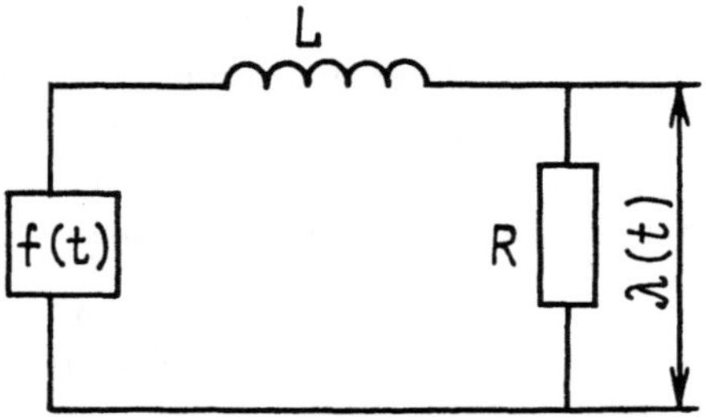

Fig. 1b

By the formulas for electric circuits that follow from the Ohm and Kirhoff laws the problem of finding the unknown voltage λ reduces to solving the ordinary differential equation

$$\frac{d\lambda}{dt} + \alpha\lambda = \gamma f(t) , \quad t \geq 0 , \tag{1.2}$$

where $\alpha = \gamma = 1/RC$ in the case a) and $\alpha = \gamma = R/C$ in the case b). When $f(t)$ is a given deterministic function of time t, this equations can be directly solved. However, in real conditions of functioning of such a circuit as an element of a radiotechnic device, the original voltage f is, as is well known, random. Therefore the solution of (1.2) becomes random. Equation (1.2) has to be treated as a stochastic one (see (Gikhman and Skorokhod 1968), (Tikhonov and Kul'man 1975)). It gives a microscopic description of the stochastic process and a macroscopic description of the same process is given in terms of the density $P(t,x)$ of the probability that the voltage λ at time instant t takes value x (with the initial distribution $P(0,x) = P_0(x)$ given). To study the voltage λ under various initial random voltages f, it turns out important and convenient to consider the case that in (1.2) instead f we have an ideal stationary stochastic process with constant spectral density of energy (equal to N_0), called *white noise.*

In such a case it can be proved (see (Tikhonov and Kul'man 1975)) that the probability density P satisfies the equation

$$\frac{\partial P}{\partial t} = \frac{1}{4}\gamma^2 N_0 \frac{\partial^2 P}{\partial x^2} + \alpha \left(\frac{\partial}{\partial x}(xP) \right) . \tag{1.3}$$

Stochastic differential equations describe a great variety of processes arising in nature and modern technics and the macro-description of such processes requires solving and investigating partial differential equations and systems and among them we often find parabolic equations and systems. We would like to indicate, for example, the monograph (Klovskij et al. 1984) which concerns mathematical models of continuous communication links in the form of stochastic differential equations and the methods of their practical use in construction of physical models of communication links and the synthesis of algorithms of optimal reception of signals.

In mathematical description of many processes in chemical and biological kinetics (see (Romanovskij et al.1984)) arises a necessity to consider systems rather than individual equations.

Example 1.2. For simplicity we assume that the space is one-dimensional, i.e. a volume where a chemical reaction takes place or live cells interact is a long narrow tube. Let u_k, $k = 1, \dots, m$, be kinetic variables depending on time t and the space variable x. In chemistry and biochemistry u_k is the concentration of reacting substances and in biology it is the biomass or the number of organisms of a given kind in the unit length of the tube. The *kinetic equation* with interaction of components and diffusion has the form

$$\frac{\partial u_k}{\partial t} = \frac{\partial}{\partial x}\left[\sum_{j=1}^{m} D_{kj}(u_1, \dots, u_m)\frac{\partial u_j}{\partial x}\right] + F_k(u_1, \dots, u_m, x)\,, \quad k = 1, \dots, m\,.$$

Here D_{kk} and D_{kj} are coefficients of diffusion and mutual diffusion and F_k are functions defining total velocities of the changes of u_k resulting from their interaction. In many cases we can assume that D_{kk} are constant quantities, $D_{kj} = 0$, $k \neq j$, and F_k do not depend on x. The above system is nonlinear. As the first step of investigations we often consider a *linearized system* which for $D_{kk} = D_k = \text{const}$ and $D_{kj} = 0$, $k \neq j$, has the form

$$\frac{\partial u_k}{\partial t} = D_k\frac{\partial^2 u_k}{\partial x^2} + \sum_{j=1}^{m} a^{kj}u_j\,, \quad j = 1, \dots, m\,. \tag{1.4}$$

Linear models may not be sufficient for satisfactory quantitative and qualitative description of the processes in question. If we consider nonlinear models, we can greatly extend the class of practical problems available to study. In the present article we expose facts of the linear theory which is sufficiently developed and, in our opinion, has possible further substantial applications.

1.2. Systems Parabolic in the Sense of Petrovskij. The initial definitions and results of the theory of quite general parabolic equations and systems belong to Petrovskij (1986, 1987). It is remarkable that numerous well-known properties of solutions of the heat equation remain valid for such systems with some natural amendments.

Everywhere in what follows we shall consider a one-dimensional variable t and an n-dimensional variable $x = (x_1, \dots, x_n)$, which are interpreted as time and space variables respectively. By Q we denote a set of points (t, x) of the space $\mathbb{R}^{n+1}$ and we also adopt the following notations: $D_t = \partial/\partial t$, $D^\alpha = D_1^{\alpha_1} D_2^{\alpha_2} \dots D_n^{\alpha_n}$, $D_j = \mathrm{i}^{-1}\partial/\partial x_j$, $|\alpha| = \alpha_1 + \dots + \alpha_n$, $\mathrm{i} = \sqrt{-1}$.

We begin with the definition of equation of arbitrary order parabolic in the sense of Petrovskij. Let $L(t,x;D_t,D)$ be a linear differential expression (in what follows we shall use the term 'differential operator' or briefly 'operator') of arbitrary order with complex coefficients depending on t and x and defined in Q. For any point $(t,x) \in Q$ the function $L(t,x;p,\sigma)$, where σ is an n-dimensional vector with coordinates $\sigma_1,\dots,\sigma_n$ and p is a scalar complex parameter, is a polynomial with respect to σ_j and p. Let b be a positive integer and suppose that the degree of the polynomial $L(t,x;p\lambda^{2b},\sigma\lambda)$ with respect to λ is equal to $2br$, where $r>0$ is an integer. We denote by L_0 the *principal part of the polynomial* L, i.e. the sum of all terms of L such that

$$L_0(t,x;p\lambda^{2b},\sigma\lambda) = \lambda^{2br} L_0(t,x;p,\sigma) .$$

Definition 1.1. The operator L is *parabolic in the sense of Petrovskij (2b-parabolic)* at the point (t,x) if for all $\sigma \in \mathbb{R}^n$ the p-zeros of the polynomial $L_0(t,x;p,\sigma)$ satisfy the inequality

$$\operatorname{Re} p(t,x;\sigma) \le -\delta(t,x)|\sigma|^{2b} , \quad \delta(t,x) > 0 . \tag{1.5}$$

The operator L is *uniformly parabolic* in the domain Q if it is parabolic at each point of Q and the inequality (1.5) is satisfied with the same positive number δ for each point $(t,x) \in Q$.

For parabolic operators we compare in strength a single differentiation with respect to time t with $2b$-fold differentiation in space variables $x_1,\dots,x_n$, i.e. the differentiation in t has the weight $2b$. It is easy to show that the weight is even as a result of the *parabolicity condition* (1.5).

And so we consider a *weighted homogeneous parabolic polynomial* $R(p,\sigma)$:

$$R(\lambda^{2b}p,\lambda\sigma) = \lambda^s R(p,\sigma) . \tag{1.6}$$

We put $\sigma = 0$ and from (1.6) we obtain

$$R(\lambda^{2b}p,0) = \lambda^s R(p,0) .$$

Hence it follows that either $s = 2br$ or $R(p,0) \equiv 0$. The latter is impossible since for $\operatorname{Re} p > 0$ this contradicts the parabolicity condition for the polynomial. Therefore, $s = 2br$ and $R(p,0) = \gamma p^r$, $\gamma \ne 0$. We can prove in a similar way that the coefficients at σ_j^{2br}, $j = 1,\dots,n$, in R are different from zero. For the parabolic operator L the numbers b and r are uniquely determined: r is the degree of the polynomial $L(t,x;p,\sigma)$ in p and $2br$ is that in σ_j, $j=1,\dots,n$.

Definition 1.2. A *matrix differential operator* $\mathcal{L}(t,x;D_t,D)$ with entries $L_{kj}(t,x;D_t,D)$, $k,j=1,\dots,m$, is *parabolic in the sense of Petrovskij* if

1) the operator $L(t,x;D_t,D) = \det\mathcal{L}(t,x;D_t,D)$ is $2b$-parabolic in the sense of Petrovskij,
2) the degree of the polynomials $L_{kj}(t,x;p\lambda^{2b},\sigma\lambda)$ with respect to λ does not exceed $2bn_j$ and

$$L_{kj}(t,x;p,\sigma) = \delta_{kj}p^{n_j} + L'_{kj}(t,x;p,\sigma)\,,$$

where L'_{kj} is a polynomial not containing p^{n_j} and δ_{kj} is the Kronecker delta.

Therefore the structure of *systems 2b-parabolic in the sense of Petrovskij* looks like that:

$$D_t^{n_k}u_k \equiv \frac{\partial^{n_k}u_k}{\partial t^{n_k}} = \sum_{j=1}^{m} \sum_{\substack{2b\alpha_0+|\alpha|\le 2bn_j \\ \alpha_0<n_j}} a^{kj}_{\alpha_0\alpha}(t,x)D_t^{\alpha_0}D^\alpha u_j + f_k(t,x)\,,$$

$$k = 1,\dots,m\,. \tag{1.7}$$

The systems parabolic in the sense of Petrovskij with $n_j = 1$, $j = 1,\dots,m$ are more extensively studied:

$$D_t u_k \equiv \frac{\partial u_k}{\partial t} = \sum_{j=1}^{m}\sum_{|\alpha|\le 2b} a^{kj}_\alpha(t,x)D^\alpha u_j + F_k(t,x)\,,$$

$$k = 1,\dots,m\,. \tag{1.8}$$

An important subclass of systems of the form (1.8) parabolic in the sense of Petrovskij are, so called, strongly parabolic systems.

Definition 1.3. A system of the form (1.8) is called *uniformly strongly parabolic* if for any vector $\eta \in \mathbb{C}^n$ and any $\sigma \in \mathbb{R}^n$

$$\operatorname{Re}\sum_{k,j=1}^{m}\sum_{|\alpha|=2b} a^{kj}_\alpha(t,x)\sigma^\alpha\eta_k\overline{\eta}_j \ge \delta|\sigma|^{2b}|\eta|^2\,,$$

where δ is a positive constant.

Strongly parabolic systems have a series of important properties which are not satisfied by general systems parabolic in the sense of Petrovskij.

In applications we usually find *equations* and *systems of the second order*

$$\frac{\partial u}{\partial t} = \sum_{\mu,\nu=1}^{n} a_{\mu\nu}(t,x)\frac{\partial^2 u}{\partial x_\mu \partial x_\nu} + \sum_{\mu=1}^{n} a_\mu(t,x)\frac{\partial u}{\partial x_\mu} + a(t,x)u\,; \tag{1.9}$$

where $a_{\mu\nu}$, a_μ $(\mu,\nu = 1,\dots,m)$, and a are quadratic matrices of order m and u is a vector of height m.

Uniform parabolicity of the system (1.9) in Q means that the λ-roots of the equation

$$\det\left(\sum_{\mu\nu=1}^{n} a_{\mu\nu}(t,x)\sigma_\mu\sigma_\nu - \lambda I\right) = 0\,, \tag{1.10}$$

where I is the unit matrix of order m, satisfy the inequality

$$\mathrm{Re}\lambda(t,x;\sigma) \geq \delta|\sigma|^2\,, \quad \delta > 0\,, \tag{1.11}$$

for any $(t,x) \in Q$ and $\sigma \in \mathbb{R}^n$.

In the case of one equation ($m = 1$) and real coefficients the inequality (1.11) takes the form

$$\sum_{\mu,\nu=1}^{n} a_{\mu\nu}(t,x)\sigma_\mu\sigma_\nu \geq \delta|\sigma|^2 \tag{1.12}$$

and *uniform parabolicity* means that the quadratic form $\sum_{\mu,\nu=1}^{n} a_{\mu\nu}(t,x)\sigma_\mu\sigma_\nu$, $\sigma \in \mathbb{R}^n$, is positive definite uniformly in $(t,x) \in Q$. For equations (1.1) and (1.3) this form is equal to $a^2|\sigma|^2$ and $(1/4)\gamma^2 N_0\sigma^2$, respectively, and so these equations are uniformly parabolic if $a^2 > 0$, $\gamma^2 > 0$, and $N_0 > 0$, which of course follows from their physical meaning.

For the system (1.4) the equation (1.10) has the form

$$\det\begin{pmatrix} D_1\sigma^2 - \lambda & 0 & \dots & 0 \\ 0 & D_2\sigma^2 - \lambda & \dots & 0 \\ \vdots & \vdots & \ddots & \vdots \\ 0 & 0 & \dots & D_m\sigma^2 - \lambda \end{pmatrix} = 0\,,$$

hence $\lambda_j = D_j\sigma^2$, $j = 1,\dots,m$, and consequently the system is uniformly parabolic since $D_j > 0$, $j = 1,\dots,m$.

In what follows we shall consider uniformly parabolic systems only, so the word 'uniformly' will be omitted.

1.3. $2\overrightarrow{b}$-parabolic Systems. The author (Ejdel'man 1960) has presented the following generalization of systems parabolic in the sense of Petrovskij. We give the weight $1/2b_1$ to the differentiation in the space variable x_1, $1/b_2$ to $x_2,\dots,1/b_n$ to x_n, relative to the single differentiation in the time variable t. By $\overrightarrow{b}$ we denote the vector $\overrightarrow{b} = (b_1,\dots,b_n)$.

Definition 1.4. An operator $\mathcal{L}(t,x;D_t,D)$ is $2\overrightarrow{b}$*-parabolic* if there are positive integers b_i, $i = 1,\dots,n$, such that

1) the functions $L_{kj}(t,x;p\lambda,\sigma_1\lambda^{1/2b_1},\dots,\sigma_n\lambda^{1/2b_n})$ with respect to λ are linear combinations of power functions with maximal degree not exceeding n_j, and

$$L_{kj}(t,x;p,\sigma) = \delta_{kj}p^{n_j} + L'_{kj}(t,x;p,\sigma)\,,$$

where L'_{kj} is a polynomial not containing p^{n_j};

2) the polynomial

$$L_0(t,x;p,\sigma) = \det\mathcal{L}_0(t,x;p,\sigma)\,,$$

where $\mathcal{L}_0$ is the matrix composed of the principal parts L^0_{kj} of the polynomials L_{kj} which are the sums of all terms of L_{kj} such that

$$L^0_{kj}(t,x;p\lambda,\sigma_1\lambda^{1/2b_1},\dots,\sigma_n\lambda^{1/2b_n}) = \lambda^{n_j} L^0_{kj}(t,x;p,\sigma_1,\dots,\sigma_n)\,,$$

has the following property: for all real $\sigma_1,\dots,\sigma_n$ the p-zeros of $L_0(t,x;p,\sigma)$ satisfy the inequality

$$\operatorname{Re} p(t,x;\sigma) \leq -\delta(\sigma_1^{2b_1} + \dots + \sigma_n^{2b_n})\,,$$

where δ is a positive constant.

We put $\widetilde{\alpha} = (\alpha_1/2b_1) + \dots + (\alpha_n/2b_n)$ and then a $2\overrightarrow{b}$*-parabolic system* can be written in the following way:

$$D_t^{n_k} u_k = \sum_{j=1}^{m} \sum_{\substack{\widetilde{\alpha}+\alpha_0 \leq n_j \\ \alpha_0 < n_j}} a^{kj}_{\alpha_0\alpha}(t,x) D_t^{\alpha_0} D^\alpha u_j + f_k(t,x)\,,$$

$$k = 1,\dots,m\,. \tag{1.13}$$

Example 1.3. A simple $2\overrightarrow{b}$*-parabolic equation* has the form

$$\frac{\partial u}{\partial t} = (-1)^{b_1-1}\frac{\partial^{2b_1} u}{\partial x_1^{2b_1}} + \dots + (-1)^{b_n-1}\frac{\partial^{2b_n} u}{\partial x_n^{2b_n}}\,.$$

Hypothetically, such equations can appear when the properties of the process with respect to various spatial variables are essentially different. However, it is more realistic to expect such a variety to appear when the description is by pseudodifferential rather than differential equations.

1.4. Systems Parabolic in the Sense of Solonnikov. The structure of systems parabolic in the sense of Petrovskij admits the following generalization. For Petrovskij's systems the degrees of generalized homogeneity of the polynomials L^0_{kj} are equal to $2bn_j$ and do not depend on k. Solonnikov (1965, 1967) gave a definition of parabolicity where the degrees of generalized homogeneity of the polynomials L^0_{kj} depend on k as well as on j. For hyperbolic systems such a generalization was proposed by Leray and for elliptic by Douglis and Nirenberg. In these definitions the *order of the operator* L_{kj} is defined as the sum of $s_k + t_j$, where s_k, t_j, $k,j = 1,\dots,m$, are some integers.

Definition 1.5. We shall call an operator $\mathcal{L}$ *parabolic in the sense of Solonnikov* if

1) the operator $L = \det\mathcal{L}$ is $2b$-parabolic in the sense of Petrovskij;
2) there exist integers s_k, t_j, $k,j = 1,\dots,m$, such that the degree of the polynomial $L_{kj}(t,x;p\lambda^{2b},\sigma\lambda)$ in λ does not exceed $s_k + t_j$ (if $s_k + t_j < 0$, then $L_{kj} \equiv 0$) and, in addition, $\sum_{k=1}^m (s_k + t_j) = 2br$, where r is the degree of the polynomial $L(t,x;p,\sigma)$ in p.

The *principal part of the operator* $\mathcal{L}$ is the operator $\mathcal{L}_0$ corresponding to the matrix whose entries are the polynomials $L^0_{kj}(t,x;p,\sigma)$ that are the principal parts of L_{kj}. These are the sums of all the terms of L_{kj} such that $L^0_{kj}(t,x;p\lambda^{2b},\sigma\lambda) = \lambda^{s_k+t_j} L^0_{kj}(t,x;p,\sigma)$.

If the degree of $L^0_{kj}(t,x;p\lambda^{2b},\sigma\lambda)$ in λ is less than s_k+t_j, then $L^0_{kj} \equiv 0$. The polynomial $L_0 = \det\mathcal{L}_0$ is the principal part of the polynomial L.

The following natural question arises: if a matrix differential operator $\mathcal{L}$ is such that the operator $L = \det\mathcal{L}$ is parabolic in the sense of Petrovskij, can we choose the numbers s_k and t_j such that the operator $\mathcal{L}$ is parabolic in the sense of Solonnikov? Note that such numbers should satisfy the following requirements: $s_k+t_j \geq \alpha_{kj}$, $\sum_{k=1}^m (s_k+t_k) = 2br$, where a_{kj} is the degree in λ of $L_{kj}(t,x;p\lambda^{2b},\sigma\lambda)$. Obviously, the numbers b and r are uniquely determined by L.

As was shown by Volevich, such numbers s_k, t_j always exist. If s_k, t_j is a choice of weights, then s_k+a and t_j-a also give an admissible choice. Therefore, s_k, t_j are usually fixed by the condition $\max s_k = 0$.

Example 1.4. We consider a system consisting of two heat equations

$$\frac{\partial u_1}{\partial t} = \Delta u_1 , \quad \frac{\partial u_2}{\partial t} = \Delta u_2 . \tag{1.14}$$

For (1.14) we have

$$\mathcal{L}(p,\sigma) = \mathcal{L}_0(p,\sigma) = \begin{pmatrix} p+|\sigma|^2 & 0 \\ 0 & p+|\sigma|^2 \end{pmatrix} .$$

The operator $\mathcal{L}$ is parabolic in the sense of Petrovskij: $s_1=s_2=0$, $t_1=t_2=2$. But this system can be also viewed upon as a parabolic in the sense of Solonnikov with $s_1=0$, $s_2=-c$, $t_1=2$, $t_2=2+c$, where c is a positive integer. So even the condition $\max_k s_k = 0$ does not guarantee the uniqueness of the choice of s_k, t_j even in the case of a simple system (1.14). However, this flexibility can be employed to extend the class of boundary-value problems available to study and to obtain theorems on well-posedness of such problems in a broad range of functional spaces.

The systems parabolic in the sense of Petrovskij form a subclass of the class of systems considered, where $s_1=\ldots=s_m=0$, $t_j=2bn_j$, $j=1,\ldots,m$.

A particular case of general parabolic systems is a class of systems defined by a Japanese mathematician Shirota (1957).

Definition 1.6. The systems of the form

$$\frac{\partial u_k}{\partial t} = \sum_{j=1}^m a_{kj}(t,x;D)u_j , \quad k=1,\ldots,m , \tag{1.15}$$

are called *parabolic in the sense of Shirota* if

1) the polynomial $\det(pI - (a_{kj}(t,x;\sigma))_{k,j=1}^m)$ is parabolic in the sense of Petrovskij;
2) there exist nonnegative integers s_k, $k = 1, \dots, m$, such that the order of the operator $a_{kj}(t,x;\sigma)$ does not exceed $s_k - s_j + 2b$.

The systems parabolic in the sense of Shirota contain, as a particular class, the systems parabolic in the sense of Petrovskij. For the systems of the form (1.8) this is obvious (all $s_k = 0$). A system of the form (1.7) is first transformed in a standard way to a system with first order derivatives in t and next it is verified to be parabolic in the sense of Shirota.

If a system is parabolic in the sense of Shirota, then the polynomials

$$L^0(t,x;p\lambda^{2b},\sigma\lambda) = p\lambda^{2b} + a_{kk}^0(t,x;\sigma\lambda) = \lambda^{s_k+t_k} L_{kk}^0(t,x;p,\sigma) ,$$

correspond to the principal parts of its diagonal operators. Hence it follows that $s_k + t_k = 2b$, $t_k = 2b - s_k$. In this way we have shown in what sense general parabolic systems contain the systems parabolic in the sense of Shirota.

1.5. Systems Parabolic in the Sense of Shilov. All definitions considered above consist of two parts:

1) a certain assumption on the structure of the systems in question which allows us to select the principal part of the operators determining the system;
2) conditions on the determinant of the corresponding principal part of the operator.

This, as we shall see in what follows, will play an important role in building a sufficiently complete theory for such systems with variable coefficients. In the definition below we do not have any condition on the structure of the system considered.

Definition 1.7. A system of the form (1.15) is called *parabolic in the sense of Shilov* if for all real σ the p-roots of the polynomial

$$L(t,x;p,\sigma) = \det(pI - (a_{kj}(t,x;\sigma)_{k,j=1}^m))$$

satisfy the inequality

$$\operatorname{Re} p(t,x;\sigma) \le -c_1|\sigma|^h + c_2 ,$$

where c_1, c_2, and h are positive constants.

Example 1.5. We consider the equation

$$\frac{\partial u}{\partial t} = \Delta u + \mathrm{i} \sum_{|\alpha| \le N} a_\alpha(t,x) D^\alpha u , \tag{1.16}$$

where $N > 2$, $a_\alpha(t,x)$ are real functions, and for $|\alpha| = N$ at least one of the coefficients is different from zero in the domain. The polynomial corresponding to (1.16) has the form

$$L(t,x;p,\sigma) = p + |\sigma|^2 - \mathrm{i} \sum_{|\alpha|\le N} a_\alpha(t,x)\sigma^\alpha .$$

In this way

$$p(t,x;\sigma) = -|\sigma|^2 + \mathrm{i} \sum_{|\alpha|\le N} a_\alpha(t,x)\sigma^\alpha ; \quad \operatorname{Re} p(t,x;\sigma) = -|\sigma|^2 .$$

Equation (1.16) is a typical example of an equation parabolic in the sense of Shilov.

1.6. Diffusion Equations with Inertia. A natural source of interesting parabolic equations is the theory of Markov stochastic processes. Passing to a macro-description of such processes with respect to the densities of transition probabilities, one obtains, under some special conditions, partial differential equations. Such equations may not be related to the parabolic equations defined above but the properties of solutions and the methods of investigation are quite close, for some natural reasons, to the properties of solutions to the heat equation. Here we define, so called, diffusion equations with inertia, and in 1.7 we consider a class of parabolic pseudodifferential equations with nonsmooth symbols. They are included into the present article to allow us to describe in a more contrasting way the results of the theory of parabolic equations and possibilities of further applications.

Starting from some problems of probability theory Kolmogorov (1934) introduced and studied an interesting equation

$$\frac{\partial u}{\partial t} - x\frac{\partial u}{\partial y} = \frac{\partial^2 u}{\partial x^2} , \tag{1.17}$$

which he called *diffusion equation with inertia.*

The *fundamental solution* (transition probability density) of (1.17) constructed by Kolmogorov has the form

$$\Gamma(t,x,y,;\tau,\xi,\eta)$$

$$= \frac{\sqrt{3}}{2\pi(t-\tau)^2} \exp\left\{ -\frac{(x-\xi)^2}{4(t-\tau)} - 3\frac{\left(y-\eta+(t-\tau)\left(\frac{x+\xi}{2}\right)\right)^2}{(t-\tau)^3} \right\} . \tag{1.18}$$

This allows us to see the similarity as well as the difference between the solution to (1.17) and the solution to the heat equation. Equation (1.16) was generalized many times and these generalizations were investigated (Il'in 1962), (Sonin 1967), (Ejdel'man and Malitskaya 1975), (Ejdel'man and Tychinskaya

1979), (Weber 1951). We note that in the Hörmander's monograph this equation is first considered as an example (Hörmander 1983, 253-254) and then an interesting generalization is given (Hörmander 1985, 470-478).

Definition 1.8. *Diffusion equation with inertia in some of the variables* is an equation of the form

$$D_t u - \sum_{j=1}^{n_1} x_j D_{y_j} u - \sum_{j=1}^{n_2} y_j D_{z_j} u = a(t, x; D_x) u \,, \tag{1.19}$$

where the operator $D_t - a(t, x; D_x)$ is uniformly parabolic in the sense of Petrovskij and $n_2 \leq n_1 \leq n$.

1.7. Some Classes of Parabolic Pseudodifferential Equations. We consider a *pseudodifferential equation* (Drin' 1986), (Kochubej 1985), (Ejdel'man and Drin' 1981)

$$D_t u + A_0 u + \sum_{k=1}^{s} A_s u = 0 \,, \tag{1.20}$$

where $A_0, A_1, \ldots, A_s$ are *pseudodifferential operators* with symbols $a_0(t, x; \sigma)$, $a_1(t, x; \sigma), \ldots, a_s(t, x; \sigma)$ homogeneous in σ. We assume that the principal symbol $a_0(t, x; \sigma)$ is homogeneous of degree $\gamma \geq 1$ and satisfies the condition (of parabolicity of (1.20))

$$\operatorname{Re} a_0(t, x; \sigma) \geq \delta |\sigma|^\gamma \,, \quad \delta > 0 \,, \tag{1.21}$$

and the symbols $a_k(t, x; \sigma)$ are homogeneous in σ of degrees γ_k, $0 < \gamma_k < \gamma$.

A simple example is a pseudodifferential operator with the symbol $|\sigma|^\gamma$, $\gamma < 2$, which is the generator of a symmetric stable stochastic process (Samko 1984). Many papers are devoted to the construction of discontinuous Markov processes by means of generating integro-differential, and in particular pseudodifferential, operators (see (Kochubej 1985) and the references therein) where either stochastic differential equations or the semigroup theory is employed. Analytic approach based on the theory of differential and pseudodifferential equations gives, as a rule, rich information on the properties of the transition probability of the process. We note that the symbols are nonsmooth for $\sigma = 0$ which excludes the possibility of applying the standard calculus of pseudodifferential operators.

We shall discuss the question of how to define such operators. We should mention that the usual formula

$$\begin{aligned} (Au)(t, x) &= \int_{\mathbb{R}^n} \mathrm{e}^{\mathrm{i}x \cdot \sigma} a(t, x; \sigma) \hat{u}(t, \sigma) d\sigma \,, \\ \hat{u}(t, \sigma) &= \frac{1}{(2\pi)^n} \int_{\mathbb{R}^n} \mathrm{e}^{-\mathrm{i}x \cdot \sigma} u(t, \sigma) d\sigma \,, \end{aligned} \tag{1.22}$$

defines a pseudodifferential operator for rapidly vanishing functions only.

However (1.20) has to be considered for bounded functions (this is concerned with the meaning of the problem and the probabilistic applications). Following Kochubej (1985), by *solution* of (1.20) we understand a bounded function $u(t,x)$ continuous on $[0,T] \times \mathbb{R}^n$ satisfying (1.20) with A_0, A_k replaced by conditionally convergent hyper-singular integral operators D^{γ}_{Ω}, $D^{\gamma_k}_{\Omega_k}$ (whose characteristics are defined in a special way by the symbols a_0, a_k) or by differential operators understood in the classical sense if the corresponding symbols are polynomials in σ. Results obtained with such an interpretation of (1.20) are interesting from the point of view of the theory of equations and stochastic processes.

We shall very briefly dwell on hyper-singular integral operators which were systematically studied by Samko (1984), (Samko et al. 1987). Suppose we have complex-valued bounded functions $f \in C(\mathbb{R}^n)$, $\Omega \in C(\mathbb{R}^n \times S^{n-1})$. The expression of the form

$$(D^{\alpha}_{\Omega} f)(x) = (1/d_{nl}(\alpha)) \int_{\mathbb{R}^n} \Omega\left(x, \frac{h}{|h|}\right) \frac{\Delta^l_h f(h)}{|h|^{n+\alpha}} dh\,, \quad x \in \mathbb{R}^n\,, \tag{1.23}$$

where $\alpha > 0$, $\Delta^l_h f(x) = \sum_{k=0}^{l} (-1)^k C^k_l f(x+kh)$, l is a natural number, and $d_{nl}(\alpha)$ is the normalizing constant, is called a *hyper-singular integral operator of order d* with characteristic Ω. A *truncated* operator is obtained by replacing the domain of integration in (1.23) by $\{h \in \mathbb{R}^n : |h| > \varepsilon\}$, $\varepsilon > 0$.

The hyper-singular integral (1.23) converges absolutely if $l > \alpha$ and the functions f have bounded derivatives up to order $[\alpha]+1$ inclusive. Under such hypotheses the conditional convergence

$$(D^{\alpha}_{\Omega} f) = \lim_{\varepsilon \to 0} (D^{\alpha}_{\Omega,\varepsilon} f)$$

takes place also in the case of $l > 2[\alpha/2]$ if the characteristic $\Omega(x,h)$ is even in h. We now consider the function $f_\sigma(x) = e^{i\sigma\cdot x}$. We have

$$(D^{\alpha}_{\Omega} f_\sigma)(x) = \widetilde{\Omega}(x;\sigma) f_\sigma(x) \tag{1.24}$$

where

$$\widetilde{\Omega}(x;\sigma) = (1/d_{nl}(\alpha)) \int_{\mathbb{R}^n} \frac{(1-e^{-i\sigma\cdot x})^l}{|h|^{n+\alpha}} \Omega\left(x, \frac{h}{|h|}\right) dh\,.$$

The function $\widetilde{\Omega}(x;\sigma)$ is called the *symbol* of the hyper-singular integral operator D^{α}_{Ω}. From (1.23) it follows that on the *Schwartz space* $S(\mathbb{R}^n)$ the operator D^{α}_{Ω} coincides with the pseudodifferential operator with symbol $\widetilde{\Omega}$. The constants $d_{nl}(\alpha)$ are chosen such that the hyper-singular integral operator are independent of l. In particular, if the characteristic is identically one, then the corresponding symbol equals $|\sigma|^{\alpha}$.

We consider a pseudodifferential operator of the form (1.22) where the symbol $a(t,x;\sigma)$ is homogeneous in σ of degree γ and satisfies the following condition

α. Let $N \geq 2n + 2[\gamma] + 1$ and suppose that for $\sigma \neq 0$, $a(t, x; \sigma)$ has N continuous derivatives in σ and for all $x \in \mathbb{R}^n$, $\sigma \in \mathbb{R}^n$, $t \in [0, T]$ we have

$$|D_0^k a(t, x; \sigma)| \leq c_N |\sigma|^{\gamma - |k|} , \quad |k| \leq N ,$$

where γ is not an integer.

To write the pseudodifferential operator (1.22) in the form of a hyper-singular integral operator we expand the symbol in the series with respect to the system $\{Y_{\nu\mu}\}$ of spherical harmonics (Samko 1984)

$$a(t, x; \sigma) = \sum_{\nu=0}^{\infty} \sum_{\mu=0}^{\delta_\nu} c_{\lambda\mu}(t, x) Y_{\nu\mu}(\sigma) , \quad |\sigma| = 1 ,$$

where δ_ν is the dimension of the space of spheric harmonics of order ν. According to (Samko 1984) the symbol of the hyper-singular integral operator $D^\alpha_{Y_{\nu\mu}}$ is equal to $|\sigma|^\gamma \lambda(\nu, \gamma) Y_{\nu\mu}(\sigma/|\sigma|)$ where $\lambda(\nu, \gamma)$ are some constants. Hence it follows that D^α_Ω has the characteristic depending on the parameter t:

$$\Omega(t, x; \sigma) = \sum_{\nu=0}^{\infty} \sum_{\mu=0}^{\delta_\nu} \frac{c_{\lambda\mu}(t, x)}{\lambda(\nu, \gamma)} Y_{\nu\mu}(\sigma) .$$

Recently, in the solid body physics the investigation of, so called, *fractal media* has intensified including, in particular, the diffusion phenomena in them. In one of the models, presented in (Nigmatullin 1986), the diffusion in a strongly porous (fractal) medium is described by an equation of the type of the heat equation but with a fractional derivative in time coordinate

$$(D_t^{(\alpha)} u)(t, x) = \frac{\partial^2 u(t, x)}{\partial x^2} , \quad 0 < \alpha < 1 . \tag{1.25}$$

The formulation of initial-boundary problems for (1.25), analogous to the problems for parabolic differential equations, makes sense if by $D^{(\alpha)}$ we understand the *regularized fractional derivative*:

$$(D^{(\alpha)} \varphi)(t) = \frac{1}{\Gamma(1 - \alpha)} \left[\frac{d}{dt} \int_0^t (t - \tau)^{-\alpha} \varphi(\tau) d\tau - t^{-\alpha} \varphi(0) \right] , \quad t \geq 0 .$$

The investigation of the equation of the form $D_t^{(\alpha)} u = Au$, where A is an elliptic operator, has been recently started by Kochubej. We note that the regularized fractional derivative $D^{(\alpha)}$ is a particular case of a generalized fractional-differential operator of Dzharbashyan and Nersesyan (1958). The Cauchy problem for ordinary fractional-differential equations connected with such operators was studied in (Dzharbashyan and Nersesyan 1968).

§2. Initial and Boundary Problems

2.1. Introduction. A differential equation alone is not sufficient to describe a concrete process. And for example, to define the temperature inside a body at any instant of time it is necessary to know the distribution of the temperature inside the body at the initial time (initial condition) and the heat conditions on the boundary $\partial\Omega$ of the body Ω (boundary condition).

Boundary conditions may be different depending on concrete physical or technical problem.

In the case of the heat equation the most popular are the following problems:

1) at each point of the boundary $\partial\Omega$ the temperature is given

$$u|_{\partial\Omega}= g_1 , \tag{1.26}$$

where g_1 is a known function of time t and a point of the surface $\partial\Omega$;

2) at each point of the surface $\partial\Omega$ the heat flow $-k\frac{\partial u}{\partial\nu}$ is given, where k is the conduction coefficient of the body and $\frac{\partial}{\partial\nu}$ denotes the differentiation along the direction normal to the surface $\partial\Omega$, i.e.

$$\left.\frac{\partial u}{\partial\nu}\right|_{\partial\Omega} = g_2 ; \tag{1.27}$$

3) on the surface of the body the heat is exchanged with the surrounding medium whose temperature u_0 is known. In a simple case we assume that the *Newton heat exchange law* is satisfied. According to this law, the amount of heat q transferred in the unit of time from the unit area of the surface $\partial\Omega$ to the surrounding medium is proportional to the difference between the temperatures:

$$q = H(u - u_0) ,$$

where H is the heat exchange coefficient. By the law of conservation of energy this amount of heat is the same as that which is exported through the unit area of the surface in the unit of time as a result of the internal heat transfer. This leads to the following condition assumed on $\partial\Omega$:

$$H(u - u_0) = -k\frac{\partial u}{\partial\nu}$$

or

$$\left[\frac{\partial u}{\partial\nu} + h(u - u_0)\right]_{\partial\Omega} = 0 , \quad h = H/k . \tag{1.28}$$

We note that the conditions (1.26)–(1.28) contain the derivatives with respect to the spatial variables of order not greater than one, i.e. the order of the boundary conditions is less than the *weighted order* of the equation. However, in applications we encounter problems for the heat equation leading to boundary conditions containing derivatives of an arbitrary order. We give as an example the problem posed and studied by Tikhonov (1950). We investigate the process of heat transfer in a homogeneous semi-infinite rod with insulated side surface under the assumption that at each instant t the temperature at all points of any cross section of the rod is the same. There are no heat sources inside the rod and to its end $x = 0$ we attach a body Ω with heat capacity c and such a great conductiveness that in practice we can assume that temperature U of Ω only depends on time t. We assume that inside Ω there are heat sources with density $f(t)$ at instant t.

The temperature $u(t, x)$ of such a rod satisfies the heat equation

$$\frac{\partial u}{\partial t} = a^2 \frac{\partial^2 u}{\partial x^2}, \quad t > 0, x > 0. \tag{1.29}$$

To find the boundary condition for $x = 0$ at instant t we take into account two circumstances:

1) for $x = 0$ the temperatures of the rod and of Ω are the same;
2) the heat flow in the cross section $x = 0$ should coincide with the amount of heat obtained by the body in time unit.

Thus we have

$$U(t) = u(t,0), \quad -k\frac{\partial}{\partial x}u(t,x)\,|_{x=0} = f(t) - c\frac{dU}{dt}, \quad t > 0.$$

Employing the fact that $u(t, x)$ is a solution to the heat equation (1.29) and the last equalities, we arrive at the boundary condition

$$\left(ca^2 \frac{\partial^2 u}{\partial x^2} - k\frac{\partial u}{\partial x} \right)\Bigg|_{x=0} = f(t), \quad t > 0,$$

which is called the *concentration of heat capacity condition* and contains a derivative of the second order.

If on the boundary we have an overfall of temperatures of Ω and the rod, i.e.

$$U(t) - u(t,0) = k_1 \frac{\partial u(t,x)}{\partial x}\Bigg|_{x=0}, \quad t > 0,$$

then the corresponding boundary condition contains derivatives of the third order. By means of a series of volumes with overfalls of temperatures between them, we can realize the schemes corresponding to boundary conditions containing derivatives of an arbitrary order.

The problem of finding a solution to the heat equation satisfying the initial condition

$$u\,|_{t=0} = \varphi(x)$$

and a certain boundary condition is called a *mixed problem* for this equation (the terms *initial-boundary, boundary-value,* or *boundary problem* are also used). In the case of boundary conditions (1.26), (1.27), (1.28) this problem is called *first, second,* and *third mixed (boundary)* problem respectively. The first problem is often called the *Dirichlet problem* and the second, the *Neuman problem.*

If we want to determine the temperature and for some reason we can neglect the influence of the boundary, then it is natural to assume that there is no boundary and we face the problem of finding a solution in the whole space from the initial condition. Such a problem is traditionally called the *Cauchy problem.* From the practical point of view such an approach is important because, as a rule, the Cauchy problem is easier to solve and investigate than the boundary problem. We note that in the questions with probabilistic origin the Cauchy problem is quite common.

2.2. Cauchy Problem. Initial Problem. For the systems parabolic in the sense of Petrovskij, Shirota, Shilov, or $2\vec{b}$-parabolic solved with respect to the highest derivatives with respect to the time variable t it is natural to formulate the *Cauchy problem* on finding a solution to the system in the band $\Pi_T = [0,T] \times \mathbb{R}^n$, satisfying in the case of the systems (1.7) and (1.13) the initial conditions

$$D_t^\mu u_j\,|_{t=0} = \varphi_j^\mu(x)\,, \quad \mu = 0, \ldots, n_j - 1\,,\ j = 1, \ldots, m\,. \tag{1.30}$$

Here it is assumed that $D_t^\mu u_j(t,x)$ converges to $\varphi_j^\mu(x)$ as $t \to 0$ in the topology of the function spaces in which we study the solvability of the problem. In a simple case we can understand (1.30) as the limit for each fixed x or the uniform limit on each compact set in $\mathbb{R}^n$

In the case of the systems parabolic in the sense of Solonnikov, the Cauchy problem may have no sense and it becomes necessary to consider a special, so called, *initial problem.*

We shall now describe it.

The initial condition will be given in the form

$$C(x; D_t, D)u\,|_{t=0} = \varphi(x)\,, \tag{1.31}$$

where C is a matrix differential operator with the entries C_{α_j}, $\alpha = 1, \ldots, m_1$, $j = 1, \ldots, m$, and $\varphi(x)$ is an arbitrary vector-function with values in $\mathbb{R}^{m_1}$. To define the structure of the matrix of initial conditions we proceed in the following way: let γ_{α_j} be the degree in λ of the polynomial $C_{\alpha_j}(x; p\lambda, \sigma\lambda^{2\beta})$ (if $c_{\alpha_j} \equiv 0$, then as γ_{α_j} we can take any integer) and we put $\rho_\alpha = \max_j(\gamma_{\alpha_j} - t_j)$. The sum $C_{\alpha_j}^0$ of all the terms of the polynomial C_{α_j} satisfying the homogeneity condition

$$C_{\alpha_j}^0(x; p\lambda, \sigma\lambda^{2\beta}) = \lambda^{\rho_\alpha + t_j} C_{\alpha_j}^0(x; p, \sigma) \tag{1.32}$$

is called the principal part of C_{α_j} and the matrix $C_0 = (C_{\alpha_j}^0)$ the *principal part of the matrix* C.

We shall explain what number m_1 should be and what condition is necessary on C for us to be able to determine uniquely the value of any derivative of any function u_j for $t = 0$ from the system

$$\mathcal{L}u = f \tag{1.33}$$

and the initial condition (1.31) by means of differentiation and solving linear algebraic systems (assuming that the solution, the coefficients of $\mathcal{L}$, and f are sufficiently smooth). We require that this condition on $\mathcal{L}$ and C be formulated in terms of $\mathcal{L}_0$ and C_0. And so, we have to study the problem

$$\mathcal{L}_0(0, x; D_t, D)u = f \ , \tag{1.34}$$

$$C_0(x; D_t, D)u\,|_{t=0} = \varphi \ . \tag{1.35}$$

In particular, taking all possible vector-functions $u(t)$ depending on t only, we obtain by (1.34), (1.35) the following problem:

$$\mathcal{L}_0(0, x; D_t, 0)u(t) = f(t, x) \ , \tag{1.36}$$

$$C_0(x; D_t, 0)u(t)\,|_{t=0} = \varphi(x) \ . \tag{1.37}$$

We note that

$$\det\mathcal{L}_0(0, x; p, 0) = L_0(0, x; p, 0) = \gamma(x)p^r \ , \qquad \text{where} \quad r = \sum_{k=1}^{m}(t_k + s_k) \ ,$$

and by the parabolicity condition we have $\gamma(x) \neq 0$ for all x for which the system is parabolic. Hence it follows that the number of conditions (1.37) is $m_1 = r$, i.e. the matrix C_0 (and the matrix C) has dimension $r \times m$.

Example 1.6. We consider a system parabolic in the sense of Solonnikov with the matrix

$$\mathcal{L}(p, \sigma) = \begin{pmatrix} p + |\sigma|^2 & p^{10} \\ 0 & p + |\sigma|^2 \end{pmatrix} \ .$$

Here we have $b = 1$, $r = 2$, $s_1 = 0$, $s_2 = -8$, $t_1 = 2$, $t_2 = 10$. Despite the fact that the system includes the derivative in t of the tenth order, the matrix C_0 has dimension 2×2.

Using the *residual method* of solving the homogeneous system

$$\mathcal{L}_0(0, x; \frac{d}{dt}, 0)u(t, x) = 0 \ ,$$

and a certain standard argument which we shall present in the next section, one can arrive at a condition which is necessary for the initial problem to have all required properties. We denote by $\hat{\mathcal{L}}_0$ the matrix associated to the matrix $\mathcal{L}_0$ ($\mathcal{L}_0\hat{\mathcal{L}}_0 = \hat{\mathcal{L}}_0\mathcal{L}_0 = \det\mathcal{L}_0 \cdot I$).

Complementarity Condition for the Initial Problem. The initial problem (1.31) for the system $\mathcal{L}u = f$ parabolic in the sense of Solonnikov satisfies the *complementarity condition* if the rows of the matrix

$$C_0(x; p, 0)\hat{\mathcal{L}}_0(0, x; p, 0)$$

are linearly independent modulo polynomial $p^r = (1/\gamma(x))L_0(0, x; p, 0)$ for any x.

We recall that linear independence of the polynomials $M_1(p), \ldots, M_l(p)$ modulo polynomial $N(p)$ means that the remainders of division of $M_k(p)$ by $N(p)$, $k = 1, \ldots, l$, are linearly independent.

We note that the Cauchy problem, as well as the initial problem for parabolic systems, are characteristic (the hyperplane $t = 0$ is a characteristic) and for the uniqueness of solution an additional condition on the behavior of solutions for $|x| \to \infty$ is necessary.

2.3. Parabolic Boundary Problems. We shall consider the solution of a general parabolic system

$$\mathcal{L}u = f \tag{1.38}$$

in a cylindric domain $Q_T = [0, T] \times \Omega$, where Ω is a domain in $\mathbb{R}^n$ with smooth boundary $\partial\Omega$ and $S_T = [0, T] \times \partial\Omega$ is the side surface of the cylinder Q_T.

For $t = 0$ we give the initial condition

$$C(t, x; D_t, D)u|_{t=0} = \varphi(x) , \quad x \in \Omega , \tag{1.39}$$

and on S_T the boundary condition

$$\mathcal{B}(t, x; D_t, D)u|_{S_T} = g , \tag{1.40}$$

where $\mathcal{B}$ is a matrix differential operator and g is an arbitrary vector given on the surface S_T.

We define the *principal part* $\mathcal{B}_l$ of the matrix $\mathcal{B}$. We denote by $B_{\beta j}$, $\beta = 1, \ldots, m_2$, $j = 1, \ldots, m$, the elements of the matrix $\mathcal{B}$ and by $r_{\beta j}$ the degree in λ of the polynomial $B_{\beta j}(t, x; p\lambda^{2b}, \sigma\lambda)$ (if $B_{\beta j} \equiv 0$, then as $r_{\beta j}$ we can take any integer). We put $\theta_\beta = \max(r_{\beta j} - t_j)$ and then $\theta_\beta + t_j \geq r_{\beta j}$.

The *principal part of the polynomial* $B_{\beta j}$ is the sum $B^0_{\beta j}$ of all its terms satisfying the generalized homogeneity condition

$$B^0_{\beta j}(t, x; p\lambda^{2b}, \sigma\lambda) = \lambda^{\theta_\beta + t_j} B^0_{\beta j}(t, x; p, \sigma) .$$

By $\mathcal{B}_0$ we denote the matrix with entries $B^0_{\beta j}$. Obviously, the choice of the principal part $\mathcal{B}_0$ depends on the choice of t_j so in general $\mathcal{B}_0$ is nonuniquely defined.

We now fix the numbers s_k, t_j, and θ_β. We take any point $(t^0, x^0) \in S_T$ and we introduce in $\mathbb{R}^n$ local coordinates $y_1, \ldots, y_n$ at the point $x^0 \in \partial\Omega$, where the y-axis is taken along the inward normal ν (Fig. 2) and in the half-plane $y_n > 0$ we consider the problem

$$\mathcal{L}_0(t^0, x^0; D_t, D_x)u = 0 , \tag{1.41}$$

$$\mathcal{B}_0(t^0, x^0; D_t, D_x)u|_{y_n=0} = g . \tag{1.42}$$

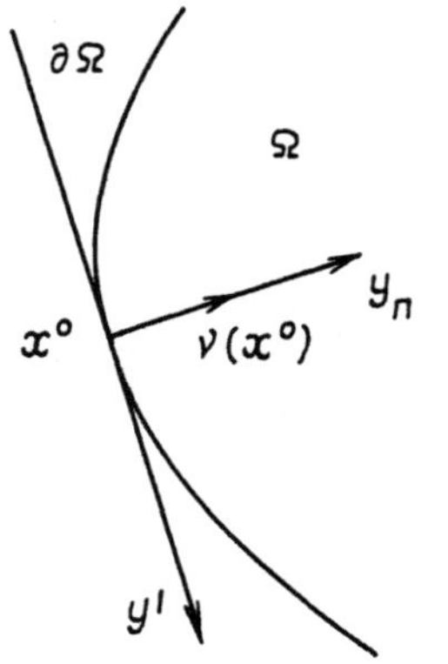

Fig. 2

Following Lopatinskij (1953) we assume that the following fundamental condition is satisfied: for any point $(t^0, x^0) \in S_T$ the problem (1.41), (1.42) can be solved for any smooth vector-function g with compact support in the class of functions having the Fourier transform with respect to the tangent variables y'.

The goal of the next steps is to give the above condition a clear algebraic form. They are based on some standard argument which is repeatedly used in the theory of boundary problems. We shall present it briefly.

We write the problem (1.41), (1.42) in new coordinates

$$\widetilde{\mathcal{L}}(D_t, D_y)u = 0 ,$$

$$\widetilde{\mathcal{B}}(D_t, D_y)u\,|_{y_n=0} = g(t, y') .$$

The operator $\mathcal{L}$ remain of course parabolic with the same constant of parabolicity δ.

Applying the Fourier transform in the tangent variable y' and the Laplace transform in the time variable t (assuming that all derivatives in t equal zero for $t = 0$) we reduce the last problem to the boundary problem for a system of ordinary differential equations on the semi-axis $y_n \geq 0$:

$$\widetilde{\mathcal{L}}\left(p, \sigma'; \frac{1}{\mathrm{i}}\frac{d}{dy_n}\right)\hat{u}(y_n) = 0 ,$$

$$\widetilde{\mathcal{B}}\left(p, \sigma', \frac{1}{\mathrm{i}}\frac{d}{dy_n}\right)\hat{u}(y_n)\,|_{y_n=0} = \hat{g}(p, \sigma') ,$$

$$|\hat{u}(y_n)| \to 0 \quad \text{as} \quad y_n \to \infty ,$$

where $\sigma' = (\sigma_1, \ldots, \sigma_{n-1})$.

And so, we have to consider the following problem

$$\mathcal{P}\left(\frac{1}{\mathrm{i}}\frac{d}{dz}\right)w(z)=0\,, \tag{1.43}$$

$$\mathcal{A}\left(\frac{1}{\mathrm{i}}\frac{d}{dz}\right)w(z)\,|_{z=0}=h\,, \tag{1.44}$$

$$|w(z)|\to 0 \quad \text{as} \quad z\to\infty\,. \tag{1.45}$$

All solutions to (1.43) form a finite-dimensional space E. By E_0 we denote the subspace of E of all solutions to (1.43) satisfying (1.45). Its dimension coincides with the number of zeros of the polynomial $P(\tau)=\det\mathcal{P}(\tau)$ with positive imaginary part (the multiplicities are taken into account). We denote these zeros by τ_s^+ and their number by r^+. We define a polynomial by

$$P^+(\tau)=\prod_{s=1}^{r^+}(\tau-\tau_s^+)\,.$$

We can easily establish that the coordinates of any solution to (1.43) from E_0 satisfy the equation

$$P^+\left(\frac{1}{\mathrm{i}}\frac{d}{dz}\right)v(z)=0\,.$$

Let $w^{(s)}(z)$, $s=1,\dots,r^+$ be a basis of E_0. Then any solution of the system (1.43) from E_0 is given by the formula

$$w(z)=\sum_{s=1}^{r^+}c_s w^{(s)}(z)\,,$$

where c_s are arbitrary constants. We shall find them so that they satisfy the boundary condition (1.44). To this end we consider the following system of algebraic equations

$$\sum_{s=1}^{r^+}\left(\sum_{j=1}^{m}A_{\beta j}\left(\frac{1}{\mathrm{i}}\frac{d}{dz}\right)w_j^{(s)}(z)\,|_{z=0}\right)c_s=h_\beta\,,\quad \beta=1,\dots,m_2\,. \tag{1.46}$$

We introduce the matrix D with entries

$$d_{\beta s}=\sum_{j=1}^{m}A_{\beta j}\left(\frac{1}{\mathrm{i}}\frac{d}{dz}\right)w_j^{(s)}(z)\,|_{z=0}\,.$$

For the system (1.46) to have a unique solution for any h_β it is necessary and sufficient that the following two conditions be fulfilled

1) $m_2=r^+$, (1.47)
2) $\det\mathcal{D}\neq 0$. (1.48)

And so we have found the answers to the questions we are concerned with: on the number of boundary conditions and on the conditions of solvability of the boundary problem we have posed. However, condition (1.48) is not expressed in terms of the original problem and some additional considerations are necessary to obtain such a formulation. This is what we are now turning to.

We employ the *residual method* (*Cauchy method*) of solving (1.43). By this method the general solution is given by the formula

$$w(z) = \int_{\gamma} \frac{\hat{\mathcal{P}}(\tau)Q(\tau)\mathrm{e}^{\mathrm{i}z\tau}}{P(\tau)}\,d\tau\,,$$

where γ is an arbitrary closed contour in the complex z-plane containing all zeros of $P(\tau)$ and $Q(\tau)$ is a vector whose coordinates are arbitrary polynomials of degrees less than the degree of $P(\tau)$.

Any vector from E_0 can be represented by the analogous formula

$$w(z) = \int_{\gamma^+} \frac{\hat{\mathcal{P}}(\tau)Q(\tau)\mathrm{e}^{\mathrm{i}z\tau})}{P^+(\tau)}\,d\tau\,, \tag{1.49}$$

where γ^+ is an arbitrary closed contour in the complex z-plane containing all zeros of $P^+(\tau)$ and the coordinates of the vector $Q(\tau)$ are arbitrary polynomials of degree $r^+ - 1$. In particular, the vector-functions $w^{(s)}(z)$ of a basis of E_0 can be written using (1.49) with some $Q^{(s)}(\tau)$

$$w^{(s)}(z) = \int_{\gamma^+} \frac{\hat{\mathcal{P}}(\tau)Q^{(s)}(\tau)\mathrm{e}^{\mathrm{i}z\tau}}{P^+(\tau)}\,d\tau\,. \tag{1.50}$$

Since $\{w^{(s)}(z)\}_{s=1,\ldots,\gamma^+}$ form a basis of E_0, for any admissible vector $Q(\tau)$ we can find constants $a_1, \ldots, a_{r^+}$ such that for any $k = 1, \ldots, m$ we have

$$\int_{\gamma^+} \frac{\sum_{j=1}^m \hat{P}_{kj}(\tau)Q_j(\tau)\mathrm{e}^{\mathrm{i}z\tau}}{P^+(\tau)}\,d\tau = \sum_{s=1}^{r^+} a_s w_k^{(s)}$$

$$= \int_{\gamma^+} \frac{\sum_{j=1}^m \hat{P}_{kj}(\tau)\sum_{s=1}^{r^+} a_s Q_j^{(s)}(\tau)\mathrm{e}^{\mathrm{i}z\tau}}{P^+(\tau)}\,d\tau\,. \tag{1.51}$$

Using (1.50) we find that the elements of the matrix $\mathcal{D}$ take the form

$$d_{\beta s} = \int_{\gamma^+} \frac{\sum_{k,j=1}^m A_{\beta k}(\tau)\hat{P}_{kj}(\tau)Q_j^{(s)}(\tau)}{P^+(\tau)}\,d\tau$$

$$= \int_{\gamma^+} \frac{\sum_{j=1}^m R_{\beta j}(\tau)Q_j^{(s)}(\tau)}{P^+(\tau)}\,d\tau\,, \tag{1.52}$$

where $R_{\beta j} = \sum_{k=1}^m A_{\beta k}(\tau)\hat{P}_{kj}(\tau)$.

We proceed to the final argument. It is based on the fact that (1.48) holds if and only if the homogeneous algebraic system

$$\sum_{\beta=1}^{r^+} e_\beta d_{\beta s} = 0 , \quad s = 1, \dots, r^+ , \tag{1.53}$$

has only zero solution.

We define the functions $R_j(\tau) = \sum_{\beta=1}^{r^+} e_\beta R_{\beta j}(\tau)$. From (1.52) and (1.53) it follows that

$$\int_{\gamma^+} \frac{\sum_{j=1}^m R_j(\tau) Q_j^{(s)}(\tau)}{P^+(\tau)} \, d\tau = 0 , \quad s = 1, \dots, r^+ , \tag{1.54}$$

which is equivalent to the fact that the equality

$$\int_{\gamma^+} \frac{\sum_{j=1}^m R_j(\tau) Q_j^{(s)}(\tau)}{P^+(\tau)} \, d\tau = 0 \tag{1.55}$$

is satisfied for *any* polynomial $Q_j(\tau)$ of degree $r^+ - 1$. The last assertion follows from (1.51) by applying the operator $\sum_{\beta=1}^{r^+} e_\beta A_{\beta k} \left(\frac{1}{i} \frac{d}{dz}\right)$, summing the result over k from 1 to m, putting $z = 0$, and using (1.54).

Since the polynomials $Q_j(\tau)$ are arbitrary, (1.55) is equivalent to the fact that for any j and any polynomial $Q(\tau)$ of degree $r^+ - 1$ we have

$$\int_{\gamma^+} \frac{R_j(\tau) Q(\tau)}{P^+(\tau)} \, d\tau = 0 ,$$

which, obviously, is equivalent to the equality

$$\sum_{\beta=1}^{r^+} e_\beta R_{\beta j}(\tau) = R_j(\tau) = p_j(\tau) P^+(\tau) , \quad j = 1, \dots, m , \tag{1.56}$$

where $p_j(\tau)$ are some polynomials. Thus we have proved that condition (1.48) is equivalent to the fact that the equality $e_\beta = 0$ follows from (1.56).

In this way we arrive at the conclusion that the solvability condition for the problem (1.43)–(1.45) can be formulated in terms of the problem itself in the following way:

The rows of the matrix

$$\mathcal{R}(\tau) = \mathcal{A}(\tau) \hat{\mathcal{P}}(\tau)$$

are linearly independent modulo polynomial $P^+(\tau)$.

We return to the problem (1.38)–(1.40). To determine the number of boundary conditions we have to find the number r^+. We have a lemma:

Lemma. *Suppose that the polynomial* $L_0(p, \sigma', \tau)$ *is* weighted-homogeneous and parabolic in the sense of Petrovskij, *i.e.*

$$L_0(p\lambda^{2b}, \sigma'\lambda, \tau\lambda) = \lambda^{2br} L_0(p, \sigma', \tau)$$

and for any real (σ', τ) all p-zeros of this polynomial satisfy

$$\operatorname{Re} p(\sigma', \tau) \leq -\delta(|\sigma'|^2 + \tau^2)^b .$$

Then for any real $\sigma_1, \ldots, \sigma_{n-1}$ and any complex p that satisfy the conditions

$$\operatorname{Re} p > -\delta_1 |\sigma'|^{2b} , \quad 0 < \delta_1 < \delta , \quad |p| + |\sigma'|^{2b} > 0 ,$$

the polynomial $L_0(p, \sigma', \tau)$ has br τ-zeros with positive imaginary part and br τ-zeros with negative.

And so for general parabolic systems we ought to impose $r^+ = br$ boundary conditions, i.e. the matrix $\mathcal{B}$ of boundary conditions has dimension $br \times m$. We now consider the problem (1.41), (1.42). An arbitrary vector $\sigma \in \mathbb{R}^n$ can be uniquely written in the form $\sigma = \xi + \tau\nu$, where ξ is a vector lying in the plane tangent to the surface $\partial\Omega$ at x^0, ν is the unit vector on the inward normal to $\partial\Omega$ at x^0, and τ is a real parameter. We denote by Γ^x the set of all points (p, ξ), where $p \in \mathbb{C}$ and ξ is tangent to $\partial\Omega$ at x^0, satisfying the conditions $|p| + |\xi|^{2b} > 0$, $\operatorname{Re} p \geq -\delta_1 |\xi|^{2b}$ with some $\delta_1 \in (0, \delta)$, where δ is the constant from the parabolicity condition of the system. We consider the τ-polynomial $L_0(t^0, x^0; p, \xi + \tau\nu)$. For $(p, \xi) \in \Gamma^{x^0}$, by the lemma L_0 has br τ -zeros with positive imaginary part (we denote them by $\tau_s^+(t^0, x^0; p, \xi)$) and br with negative. We put

$$L_0^+(t^0, x^0; p, \xi, \tau) = \prod_{s=1}^{br} (\tau - \tau_s^+(t^0, x^0; p, \xi)) .$$

The *solvability condition (Lopatinskij condition)* that was obtained above for the boundary problem can be given a concrete form in the case in question as follows:

The Condition for the Complementary Boundary Problem (1.40) for the System (1.38).

For any $(t^0, x^0) \in S_T$ and $(p, \xi) \in \Gamma^{x^0}$ the columns of the matrix

$$\mathcal{R}(t^0, x^0; p, \xi, \tau) = \mathcal{B}_0(t^0, x^0; p, \xi + \tau\nu) \hat{\mathcal{L}}_0(t^0, x^0; p, \xi + \tau\nu) , \tag{1.57}$$

regarded as τ-polynomials are linearly independent modulo polynomial

$$L_0^+(t^0, x^0; p, \xi, \tau).$$

This important condition can be also given another form. Consider a matrix $\mathcal{R}'(t^0, x^0; p, \xi, \tau)$ whose entries are the remainders of division of the elements of $\mathcal{R}$ by L_0^+. The elements $R'_{\beta j}$ of $\mathcal{R}'$ can be written in the form

$$R'_{\beta j}(t^0, x^0; p, \xi, \tau) = \sum_{l=1}^{br} d_{\beta j}^{(l)}(t^0, x^0; p, \xi) \tau^{l-1} , \quad \beta, j = 1, \ldots, br ,$$

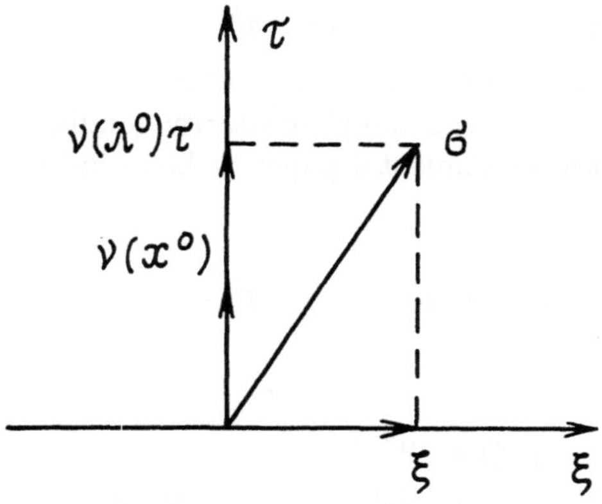

Fig. 3

and we consider the matrix $\mathcal{D}$ of dimension $br \times (br)^2$ composed of $d^{(l)}_{\beta j}$, $\beta, j = 1, \ldots, br$.

The complementarity condition is equivalent to the following hypothesis: for any $(t^0, x^0) \in S_T$ and $(p, \xi) \in \Gamma^{x^0}$ the rank of the matrix $\mathcal{D}(t^0, x^0; p, \xi)$ is equal to br or

$$\sum_j |\Delta_j(t^0, x^0; p, \xi)| > 0 \,, \tag{1.58}$$

where Δ_j are all possible minors of order br of the matrix $\mathcal{D}$.

Definition 1.9. The problem (1.38), (1.39), (1.40) is called a *general parabolic mixed (boundary) problem* if

1) the system (1.38) is parabolic in the sense of Solonnikov,
2) the initial problem (1.39) satisfies the complementarity condition,
3) the boundary problem (1.40) satisfies the complementarity condition.

In particular, for the systems parabolic in the sense of Petrovskij, the parabolicity of the mixed problem means that for $t = 0$ we have to prescribe the Cauchy data and require that the complementarity condition for boundary conditions be satisfied.

2.4. Particular Cases. Examples. The boundary problems for the systems (1.8) parabolic in the sense of Petrovskij with derivatives in time t of the first order deserve a special attention. In the matrix notation this problem takes the form:

$$D_t u = \sum_{|\alpha| \le 2b} a_\alpha(t, x) D^\alpha u + f(t, x) \equiv A(t, x; D)u + f(t, x) \,, \tag{1.59}$$

$$u|_{t=0} = \varphi(x) \,, \quad x \in \Omega \,, \tag{1.60}$$

$$\mathcal{B}(t, x; D_t, D)u|_{S_T} = g(t, x)\ , \quad (t, x) \in S_T\ , \tag{1.61}$$

where the matrix operator $\mathcal{B}$ has dimension $bm \times m$.

In the case in question the complementarity condition can be written in the form given in the fundamental paper of Lopatinskij (1953), (Ejdel'man 1964):

$$\operatorname{rank} \int_{\gamma^+} \mathcal{B}_0(t^0, x^0; p, \xi + \tau\nu)(pI - A_0(t^0, x^0, \xi + \tau\nu))^{-1}(I, \tau I, \ldots, \tau^{2b-1}I)d\tau$$
$$= bm \tag{1.62}$$

for any $(t^0, x^0) \in S_T$, $(p, \xi) \in \Gamma^{x^0}$.

We shall now write the *complementarity condition in the case of one equation* $(m = 1)$ *and the half-space* $x_n > 0$. To this end we need to perform the following operations:

1. Write down the polynomial $L_0(p, \sigma', \tau) = -\sum_{|\alpha|=2b} a_\alpha \sigma'^{\alpha'} \tau^{\alpha_n} + p$.
2. Find its τ-zeros $\tau_s^+(p, \sigma')$, $s = 1, \ldots, b$, with positive imaginary parts and write down the polynomial $L_0^+(p, \sigma', \tau) = \prod_{s=1}^b (\tau - \tau_s^+(p, \sigma'))$.
3. Write down the τ-polynomials $b_\beta^0(p, \sigma', \tau)$, $\beta = 1, \ldots, b$, corresponding to the principal parts of the boundary operators b_β.
4. Find the remainders $R_\beta(p, \sigma', \tau) = \sum_{l=1}^b d_\beta^{(l)}(p, \sigma')\tau^{l-1}$, $\beta = 1, \ldots, b$, of the division of τ-polynomials $b_\beta^0(p, \sigma', \tau)$ by τ-polynomial $L_\beta^+(p, \sigma', \tau)$.
5. Find the determinant $\Delta(p, \sigma) = \det\{(d_\beta^{(l)}(p, \sigma'))_{\beta, l=1}^b\}$.
6. Verify whether the complementarity condition is satisfied: for any

$$(p, \sigma') \in \Gamma^0$$
$$= \{(p, \sigma') : p \in \mathbb{C}, \sigma' \in \mathbb{R}^{n-1}; \operatorname{Re} p \geq -\delta_1 |\sigma'|^{2b}, |p| + |\sigma'|^{2b} > 0\}\ ,$$

$0 < \delta_1 < \delta$, the determinant $\Delta(p, \sigma')$ is nonzero.

Example 1.7. We shall analyse the validity of the complementarity conditions for model problems for the heat conduction theory

$$\frac{\partial u}{\partial t} = \Delta u\ , \quad u|_{t=0} = \varphi(x)\ ,$$

1) *Dirichlet problem:* $u|_{x_n=0} = g_1(t, x')$;
2) *Neuman problem:* $D_{x_n} u|_{x_n=0} = g_2(t, x')$;
3) *The condition with directional derivative:* $\sum_{j=1}^n b_j D_{x_j} u|_{x_n=0} = g_3(t, x')$;
4) *Tikhonov problem:*

$$\sum_{2b\alpha_0 + |\alpha'| + \alpha_n \leq r} b_{\alpha\alpha'\alpha_n} D_t^{\alpha_0} D_{x'}^{\alpha'} D_{x_n}^{\alpha_n} u|_{x_n=0} = g_4(t, x')\ ;$$

5) $D_t u|_{x_n=0} = g_5(t, x')$.

In this case $\mathcal{L}_0(p,\sigma',\tau) = p + a^2(\sigma'^2+\tau^2)$, $\delta = a^2$, $\tau_1^+ = \mathrm{i}\sqrt{a^{-2}p+|\sigma'|^2}$, where for the two-valued algebraic function we take the branch with positive real part,

$$\Gamma^0 = \{(p,\sigma') : \sigma' \in \mathbb{R}^{n-1}, \mathrm{Re}\, p > -(a^2-\varepsilon)|\sigma'|^2, 0<\varepsilon<a^2\}\,,$$

$B_0(p,\sigma',\tau)$ are respectively equal to

$$1\,,\tau\,,\sum_{j=1}^{n-1} b_j\sigma_j + b_n\tau\,, \quad \sum_{2b\alpha_0+|\alpha'|+\alpha_n=r} b_{\alpha\alpha'\alpha_n} p^{\alpha_0}\sigma'^{\alpha'}\tau^{\alpha_n}\,, p,$$

and $R'(p,\sigma',\tau)$ are equal to

$$1\,, \mathrm{i}\sqrt{a^{-2}p+|\sigma'|^2}\,, \sum_{j=1}^{n-1} b_j\sigma_j + \mathrm{i}b_n\sqrt{a^{-2}p+|\sigma'|^2}\,,$$

$$\sum_{2b\alpha_0+|\alpha'|+\alpha_n=r} b_{\alpha_0\alpha'\alpha_n} p^{\alpha_0}\sigma'^{\alpha'}(\mathrm{i}\sqrt{a^{-2}p+|\sigma'|^2})^{\alpha_n} = R_4'(p,\sigma')\,, p\,.$$

Therefore, for the Dirichlet and Neuman problems the complementarity conditions are satisfied, for the problem with directional derivative this condition is satisfied if $b_n \neq 0$ and is not satisfied if $b_n = 0$ (when the direction of differentiation lies in the tangent plane $x_n = 0$), for the Tikhonov problem complementarity means that $R_4'(p,\sigma') \neq 0$ in Γ^0, and finally for the problem 5) it is satisfied for $n = 1$ since there are no conditions on p and for $n > 1$ it is clearly not satisfied (for $p = 0$).

Example 1.8. A simple system of the theory of heat and mass exchange has the form

$$D_t u - M\Delta u = f\,, \tag{1.63}$$

where M is the square matrix of rank 2 with constant real entries and characteristic numbers λ_1 and λ_2 such that $0 < \lambda_1 < \lambda_2$. Such a system is parabolic in the sense of Petrovskij with the parabolicity constant $\delta = \lambda_1$. By means of a nondegenerate linear transformation that reduces the matrix to the diagonal form, the system (1.63) can be converted into a system consisting of two separate heat equations. We obtain

$$D_t u_j - \lambda_j \Delta u_j = f_j\,, \quad j = 1, 2\,. \tag{1.64}$$

In the domain $\Pi^+ = \{(t,x',x_n) : t > 0, x \in \mathbb{R}^{n-1}, x_n > 0\}$ we consider the problem of finding a solution $u(t,x) = (u_1(t,x), u_2(t,x))$ with the initial conditions

$$u_j|_{t=0} = \varphi_j\,, \quad j = 1, 2\,, \tag{1.65}$$

and the boundary conditions (arising form applications)

$$\sum_{k=1}^{2} \alpha_k D_{x_k} u|_{x_n=0} = g_1\,, \quad \sum_{k=1}^{2} \beta_k u_k|_{x_n=0} = g_2\,, \tag{1.66}$$

where α_k and β_k, $k = 1, 2$, are given real numbers.

We turn to finding complementarity conditions for the problem (1.64)–(1.66). We successively have

$$\mathcal{L}_0(p,\sigma',\tau) = \begin{pmatrix} p+\lambda_1(|\sigma'|^2+\tau^2) & 0 \\ 0 & p+\lambda_2(|\sigma'|^2+\tau^2) \end{pmatrix} ;$$

$$\mathcal{B}_0(p,\sigma',\tau) = \begin{pmatrix} \alpha_1\tau & \alpha_2\tau \\ \beta_1 & \beta_2 \end{pmatrix} ,$$

$$\tau_j^+(p,\sigma') = \mathrm{i}\sqrt{(p/\lambda_j)+|\sigma'|^2} ; \quad L_0^+(p,\sigma',\tau) = (\tau-\tau_1^+)(\tau-\tau_2^+) ;$$

$$\begin{aligned} \mathcal{R}(p,\sigma',\tau) &= \mathcal{B}_0(p,\sigma',\tau)\hat{\mathcal{L}}_0(p,\sigma',\tau) \\ &= \begin{pmatrix} \alpha_1\lambda_2\tau(\tau^2-\tau_2^{+2}) & \alpha_2\lambda_1\tau(\tau^2-\tau_1^{+2}) \\ \beta_1\lambda_2(\tau^2-\tau_2^{+2}) & \beta_2\lambda_1(\tau^2-\tau_1^{+2}) \end{pmatrix} \cdot \end{aligned}$$

The matrix $\mathcal{R}'(p,\sigma',\tau)$ composed of the remainders from the division of the elements of $\mathcal{R}(p,\sigma',\tau)$ by $L_0^+(p,\sigma',\tau)$ has the form

$$\mathcal{R}'(p,\sigma',\tau) = (\tau_1^+ + \tau_2^+)\begin{pmatrix} \alpha_1\lambda_2\tau_1^+(\tau-\tau_2^+) & \alpha_2\lambda_1\tau_2^+(\tau-\tau_1^+) \\ \beta_1\lambda_2(\tau-\tau_2^+) & \beta_2\lambda_1(\tau-\tau_1^+) \end{pmatrix} .$$

The complementarity condition reads

$$\operatorname{rank}\left((\tau_1^+ + \tau_2^+)\begin{pmatrix} \alpha_1\lambda_2\tau_1^+ & \alpha_2\lambda_1\tau_2^+ & -\alpha_1\lambda_2\tau_1^+\tau_2^+ & -\alpha_2\lambda_1\tau_1^+\tau_2^+ \\ \beta_1\lambda_2 & \beta_2\lambda_1 & -\beta_1\lambda_2\tau_2^+ & -\beta_2\lambda_1\tau_1^+ \end{pmatrix}\right) = 2 ,$$

that is

$$\begin{vmatrix} \alpha_1\tau_1^+ & \alpha_2\tau_2^+ \\ \beta_1 & \beta_2 \end{vmatrix} \neq 0$$

or

$$\alpha_1\beta_2\sqrt{(p/\lambda_1)+|\sigma'|^2} - \alpha_2\beta_1\sqrt{(p/\lambda_2)+|\sigma'|^2} \neq 0 , \quad (p,\sigma') \in \Gamma^0 \qquad (1.67)$$

Condition (1.67) is obviously satisfied if $\alpha_1\beta_2 = 0$ or $\alpha_2\beta_1 = 0$ but not simultaneously. The analysis of (1.67) shows that in the case $\alpha_1\alpha_2\beta_1\beta_2 \neq 0$ the complementarity condition is satisfied if

$$\frac{\alpha_2\beta_1}{\alpha_1\beta_2} \overline{\in} \left[\sqrt{\gamma\frac{1-\delta_1/\lambda_1}{\gamma-\delta_1/\lambda_1}}, \sqrt{\gamma}\right) ,$$

where $\gamma = \lambda_2/\lambda_1$ and δ_1 is the number from the definition of Γ^0.

We turn our attention to the fact that despite the simplicity in appearance of the example, a difficult and important problem of finding conditions of physical constants such that the boundary problem is well-posed can be explicitly solved. In a general case the complementarity condition concerned with the parameters $(p,\sigma') \in \Gamma^{x^0}$ is, putting it mildly, difficult to verify.

Example 1.9. We consider the *Dirichlet problem* for an equation of the form (1.59)

$$D_{x_n}^{\beta-1} u|_{x_n=0} = g_\beta(t,x') , \quad \beta = 1,\ldots,b .$$

Here $b_\beta^0(p,\sigma',\tau) \equiv \tau^{\beta-1}$, $\beta = 1,\ldots,b$. Since $b_\beta^0(p,\sigma',\tau) \equiv R_\beta^0(p,\sigma',\tau) = \tau^{\beta-1}$ and these functions are linearly independent, the complementarity condition is satisfied for any equation of the form (1.59) parabolic in the sense of Petrovskij.

Example 1.10. We consider the *Dirichlet problem for the parabolic system* (1.59) with one spatial coordinate

$$D_t u = A_0 D_x^{2b} u \; ; \quad u|_{t=0} = \varphi \; ; \quad D_x^{\beta-1} u|_{x=0} = g_\beta \, , \quad \beta = 1,\ldots,b .$$

The verification of the complementarity condition reduces to proving that the problem

$$p\hat{u} = A_0 D_x^{2b}\hat{u} \; ; \quad \hat{u}^{(\beta-1)}(0) = 0 \, , \quad \beta = 1,\ldots,b \; ;$$
$$|\hat{u}(p,x)| \longrightarrow 0 \quad \text{as} \quad x \longrightarrow \infty$$

has trivial solution only. By a linear nonsingular transform, preserving the boundary Dirichlet conditions and the behavior at infinity, the matrix of the new problem is Jordanian:

$$\hat{v}(x) = T\hat{u}(x) \; ; \quad p\hat{v} = A_0' D_x^{2b}\hat{v} \; ; \quad \hat{v}^{(\beta-1)}(0) = 0 \, , \quad \beta = 1,\ldots,b \; ;$$
$$|\hat{v}(p,x)| \longrightarrow 0 \quad \text{as} \quad x \longrightarrow \infty .$$

We write down the equations of the system:

$$p\hat{v}_1 = \lambda_1 D_x^{2b}\hat{v}_1 \, , \quad p\hat{v}_2 = \lambda_2 D_x^{2b}\hat{v}_2 + \hat{v}_1, \ldots .$$

Applying successively the complementarity condition for a single parabolic equation we obtain what is required.

Example 1.11. We consider the Dirichlet problem for a system of the form (1.59) of two equations of the second order with two spatial variables with the matrix $A_0(\sigma,\tau)$ of the form:

$$A_0(\sigma,\tau) = \begin{pmatrix} a\sigma^2 - (\tau - \mathrm{i}\sqrt{1+a}\sigma)^2 & -4(1+a)\sigma^2 \\ (\tau - \mathrm{i}\sqrt{1+a}\sigma)^2 & a\sigma^2 - (\tau + \mathrm{i}\sqrt{1+a}\sigma)^2 - 4(1+a)\sigma^2 \end{pmatrix} .$$

It can be directly verified that $\det(pI - A_0(\sigma,\tau)) = (p+\sigma^2+\tau^2)^2$ and that for $p = a\sigma^2$, $\sigma > 0$, the system of ordinary differential equations

$$\begin{cases} \{p - a\sigma^2 + (D_{x_2} - \mathrm{i}\sqrt{1+a}\sigma)^2\}\hat{u}_1 + 4(1+a)\sigma^2\hat{u}_2 = 0 \\ \\ -(D_{x_2} - \mathrm{i}\sqrt{1+a}\sigma)^2\hat{u}_1 \\ \qquad + \{p - a\sigma^2 + 4(1+a)\sigma^2 + (D_{x_2} + \mathrm{i}\sqrt{1+a}\sigma)^2\}\hat{u}_2 = 0 \end{cases}$$

has nontrivial solution $\hat{u}(x_n) = (x_2 \exp\{-\sqrt{1+a\sigma x_2}\}, 0)$ satisfying homogeneous Dirichlet condition. Thus, if the number n of spatial variables is greater than one, then there exist systems parabolic in the sense of Petrovskij for which the first boundary value problem is ill-posed.

If the Dirichlet problem for the system (1.59), parabolic in the sense of Petrovskij, satisfies the complementarity condition, then the matrix (polynomial in τ)

$$\mathcal{L}_0(t^0, x^0; p, \xi, \tau) = pI - A_0(t^0, x^0, \xi + \tau\nu)$$

admits the following factorization

$$\mathcal{L}_0(t^0, x^0; p, \xi, \tau) = \mathcal{L}_0^-(t^0, x^0; p, \xi, \tau)\mathcal{L}_0^+(t^0, x^0; p, \xi, \tau) \,, \tag{1.68}$$

where $\mathcal{L}_0^+$ and $\mathcal{L}_0^-$ are matrices polynomial in τ such that the τ-zeros of $\det\mathcal{L}_0^+$ coincide with the τ-zeros of $\mathcal{L}_0$ with $\mathrm{Im}\tau > 0$ and the τ-zeros of $\det\mathcal{L}_0^-$ with the τ-zeros of $\mathcal{L}_0$ with $\mathrm{Im}\tau < 0$.

The factorization (1.68) allows us to give the complementarity condition in the form (1.62) a more concrete shape:

for any $(t^0, x^0) \in S_T$, $(p, \xi) \in \Gamma^{x^0}$

$$\det \int_{\gamma^+} \mathcal{B}_0(t^0, x^0; p, \xi + \nu\tau)(\mathcal{L}_0^+(t^0, x^0; p, \xi, \tau))^{-1}(I, \tau I, \ldots, \tau^{b-1}I)d\tau \neq 0 \,. \tag{1.69}$$

Example 1.12. We consider a system parabolic in the sense of Petrovskij with two spatial coordinates, consisting of two separate heat equations

$$\mathcal{L}_0(p, \sigma, \tau) = \begin{pmatrix} p + \sigma^2 + \tau^2 & 0 \\ 0 & p + \sigma^2 + \tau^2 \end{pmatrix} .$$

We have already met this system in Example 1.4. Here $b = 1$, $m = r = 2$. We can treat it as parabolic in the sense of Solonnikov with $s_1 = -c$, $s_2 = 0$, $t_1 = 2 + c$, $t_2 = 2$ $(c > 0)$; $s_1 = 0$, $s_2 = -c$, $t_1 = 2$, $t_2 = 2 + c$ $(c \geq 0)$.

Let $\mathcal{B}(p, \sigma, \tau) = \begin{pmatrix} 1 & 0 \\ \sigma & 1 \end{pmatrix}$. We can indicate the numbers θ_1 and θ_2 for this operator (the weights of boundary conditions): if $t_1 = 2 + c$, $t_2 = 2$ $(c > 0)$, then $\theta_1 = -2 - c$, $\theta_2 = -2$, and if $t_1 = 2$, $t_2 = 2 + c$ $(c > 0)$, then $\theta_1 = -2$, $\theta_2 = -1$.

In the former case $\theta_1 + t_1 = 0$, $\theta_1 + t_2 = -c$, $\theta_2 + t_1 = c$, $\theta_2 + t_2 = 0$ and therefore

$$\mathcal{B}_0^{(1)}(p, \sigma, \tau) = \begin{pmatrix} 1 & 0 \\ \sigma & 1 \end{pmatrix} \quad \text{for } c = 1 \,, \quad \mathcal{B}_0^{(2)}(p, \sigma, \tau) = \begin{pmatrix} 1 & 0 \\ 0 & 1 \end{pmatrix} \quad \text{for } c > 1 \,.$$

In the latter case $\theta_1 + t_1 = 0$, $\theta_1 + t_2 = c$, $\theta_2 + t_1 = 1$, $\theta_2 + t_2 = 1 + c$ $(c \geq 0)$. Therefore

$$\mathcal{B}_0^{(3)}(p, \sigma, \tau) = \begin{pmatrix} 1 & 0 \\ \sigma & 0 \end{pmatrix} ,$$

the rows of this matrix are linearly dependent, and so the rows of the matrix $\mathcal{B}_0^{(3)}\hat{\mathcal{L}}_0$ are linearly dependent. Thus the complementarity condition is not satisfied in this case. In particular, it is not satisfied if we consider $\mathcal{L}_0$ as an operator parabolic in the sense of Petrovskij ($s_1 = s_2 = 0$, $t_1 = t_2 = 2$). Therefore, having introduced the structure of general parabolic systems we can find new interesting well-posed boundary-value problems also for simple systems which are important from the practical point of view.

2.5. Parabolic Conjugation Problem. We move to the presentation of some simple generalizations of the linear parabolic boundary-value problems defined above. The first is, so called, parabolic conjugation problem. For simplicity we shall consider the case of systems parabolic in the sense of Petrovskij of the first order in time variable t. The general case is quite analogous.

In $\mathbb{R}^n$ we consider a domain Ω^0 with smooth boundary $\partial\Omega^0$. The domain Ω^0 is divided into two subdomains Ω^1 and Ω^2 by a hypersurface $\partial\Omega^1$, $\overline{\Omega}^1 \cup \Omega^2 = \Omega^0$, $\Omega^1 \cap \Omega^2 = \emptyset$, $\partial\Omega^1 \cap \partial\Omega^0 = \emptyset$, $\partial\Omega^2 = \partial\Omega^0 \cup \partial\Omega^1$.

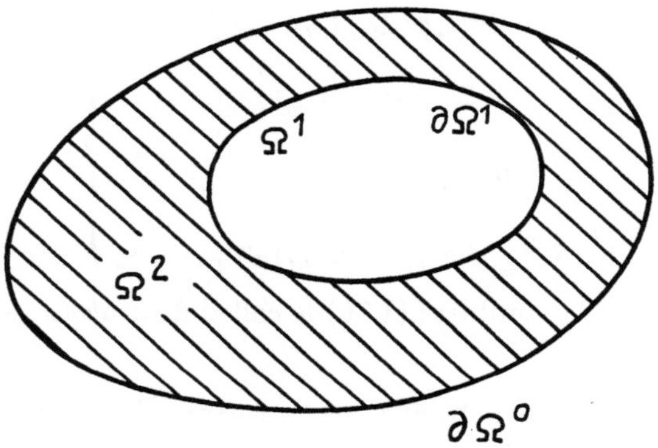

Fig. 4

We denote $Q^\mu = (0,T] \times \Omega^\mu$, $S^\mu = (0,T] \times \partial\Omega^\mu$, $\mu = 0,1,2$.

We formulate the *conjugation problem.* We need to find a vector-function $u(t,x)$ defined in Q^0 such that

$$u(t,x) = \begin{cases} u^1(t,x) & (t,x) \in Q^1\,, \\ u^2(t,x) & (t,x) \in Q^2\,. \end{cases} \tag{1.71}$$

The vector-function $u^\mu = (u_1^\mu, \dots, u_{m_\mu}^\mu)$ is a solution to the system

$$\sum_{j=1}^{m_\mu} L_{kj}^\mu(t,x;D_t,D)u_j^\mu(t,x) = f_k^\mu(t,x)\,, \quad k = 1,\dots,m_\mu\,, \tag{1.72}$$

in the domain Q^μ, $\mu = 1, 2$, and for $t = 0$ it satisfies the initial conditions

$$u_i^\mu(t,x)\,|_{t=0} = \varphi_i^\mu(x)\,, \qquad x \in \Omega^\mu\,, \quad i = 1, \dots, m_\mu\,. \tag{1.73}$$

On the surface S^0 the vector-function $u^2(t,x)$ satisfies the boundary conditions

$$\sum_{j=1}^{m_2} B_{\beta j}(t,x;D_t,D)u_j^2(t,x)\,|_{S^0} = g_\beta^2(t,x)\,, \quad \beta = 1, \dots, m_2\,. \tag{1.74}$$

Finally, on the common boundary S^1 of the cylinders Q^1 and Q^2 we impose the *conjugation conditions*:

$$\left.\left(\sum_{j=1}^{m_1} B_{\beta j}^1(t,x;D_t,D)u_j^1 + \sum_{j=1}^{m_2} B_{\beta j}(t,x;D_t,D)u_j^2\right)\right|_{S^1} = g_\beta^1(t,x)\,, \tag{1.75}$$

$\beta = 1, \dots, m_1 + m_2$. The sum $[v^1(t,x) + v^2(t,x)]\,|_{S^1}$ in (1.75) is understood as the sum of the limit values as we approach S^1 from Q^1 and Q^2 respectively.

We write the problem (1.72)-(1.75) in a more compact form:

$$\left.\begin{aligned} \mathcal{L}^\mu(t,x;D_t,D)u^\mu &= f^\mu\,, \\ u^\mu\,|_{t=0} &= \varphi^\mu\,, \\ \mathcal{B}(t,x;D_t,D)u^2\,|_{S^0} &= g^2\,, \\ \textstyle\sum_{\mu=1}^2 \mathcal{B}^\mu(t,x;D_t,D)u^\mu\,|_{S^1} &= g^1\,. \end{aligned}\right\} \tag{1.76}$$

Definition 1.10. The problem (1.76) is called a *parabolic conjugation problem* if

1) the systems (1.72) are parabolic in the sense of Petrovskij of the first order in t (with the common parabolicity constant δ) in Q^1 and Q^2 respectively,
2) the boundary conditions (1.74) satisfy the complementarity condition with respect to the operator $\mathcal{L}^2$,
3) the boundary operators $(\mathcal{B}^1, \mathcal{B}^2)$ satisfy the compatible complementarity condition (the condition of compatible covering) with respect to the operators $(\mathcal{L}^1, \mathcal{L}^2)$.

A new notion appears in the above definition, the *condition of compatible covering*, and we turn to the precise definition.

We denote by $\mathcal{L}_0^\mu$, $\mathcal{B}_0^\mu$ the principal parts of the corresponding operators and we put $L_0^\mu(t,x;p,\sigma) = \det\mathcal{L}_0^\mu(t,x;p,\sigma)$, $\mu = 1, 2$. Let (t^0, x^0) be an arbitrary point of S^1, let ν^1 be a vector inward normal to $\partial\Omega^1$ relative to Ω^1, and let ν^2 be the same but relative to Ω^2. We write down the factorizing identities:

$$L_0^\mu(t,x;p,\xi + \tau\nu^\mu) = L_0^{\mu+}(t,x;p,\xi,\tau)L_0^{\mu-}(t,x;p,\xi,\tau)\,,$$

where the τ-zeros of $L_0^{\mu+}$ and $L_0^{\mu-}$ coincide with the τ-zeros of L_0^μ lying respectively in the upper and lower half-planes of the complex τ-plane, $\Gamma_1^{x^0} = \{(p,\xi) : p \in \mathbb{C}, \xi$ is an arbitrary vector in the plane tangent to $\partial\Omega^1$ at x^0, $\mathrm{Re}\, p > -\delta_1|\xi|^{2b}\}$.

Complementarity Condition of the Conjugation Problem (compatible covering condition). The operators $\{\mathcal{B}^1, \mathcal{B}^2\}$ satisfy the *compatible covering condition* with respect to the operators $\{\mathcal{L}^1, \mathcal{L}^2\}$ if for any $(t^0, x^0) \in S^1$ and any $(p,\xi) \in \Gamma_1^{x^0}$ the remainders of the division of $\mathcal{B}_0^\mu(t^0, x^0; p, \xi + \nu^\mu\tau)\hat{\mathcal{L}}_0^\mu(t^0, x^0; p, \xi + \nu^\mu\tau)$ by $L_0^{\mu+}(t^0, x^0; p, \xi, \tau)$, $\mu = 1, 2$, form a matrix

$$(\mathcal{R}^1(t^0, x^0; p, \xi, \tau), \mathcal{R}^2(t^0, x^0; p, \xi, \tau))$$

whose rows, as τ-polynomials, are linearly independent.

We shall briefly explain how the compatible covering condition follows from the usual complementarity condition. We consider a model conjugation problem where on both sides of the hyperplane we have systems of differential equations

$$\mathcal{L}_0^1 u^1 = f^1\,, \quad x_n > 0\,; \qquad \mathcal{L}_0^2 u^2 = f^2\,, \quad x_n < 0\,, \tag{1.77}$$

zero initial conditions

$$u^1\,|_{t=0} = 0\,, \quad x_n > 0\,; \qquad u^2\,|_{t=0} = 0\,, \quad x_n < 0\,, \tag{1.78}$$

and conjugation conditions

$$\mathcal{B}_0^1(D_t, D_{x'}, D_{x_n})u^1\,|_{x_n=+0} + \mathcal{B}_0^2(D_t, D_{x'}, D_{x_n})u^2\,|_{x_n=-0} = g^1(t, x')\,. \tag{1.79}$$

By means of the reflection $u^2(t, x', x_n) = u^2(t, x', -y_n) \equiv \widetilde{u}^2(t, x', y_n)$ the problem (1.77)–(1.79) is converted into a usual boundary-value problem with $x_n > 0$ for the system (we maintain the old notation)

$$\begin{pmatrix} \mathcal{L}_0^1 & 0 \\ 0 & \mathcal{L}_0^2 \end{pmatrix} \begin{pmatrix} u^1 \\ u^2 \end{pmatrix} = \begin{pmatrix} f^1 \\ f^2 \end{pmatrix} \tag{1.80}$$

with diagonal matrix and the boundary conditions

$$(\mathcal{B}_0^1(D_t, D_{x'}, D_{x_n})u^1 + \mathcal{B}_0^2(D_t, D_{x'}, -D_{x_n})u^2)\,|_{x_n=+0} = g^1\,. \tag{1.81}$$

If we write down the complementarity condition for the problem (1.80), (1.81), then we obtain what is required, i.e. the compatible covering condition.

Example 1.13. We study the spreading of heat in a composite body Ω^0 whose parts Ω^μ, $\mu = 1, 2$, have densities ρ_μ, heat capacities c_μ, heat conductivity coefficients k_μ, and we put $a_\mu^2 = \rho_\mu c_\mu / k_\mu$. The mutual position of the parts Ω^1 and Ω^2 of the body satisfies the condition presented at the beginning of this section. The unknown temperature $u(t,x)$ in Ω^0 is defined by (1.71) and the functions $u^\mu(t,x)$ satisfy the heat equations

$$D_t u^\mu = a_\mu^2 \Delta u^\mu \,, \qquad (t,x) \in Q^\mu \,, \quad \mu = 1,2 \,, \tag{1.82}$$

and initial conditions

$$u^\mu |_{t=0} = \varphi^\mu(x) \,, \qquad x \in \Omega^\mu \,, \quad \mu = 1,2. \tag{1.83}$$

On the boundary $\partial\Omega^1$ separating the parts Ω^1 and Ω^2 the temperatures and heat flows should coincide:

$$u^1 |_{\partial\Omega^1} = u^2 |_{\partial\Omega^1} \,, \quad k_1 \frac{\partial u^1}{\partial \nu^1} |_{\partial\Omega^1} = k_2 \frac{\partial u^2}{\partial \nu^2} |_{\partial\Omega^1} \,. \tag{1.84}$$

This is a natural conjugation problem. We write down the compatible covering condition in the model case when $\partial\Omega^1$ is the plane $x_3 = 0$, the domain Ω^1 coincides with the half-plane $x_3 < 0$, and Ω^2 with the half-plane $x_3 > 0$. Conditions (1.84) take now the form

$$u^1(x_1,x_2,0) = u^2(x_1,x_2,0) \,, \quad k_1 \frac{\partial u^1}{\partial x_3}\bigg|_{x_3=0} = -k_2 \frac{\partial u^2}{\partial x_3}\bigg|_{x_3=0} \,.$$

In the case in question we have

$$L_0^{\mu +}(p,\sigma',\tau) = \tau - \tau_\mu^+ \,, \qquad \tau_\mu^+ = \mathrm{i}\sqrt{\frac{p}{a_\mu^2} + \sigma_1^2 + \sigma_2^2} \,, \quad \mu = 1,2 \,;$$

$$\mathcal{B}^1 = \begin{pmatrix} 1 \\ ik_1\tau \end{pmatrix} , \quad \mathcal{B}^2 = \begin{pmatrix} -1 \\ ik_2\tau \end{pmatrix} , \qquad \mathcal{R} = (\mathcal{R}^1, \mathcal{R}^2) = \begin{pmatrix} 1 & -1 \\ ik_1\tau_1^+ & ik_2\tau_2^+ \end{pmatrix} ,$$

$$\det\mathcal{R} = i(k_1\tau_1^+ + k_2\tau_2^+) \neq 0 \,, \quad k_1 > 0 \,, k_2 > 0 \,.$$

And so the compatible covering condition is satisfied for all physical values of parameters.

2.6. Nonlocal Parabolic Boundary-Value Problems. We move to the discussion of one more possible generalization of the class of boundary-value problems defined above which, from the general point of view, can be naturally called local boundary-value problems. Here we shall consider nonlocal problems in the simplest possible variant. We keep the assumptions of 2.5 on the structure of the domain Ω^0 and in addition we require that the surfaces $\partial\Omega^0$ and $\partial\Omega^1$ be diffeomorphic, i.e. there exist a one-to-one mapping of $\partial\Omega^0$ to $\partial\Omega^1$: $y = \alpha(x)$, $x \in \partial\Omega^1$, $y \in \partial\Omega^0$. If ν^0 and ν^1 are the unit vectors of inward normal directions to $\partial\Omega^0$ and $\partial\Omega^1$, then for sufficiently small ε the mapping $y + \varepsilon\nu^0 \mapsto x + \varepsilon\nu^1$ is a diffeomorphism of a ε-neighborhood of $\partial\Omega^0$ (we denote it by $\partial\Omega^0_\varepsilon$) on $\partial\Omega^1_\varepsilon$. If on $S^0_\varepsilon = (0,T] \times \partial\Omega^0_\varepsilon$ we have a function $u(t,y)$ and a differential operator $\mathcal{L}(t,y;D_t,D_y)$ with smooth coefficients, then the function $u(t,\alpha(x))$ and the operator $\mathcal{L}(t,\alpha(x);D_t,ID_x)$, where I is the transposed Jacoby matrix of the mapping $\alpha^{-1}(y)$, are defined on S^1_ε.

We shall consider a problem of finding a function $u(t,x)$ defined by (1.71) satisfying equations (1.72), initial conditions (1.73), but condition (1.74) on S will be replaced by a nonlocal boundary condition defined below.

Suppose that on $S^1_\varepsilon \cap Q^1$ and $S^1_\varepsilon \cap Q^2$ we have matrix differential operators $\mathcal{B}^1(t, x; D_t, D_x)$ and $\mathcal{B}^2(t, x; D_t, D_x)$ of dimensions $r \times m_1$ and $r \times m_2$, respectively. Suppose that on $S^0_\varepsilon \cap Q^0$ we have a differential operator $\check{\mathcal{B}}^3(t, x; D_t, D_y)$ which by the diffeomorphism $y = \alpha(x)$ is converted into an operator $\mathcal{B}^3(t, x; D_t, D_x) \equiv \check{\mathcal{B}}^3(t, \alpha(x); D_t, ID_x)$ defined on $S^1_\varepsilon \cap \Omega^1$.

In this transformation the operator $\mathcal{L}^2(t, y; D_t, D_y)$ defined on $S^0_\varepsilon \cap Q^0$ will be converted into the operator $\mathcal{L}^3(t, x; D_t, D_x) \equiv \mathcal{L}^2(t, \alpha(x); D_t, ID_x)$ acting in $S^1_\varepsilon \cap Q^1$. The inward normal to $S^1_\varepsilon \cap Q^1$ at $x \in S^1$ will be denoted by ν_3.

A *nonlocal boundary condition (nonlocal conjugation condition)* is given as follows:

$$(\mathcal{B}^1 u^1 + \mathcal{B}^2 u^2 + \mathcal{B}^3 u^3)|_{S^1} = g^1 , \tag{1.85}$$

where $\mathcal{B}^1$, $\mathcal{B}^2$, $\mathcal{B}^3$ are matrix operators of dimensions $t \times m_1$, $t \times m_2$, $t \times m_2$, respectively, $r = b(m_1 + 2m_2)$, and g^1 is a vector-function of dimension r defined on S^1.

To find the complementarity condition for the problem (1.85) we transform this problem by the reflection into an ordinary boundary problem and next we write down the usual complementarity condition.

We denote by $\mathcal{L}^\mu_0$, $\mathcal{B}^\mu_0$, $\mu = 1, 2, 3$, the principal parts of the corresponding operators. We fix a point $(t^0, x^0) \in S^1$. We consider

$$\mathcal{B}^\mu_0(t^0, x^0; p, \xi + \nu^\mu \tau) , \quad \mathcal{L}^\mu_0(t^0, x^0; p, \xi + \nu^\mu \tau) ,$$

$$L^\mu_0(t^0, x^0; p, \xi + \nu^\mu \tau) = L^{\mu+}_0 \cdot L^{\mu-}_0 , \quad \mu = 1, 2, 3 .$$

Nonlocal Compatible Covering Condition (nonlocal complementarity condition). The operators $\{\mathcal{B}^1, \mathcal{B}^2, \mathcal{B}^3\}$ satisfy the *nonlocal compatible covering condition* with respect to the operators $\{\mathcal{L}^1, \mathcal{L}^2, \mathcal{L}^3\}$ if for any $(t^0, x^0) \in S^1$ and any $(p, \xi) \in \Gamma^{x^0}_1$ the remainders $\mathcal{R}^\mu(t^0, x^0; p, \xi, \tau)$ of division of the elements of the matrices

$$\mathcal{B}^\mu_0(t^0, x^0; p, \xi + \nu^\mu \tau) \check{\mathcal{L}}^\mu_0(t^0, x^0; p, \xi + \nu^\mu \tau)$$

by the τ-polynomials $L^{\mu+}_0(t^0, x^0; p, \xi, \tau)$, $\mu = 1, 2, 3$, form a matrix

$$(\mathcal{R}^1(t^0, x^0; p, \xi, \tau), \mathcal{R}^2(t^0, x^0; p, \xi, \tau), \mathcal{R}^3(t^0, x^0; p, \xi, \tau)) ,$$

whose rows are linearly independent in τ.

Definition 1.11. The problem (1.71)–(1.73), (1.85) is called a *nonlocal parabolic boundary-value problem* if:

1) the systems $\mathcal{L}^1$ and $\mathcal{L}^2$ are parabolic in the sense of Petrovskij of the first order in t with common parabolicity constant δ in Q^1 and Q^2 respectively;
2) the boundary operators $\{\mathcal{B}^1, \mathcal{B}^2, \mathcal{B}^3\}$ satisfy on S^1 nonlocal complementarity condition with respect to $\{\mathcal{L}^1, \mathcal{L}^2, \mathcal{L}^3\}$.

The parabolic conjugation problems formally are particular cases of nonlocal parabolic problems. However, we must not forget that in the conjugation problems we do not have the condition on the diffeomorphism of the boundaries and so the external and internal boundary surfaces can be 'built up' in a completely different way. We note that nonlocal problems for general parabolic systems were defined and studied in (Ejdel'man and Zhitarashu 1970).

§3. Functional Spaces

3.1. Introduction. Parabolic boundary-value problems defined in the previous section turn out to be well-posed in a broad class of functional spaces of smooth as well as generalized functions. And this is the parabolicity condition which is necessary and sufficient for that well-posedness. For a substantial description of these results we need a developed theory of various spaces of functions of many variables taking into account the fact that time and space variables have different weights and therefore the functional spaces should possess a special anisotropy property. The necessary information about this theory can be found in (Agranovich and Vishik 1964), (Volevich and Paneyah 1965), (Maz'ya 1985), and (Slobodetskij 1958). In this section we briefly dwell on two classes of functions that can be simply described: Banach spaces of Hölder functions, which are convenient since all definitions are classical, and Hilbert spaces of type $W_{2\ t,x}^{l/2b,l}$, where the problems can be successfully investigated by means of the Fourier integral. Regretfully, an important L_p-theory of such problems will be not considered here. For this we refer the reader to fundamental papers of Solonnikov (Ladyzhenskaya et al. 1967), (Solonnikov 1965), and the extension to nonlocal problems is contained in (Ejdel'man and Zhitarashu 1970).

3.2. Spaces of Hölder Functions. Let l be a positive number and $[l]$ its integer part. We denote by $C^l(\overline{\Omega})$ the linear space of continuous functions $u(x)$ with all derivatives up to order $[l]$ continuous in Ω with finite values of the expressions

$$\|u, \Omega\|_l = \sum_{|\alpha| \le [l]} \sup_{x \in \Omega} |D_x^\alpha u(x)| + \sum_{|\alpha|=[l]} \sup_{\{x,y\} \subset \overline{\Omega}} \left\{ \frac{\Delta_x^y D_x^\alpha u(x)}{|x-y|^{l-[l]}} \right\}, \tag{1.86}$$

where $\Delta_x^y = f(y) - f(x)$.

We introduce the linear space $C^l(\overline{Q}_T) \equiv C_{t,x}^{l/2b,l}(\overline{Q}_T)$ of functions $v(t,x)$ with continuous derivatives $D_t^{\alpha_0} D_x^\alpha v(t,x)$ in $\overline{Q}_T$, $2b\alpha_0 + |\alpha| \le [l]$, and finite values of the expressions

$$\|v, Q_T\|_l = \sum_{2b\alpha_0+|\alpha|\le[l]} \sup_{(t,x)\in\overline{Q}_T} |D_t^{\alpha_0} D_x^\alpha v(t,x)|$$

$$+ \sum_{2b\alpha_0+|\alpha|=[l]} \sup_{\{(t,x),(t,y)\}\subset\overline{Q}_T} \left\{ \frac{\Delta_x^y D_t^{\alpha_0} D_x^\alpha v(t,x)}{|x-y|^{l-[l]}} \right\}$$

$$+ \sum_{0<l-2b\alpha_0-|\alpha|\le 2b} \sup_{\{(t,x),(\tau,y)\}\subset\overline{Q}_T} \left\{ \frac{|\Delta_\tau^t D_t^{\alpha_0} D_x^\alpha v(t,x)|}{|t-\tau|^{(l-2b\alpha_0-|\alpha|)/2b}} \right\}, \tag{1.87}$$

where $\Delta_\tau^t f(t,x) = f(t,x) - f(\tau,x)$. Formulas (1.86) and (1.87) define norms in the linear spaces $C^l(\Omega)$ and $C^l(Q_T)$ with respect to which they are Banach spaces.

We now define the Hölder spaces $C^l(S_T)$ of functions defined on the lateral surface S_T of the cylinder Q_T. We assume that at each point of the boundary $\partial\Omega$ there exists the tangent plane. We fix a point y on $\partial\Omega$ and we consider a local system of coordinates with the origin at y, i.e. a rectangular Cartesian system of coordinates $\hat{x}_1, \dots, \hat{x}_n$ where the axis $\hat{x}_n$ has the direction of the inward normal to $\partial\Omega$ at y and the remaining lie in the tangent plane. We assume that for every $y \in \partial\Omega$ there is a neighborhood U_y such that the part $\partial\Omega \cap U_y$ of the boundary is given by the equation

$$\hat{x}_n = F(\hat{x}'), \quad \hat{x}' \in B = \{\hat{x}' : \hat{x}' \in \mathbb{R}^{n-1}, |\hat{x}'| < d\},$$

where $F \in C^l(B)$, $\|F; B\|_l \le c$, and the constants d and c do not depend on $y \in \partial\Omega$.

We consider the mapping

$$\hat{x} \mapsto \overline{x}, \quad \overline{x}_i = \hat{x}_i, \ i = 1, \dots, n-1, \quad \overline{x}_n = \hat{x}_n - F(\hat{x}').$$

Let the mapping

$$x \mapsto \overline{x} \tag{1.88}$$

be the superposition of the mappings $\hat{x} \mapsto \overline{x}$ and $x \mapsto \hat{x}$. It maps $\partial\Omega \cap U_y$ into the ball B in the hyperplane $\overline{x}_n = 0$.

We consider a function v defined on S_T. For every fixed point $y \in \partial\Omega$ the function $v(t,x)$, $(t,x) \in [0,T] \times (\partial\Omega \cap U_y)$ composed with the mapping (1.88) becomes the function $\overline{v}(t,\overline{x}')$, $(t,\overline{x}') \in [0,T] \times B$. Suppose that $\partial\Omega$ is a surface of class C^l. We say that $v \in C^l(S_T)$ if for all $y \in \partial\Omega$ the function $\overline{u}_y$ lies in $C^l([0,T] \times B)$ and the number

$$\|u; S_T\|_l = \sup_y \|\overline{u}_y; (0,T] \times B\|_l$$

is finite; thus a norm is defined on the Banach space $C^l(S_T)$.

3.3. Spaces $\mathcal{H}^l$. This is a broad class of isotropic and anisotropic Hilbert spaces of the Sobolev-Slobodetskij spaces type, consisting of smooth as well as generalized functions. There is substantial literature devoted to the subject (Agranovich and Vishik 1964), (Volevich and Paneah 1965), and (Slobodetskij 1958). In the simplest and most accessible variant all we directly need is presented for the smooth case in the paper by Agranovich and Vishik (1964), which is important for the theory of parabolic and elliptic boundary problems.

We very briefly present some basic definitions and facts of this theory in the smooth case, to at least indicate how attractive and convenient they are for the theory of boundary problems.

The *space* $H^l(\mathbb{R}^n)$ is defined as the completion of the linear space $C_0^\infty(\mathbb{R}^n)$ of smooth complex-valued functions $u(x)$ with compact supports, with respect to the norm

$$\|u;\mathbb{R}^n\|_l = \left(\int (1+|\sigma|^{2l})|(F_n u)(\sigma)|^2 d\sigma\right)^{\frac{1}{2}},$$

where $(F_n u)(\sigma) = (2\pi)^{-n/2}\int \exp\{-ix\cdot\sigma\}u(x)dx$; here and in what follows $\int_{\mathbb{R}^n}$ is briefly denoted by $\int$. We note that by virtue of the Parseval identity this norm for integer l is equivalent to the norm

$$\left(\sum_{|\alpha|\le l}\int |D^\alpha u(x)|^2 dx\right)^{\frac{1}{2}} \tag{1.89}$$

and for fractional l to the norm

$$\left(\sum_{|\alpha|\le l}\int |D^\alpha u(x)|^2 dx + \sum_{|\alpha|=l}\int\int \frac{|D^\alpha u(x) - D^\alpha u(y)|^2}{|x-y|^{n+2l-2[l]}}dxdy\right)^{\frac{1}{2}}. \tag{1.90}$$

The space $H^0(\mathbb{R}^n)$ obviously coincides with $L_2(\mathbb{R}^n)$.

By $\mathbb{R}^n_+$ we denote the half-space in $\mathbb{R}^n$ defined by the inequality $x_n \ge 0$. For integer l the *space* $H_l(\mathbb{R}^n_+)$ is defined as the completion of the space of restrictions to $\mathbb{R}^n_+$ of the functions from $C_0^\infty(\mathbb{R}^n)$, with respect to the norm

$$\|u;\mathbb{R}^n_+\| = \left(\sum_{|\alpha|\le l}\int_{\mathbb{R}^n_+} |D^\alpha u(x)|^2 dx\right)^{\frac{1}{2}}, \tag{1.91}$$

or, with respect to an equivalent norm

$$\left(\sum_{j+m\le l}\int_{\mathbb{R}^{n-1}} |\sigma'|^{2j}d\sigma' \int_0^\infty |D^m_{x_n}(F_{n-1}u)(\sigma',x_n)|^2 dx_n\right)^{\frac{1}{2}}. \tag{1.92}$$

For any function $u(x)$ in $C_0^\infty(\mathbb{R}^n)$ and any $l \ge \frac{1}{2}$ we have the inequality

$$\|u(x',0);\mathbb{R}^{n-1}\|_{l-\frac{1}{2}} \le C\|u(x);\mathbb{R}^n_+\|_l\,, \tag{1.93}$$

where obviously the constant C does not depend on $u(x)$.

Inequality (1.93) allows us to define the limit values $u(x',0) \in H^{l-\frac{1}{2}}(\mathbb{R}_{n-1})$ for any $u(x) \in H^l(\mathbb{R}^n_+)$ by the limit passage. Inequality (1.93) remains true for all $u(x) \in H^l(\mathbb{R}^n_+)$. Therefore, for any function in $H^l(\mathbb{R}^n_+)$, $l > \frac{1}{2}$ its *trace* $u(x',0)$ on the hyperplane $x_n = 0$ belongs to $H^{l-\frac{1}{2}}(\mathbb{R}^{n-1})$. The converse is also true and allows us to 'reconstruct' functions from their traces so that the results belong to the corresponding function spaces and (1.93) is preserved.

Now let Ω be a bounded domain in $\mathbb{R}^n$ with smooth boundary $\partial\Omega$ satisfying the properties given in 3.2. For integer $l \geq 0$ we define the space $H^l(\Omega)$ as the completion of the space $C^\infty(\overline{\Omega})$, of complex-valued functions that are smooth in the closure $\overline{\Omega}$ of Ω, with respect to the norm

$$\|u; \Omega\|_l = \left(\sum_{|\alpha| \leq l} \int_\Omega |D^\alpha u(x)|^2 dx \right)^{\frac{1}{2}} . \tag{1.94}$$

The functions in $H^l(\Omega)$ have *generalized derivatives in the sense of Sobolev* of order $|\alpha| \leq l$ belonging to $H^{l-|\alpha|}(\Omega)$.

We require that the boundary $\partial\Omega$ can be covered by a finite number of domains $U_1, \dots, U_r$ in $\mathbb{R}^n$ such that $\partial\Omega$ can be locally rectified in each U_j. We also assume that for any finite partition $\{\varphi_\mu\}$ of unity in a neighborhood of Ω if the support of φ_μ intersects $\partial\Omega$, then it is contained in one of the domains U_j. We recall that the *support* of a function is the closure of the set of all points where it is nonzero, and a *partition of unity* is a collection of smooth functions φ_μ with compact supports, such that $\sum_\mu \varphi_\mu(x) = 1$ for any x in the domain.

Let φ_μ, $\mu = 1, \dots, s$, be those functions of the partition of unity whose supports intersect $\partial\Omega$. Their restrictions to $\partial\Omega$ will also be denoted by φ_μ. For $l \geq 0$ we define the *space* $H^l(\partial\Omega)$ as the completion of the space $C^\infty(\partial\Omega)$ of smooth functions $u(x')$ on $\partial\Omega$ with respect to the norm

$$\|u; \partial\Omega\|_l' = \left(\sum_{\mu=1}^{s} \|\varphi_\mu u; \mathbb{R}^{n-1}\|_l'^2 \right)^{\frac{1}{2}} , \tag{1.95}$$

where the norm $\|\varphi_\mu u, \mathbb{R}^{n-1}\|_l'$ is taken in local coordinates. The space $H^l(\partial\Omega)$ does not in fact depend on the choice of (sufficiently fine) partition of unity and local coordinates.

For $l > \frac{1}{2}$ the functions in $H^l(\Omega)$ have boundary values in $H^{l-\frac{1}{2}}(\partial\Omega)$ and an inequality of type (1.93) holds:

$$\|u; \partial\Omega\|_{l-\frac{1}{2}}' \leq C\|u; \Omega\|_l . \tag{1.96}$$

We now briefly describe anisotropic spaces of functions $v(t, x)$ of $n+1$ variables. Suppose that $\Omega \subset \mathbb{R}^n$ satisfies above conditions. Put $Q_T = [0, T] \times \Omega$ (it is a cylinder in $\mathbb{R}^{n+1}$) and $S_T = [0, T] \times \partial\Omega$. We consider the *spaces* $\mathcal{H}^{k,l}(Q_T) = W^{k,l}_{t,x,2}(Q_T)$ and $\mathcal{H}^{k,l}(S_T) = W^{k,l}_{t,x,2}(S_T)$. Roughly speaking, the functions $v(t, x)$ in these spaces have smoothness of order k in the time variable t and smoothness of order l in the space variables $x_1, \dots, x_n$. The norms in these spaces are defined by

$$\|v; Q_T\|_{k,l} = \left(\int_0^T \|v(t, x); \Omega\|_l^2 dt + \int_\Omega \|v(t, x); [0, t]\|_k^2 dx \right)^{\frac{1}{2}} , \tag{1.97}$$

$$\|v; S_T\|_{k,l}$$

$$= \left(\int_0^T \|v(t,x); \partial\Omega\|_l'^2 dt + \sum_j \int_{\partial\Omega} \|\varphi_j(x')v(t,x'); [0,T]\|_k^2 dx' \right)^{\frac{1}{2}} . \quad (1.98)$$

Spaces $\mathcal{H}^{k,l}(Q_T)$ are defined as the completions with respect to the norms (1.97) of the space of smooth functions. This definition is equivalent to the following: $\mathcal{H}^{k,l}(Q_T)$ consists of all $v(t,x)$ that belong to $H^k([0,T])$ for almost all $x \in \Omega$ and to $H^l(\Omega)$ for almost all $t \in [0,T]$, and have finite norm (1.97). Here the generalized derivatives $D_x^\alpha v$ in Ω, $|\alpha| \le l$, and the generalized derivatives $D_t^{\alpha_0} v$ in $[0,T]$, $\alpha_0 \le k$, are the corresponding derivatives in Q_T. The spaces $\mathcal{H}^{k,l}(S_T)$ can be defined analogously but here the generalized derivatives in x' are considered within the domains of a system of local coordinates. In the theory of parabolic boundary-value problems the anisotropic spaces $\mathcal{H}^{\frac{l}{2b},l}(Q_T)$, $\mathcal{H}^{\frac{\lambda}{2b},\lambda}(S_T)$ are used, where $2b$ is the parabolic weight.

We give some important properties of functions from these spaces assuming that $l+1$, $\lambda + \frac{1}{2}$ are positive integers.

1) Any function $v(t,x) \in \mathcal{H}^{l/2b,l}(Q_T)$ has the generalized derivatives $v_{\alpha_0,\alpha}(t,x) = D_t^{\alpha_0} D_x^\alpha v \in \mathcal{H}^{r/2b,r}(Q_T)$, $r = l - 2b\alpha_0 - |\alpha|$, satisfying the inequality

$$\|v_{\alpha_0,\alpha}(t,x); Q_T\|_{r/2b,r} \le \|v(t,x); Q_T\|_{l,l/2b} .$$

2) If $r \ge 1$, then $v_{\alpha_0,\alpha}(t,x)$ has the boundary values $v_{\alpha_0,\alpha}(t,x')$ in $\mathcal{H}^{\lambda/2b,\lambda}(S_T)$, $\lambda = r - \frac{1}{2}$, and we have

$$\|v_{\alpha_0,\alpha}(t,x'); S_T\|_{\lambda/2b,\lambda} \le \|v_{\alpha_0,\alpha}(t,x); Q_T\|_{r/2b,r} .$$

The functions $v_{\alpha_0,\alpha}(t,x')$ clearly make sense if $v(t,x)$ is a smooth function in $\overline{Q}_T$. In general $v_{\alpha_0,\alpha}(t,x)$ can be approximated in the metric of $\mathcal{H}^{r/2b,b}(Q_T)$ by functions $v_{\alpha_0,\alpha}^{(N)}(t,x)$ smooth in $\overline{Q}_T$, and then $v_{\alpha_0,\alpha}(t,x') = \lim_{N\to\infty} v_{\alpha_0,\alpha}^{(N)}(t,x')$ in $\mathcal{H}^{\lambda/2b,\lambda}(Q_T)$.

3) If $r > b$, then the function $v_{\alpha_0,\alpha}(t,x)$ has the trace $v_{\alpha_0,\alpha}(t_0,x)$ on the section of Q_T by the hyperplane $t = t_0$, $t_0 \in [0,T]$. Moreover, $v_{\alpha_0,\alpha}(t_0,x) \in H^{r-b}(\Omega)$ and

$$\|v_{\alpha_0,\alpha}(t_0,x); \Omega\|_{r-b} \le C\|v_{\alpha_0,\alpha}(t,x); Q_T\|_{r/2b,r} .$$

In particular, if $t = 0$ we can say that the function $v(t,x)$ in $\mathcal{H}^{l/2b,l}(Q_T)$ has well-defined initial values

$$v|_{t=0} = \varphi_0(x), \ldots, D_t^{k_0-1} v|_{t=0} = \varphi_{k_0-1}(x) , \quad (1.99)$$

where

$$k_0 = \begin{cases} [l/2b] & \text{if } l/2b - [l/2b] \le 1/2 , \\ [l/2b]+1 & \text{if } l/2b - [l/2b] > 1/2 . \end{cases}$$

Moreover, $\varphi_j(x) \in H^{l-2b_j-b}(\Omega)$ and

$$\|\varphi_j; \Omega\|_{l-2b_j-b} \leq C\|v; Q_T\|_{l/2b,l} \,.$$

4) The previous result admits the following conversion: if the functions $\varphi_j(x) \in H^{l-2b_j-b}(\Omega)$, $j = 0, \dots, k_0 - 1$, are given, then there is a function $v(t,x) \in \mathcal{H}^{l/2b,l}(Q_T)$ satisfying both (1.99) and the inequality

$$\|v; Q_T\|_{l/2b,l} \leq C \sum_{j=0}^{k_0-1} \|\varphi_j; \Omega\|_{l-2b_j-b} \,.$$

5) We consider one more way of introducing the norm in the space of traces of functions from $\mathcal{H}^{l/2b,l}(Q_T)$, which is used in the theory of boundary-value problems. In $\mathcal{H}^{\lambda/2b,\lambda}(S_T)$ we can introduce the norm

$$< v; S_T >'_{\lambda/2b,\lambda} = \inf \|w; S_{[0,\infty)}\|_{\lambda/2b,\lambda} \,,$$

where the infimum is taken over all functions $w \in \mathcal{H}^{\lambda/2b,\lambda}(S_{[0,\infty)})$ that coincide with $v(t,x)$ almost everywhere on S_T. This norm is equivalent to the norm $\|v; S_T\|_{\lambda/2b,\lambda}$.

Chapter 2
Fundamental Solutions. Cauchy Problem

§1. Fundamental Solutions

1.1. Introduction. In mathematical physics, probability theory, and their numerous applications an exceptional role is played by the function

$$\Gamma_0(t-\tau, x-\xi) = \frac{1}{(2a\sqrt{\pi(t-\tau)})^n} \exp\left\{-\frac{|x-\xi|^2}{4a^2(t-\tau)}\right\} , \tag{2.1}$$

$0 \leq \tau < t$, $\{x,\xi\} \subset \mathbb{R}^n$. By means of (2.1) the solution to the Cauchy problem for the heat equation

$$\frac{\partial u}{\partial t} = a^2 \Delta u + f(t,x) \,, \tag{2.2}$$

$$u\,|_{t=0} = \varphi(x) \tag{2.3}$$

is expressed by the formula

$$u(t,x) = \int \Gamma_0(t, x-\xi)\varphi(\xi)d\xi + \int_0^t d\tau \int \Gamma_0(t-\tau, x-\xi) f(\tau,\xi)d\xi \,, \tag{2.4}$$

if, for example, $\varphi(x)$ is a bounded continuous function defined in $\mathbb{R}^n$ and $f(t,x)$ is defined in a band $\Pi_T = [0,T] \times \mathbb{R}^n$ and is continuous, bounded, and satisfies the Hölder condition with respect to the space variables $x_1, \dots, x_n$. By *fundamental solution* $Z(t, x; \tau, \xi)$ to the Cauchy problem for the system

$$\mathcal{L}(t, x; D_t, D)u \equiv D_t u - A(t, x; D)u = 0 \tag{2.5}$$

we understand a square matrix of dimension m such that the solution to the Cauchy problem

$$u|_{t=0} = \varphi(x) \tag{2.6}$$

for (2.5) is given by the formula

$$u(t, x) = \int Z(t, x; \tau, \xi)\varphi(\xi)d\xi \,. \tag{2.7}$$

The matrix $Z(t, x; \tau, \xi)$ is a solution of the system (2.5) for $t > \tau$ and satisfies the initial condition

$$Z|_{t=\tau} = \delta(x - \xi)I \,, \tag{2.8}$$

where $\delta(x)$ is the Dirac delta-function.

Fundamental solutions to the Cauchy problem are constructed and investigated for systems of a general form parabolic in the sense of Petrovskij. But here we restrict ourselves to the systems of the form (2.5) since the results for them have the most complete character, can be relatively simply formulated, and in addition the equations and systems encountered in applications quite often have the form (2.5). Here we only consider fundamental solutions to the Cauchy problem and we shall briefly call them fundamental solutions.

The fundamental solutions of parabolic systems defined by Petrovskij maintain surprisingly many properties of the function (2.1) with only some natural modifications. We move to their construction and the analysis of their properties. We note that the proofs of a series of basic results on fundamental solutions are contained in the monographs (Ladyzhenskaya et al. 1967), (Ejdel'man 1964), (Friedman 1964).

1.2. Systems Parabolic in the Sense of Petrovskij with Bounded Coefficients. We consider a system uniformly parabolic in the sense of Petrovskij

$$\mathcal{L}(t, x; D_t, D)u \equiv D_t u - \sum_{|\alpha| \le 2b} a_\alpha(t, x) D^\alpha u = 0 \,, \tag{2.9}$$

where $a_\alpha(t, x)$ are quadratic matrices of dimension m and $u(t, x)$ is a vector-function with m coordinates.

We begin with the construction and investigation of fundamental solution to the model system with constant coefficients

$$\mathcal{L}_0(D_t, D)u \equiv D_t u - \sum_{|\alpha| = 2b} a_\alpha D^\alpha u = (D_t - A_0(D))u = 0 \,. \tag{2.10}$$

Using the Fourier transform in space variables $x_1, \dots, x_n$ and the fact that the Fourier transform of $\delta(x)$ is the constant function equal to 1, we obtain the equality

$$\Gamma_0(t,x) = \frac{1}{(2\pi)^n} \int e^{ix\cdot\sigma} e^{A_0(\sigma)t} d\sigma \,. \tag{2.11}$$

We introduce new variables of integration and we rewrite (2.11) in the form

$$\Gamma_0(t,x) = \frac{1}{(2\pi)^n t^{n/2b}} \int e^{ixt^{-1/2b}\cdot\alpha} e^{A_0(\alpha)} d\alpha \,.$$

The matrix-function $\exp\{A_0(\gamma)\}$ is an entire analytic function of complex variables $\gamma_1, \ldots, \gamma_n$ (analytic in $\mathbb{C}^n$) and by the parabolicity condition and the homogeneity of the matrix $A_0(\gamma)$, we have the estimate

$$\| \exp\{A_0(\gamma)\} \| \le C \exp\{-c_1 |\mathrm{Re}\gamma|^{2b} + c_2 |\mathrm{Im}\gamma|^{2b}\} \,, \tag{2.12}$$

where here and in what follows C and c denote positive constants which we shall equip with various indices if necessary. Using the Gel'fand-Shilov theorem (Gel'fand and Shilov 1958a) on the Fourier transforms of entire analytic functions satisfying an inequality of the type (2.12) we obtain the following result.

Theorem 2.1. *A fundamental solution $\Gamma_0(t,x)$ to the system* (2.10), *considered as a matrix-function of arguments $x_1 t^{-1/2b}, \ldots, x_n t^{-1/2b}$, is an entire analytic function of the growth order $q = 2b/(2b-1)$ for complex values of the arguments and the same order of decay for real values. All its derivatives satisfy the estimates*

$$|D^\alpha \Gamma_0(t_0 x + iy)| \le C_\alpha t^{-\frac{n+|\alpha|}{2b}} \exp\{(-c_3 |x|^q + c_4 |y|^q) t^{1-q}\} \,. \tag{2.13}$$

Thus the internal properties of $\Gamma_0(t,x)$ are completely analogous to the properties of the function (2.1) which for comparison we rewrite in the following way

$$\frac{1}{(2a\sqrt{\pi t})^n} \exp\{-\frac{1}{4a^2} \sum_{j=1}^n (x_j t^{-1/2})^2\} \,.$$

Example 2.1. We find the fundamental solution to the parabolic equation

$$D_t u + (-1)^b \Delta^b u = 0 \,. \tag{2.14}$$

It is given by the integral

$$\Gamma_0(t,x) = \frac{1}{(2\pi)^n} \int \exp\{ix\cdot\sigma - |\sigma|^{2b} t\} d\sigma \,.$$

Passing to the spherical coordinates and using some identities from the theory of Bessel functions we arrive at the formula

$$\Gamma_0(t,x) = \frac{1}{(2\pi)^{n/2} t^{n/2b}} \int_0^\infty e^{-r^{2b}} r^{n/2} J_{(n-2)/2}\left(r \frac{|x|}{t^{1/2b}}\right) dr \,, \tag{2.15}$$

where $J_{(n-2)/2}$ is the Bessel function of the first kind of order $(n-2)/2$.

Despite the fact that the last integral can be explicitly computed for $b = 1$ only, the formula (2.15) is useful, for example, in the study of stabilization of the solution to the Cauchy problem.

We note that the monograph of Fedoryuk (1987) contains results of a precise study by the saddle-point method of asymptotic behavior of $\Gamma_0(t, x)$ as $|x| \to \infty$ (in the case of one equation).

To construct fundamental solutions to the system (2.9) with variable coefficients the *Levy method* is usually used. It is based on the beforehand study of fundamental solutions to systems with 'frozen' coefficients (taken at parametric points (τ, ξ) or (t, ξ)) and the theory of so called parabolic potentials. This method is presented in detail in (Il'in et al. 1962), (Ejdel'man 1964), (Friedman 1964). If we apply the Levy method in the version presented by the author in (Ejdel'man 1964), then we obtain

Theorem 2.2. *Suppose that the coefficients $a_\alpha(t, x)$ of the system (2.9) parabolic uniformly in the band $\overline{\Pi}_T = [0, T] \times \mathbb{R}^n$ satisfy the conditions*

α_1. *$a_\alpha(t, x)$ are continuous in $\overline{\Pi}_T$ and the continuity in t of $a_\alpha(t, x)$ with $|\alpha| = 2b$ is uniform with respect to x;*

α_2. *$a_\alpha(t, x)$ is bounded in $\overline{\Pi}_T$ by a constant M_0;*

α_3. *$a_\alpha(t, x)$ satisfies the Hölder condition with exponent $\lambda \in (0, 1]$ with respect to space variables $x_1, \dots, x_n$, i.e.*

$$\|\Delta_x^y a_\alpha(t, x)\| \leq |x - y|^\lambda \qquad \forall (t, x), (t, y) \in \overline{\Pi}_T .$$

Then in the class of bounded functions there is a unique fundamental solution $Z(t, x; \tau, \xi)$ of (2.9) which satisfies the estimates

$$|D_x^\alpha Z(t, x; \tau, \xi)| \leq C_\alpha (t - \tau)^{-\frac{n+|\alpha|}{2b}} \exp\{-c\rho(t, x; \tau, \xi)\} , \quad |\alpha| \leq 2b , \tag{2.16}$$

$$\begin{gathered} |\Delta_x^y D_x^\alpha Z(t, x; \tau, \xi)| \\ \leq C_\alpha |x - \xi|^\lambda (t - \tau)^{-\frac{n+2b+\lambda}{2b}} [\exp\{-c\rho(t, x; \tau\xi)\} + \exp\{-c\rho(t, y; \tau\xi)\}] , \\ |\alpha| = 2b , \end{gathered} \tag{2.17}$$

where the constants C_α depend on n, T, δ, M_0, λ, L, the constant c depends only on the parabolicity constant δ, and $\rho(t, x; \tau\xi) = |x - \xi|^q (t - \tau)^{1-q}$.

We note that the symbol $|\cdot|$ denotes the modulus of a function, the length of a vector, or the norm of a matrix, depending on what is inside. It is important to underline that the estimates (2.16) and (2.13) fully coincide by appearance (for real arguments) but the constants C_α in (2.13) depend only on M_0 and the parabolicity constant δ. In particular, the estimates (2.13) are satisfied in the half-space $t > 0$ of $\mathbb{R}^{n+1}$. In general in the theory we present, it is essential to clarify what the constants in (2.16) and (2.17) depend on.

The fundamental solution $Z(t, x; \tau\xi)$ has a series of simple and useful properties. We present now the two most important assuming that conditions

α_1–α_3 are satisfied. Various properties are presented in (Il'in et al. 1962), (Ejdel'man 1964), (Friedman 1964).

Property. 2.1. The fundamental solution $Z(t,x;\tau,\xi)$ satisfies the *convolution formula*

$$Z(t,x;\tau,\xi) = \int Z(t,x;\beta,y)Z(\beta,y;\tau,\xi)dy\ , \quad \forall\beta \in (\tau,t)\ . \tag{2.18}$$

In the theory of Markov processes equations (2.18) is called the *Chapman–Kolmogorov equation* for densities of transition probabilities of a random process.

We note that the equality (2.18) for the function (2.1) can be directly proved, which is a simple and useful exercise.

To formulate the next important property we need an additional condition.

α_4. There exist derivatives $D^j a_\alpha(t,x)$ for all $j < d$, satisfying conditions $\alpha_1 - \alpha_3$.

Under this condition for the system (2.9) we can consider *the Lagrange adjoint system* which has the form:

$$\mathcal{L}^*(\tau,\xi;D_\tau,D_\xi)v(\tau,\xi) = -D_\tau v - \sum_{|\alpha|\le 2b}(-1)^{|\alpha|}D_\xi^\alpha(a_\alpha^*(\tau,\xi)v) = 0\ , \tag{2.19}$$

where the matrix a^* is Hermitian adjoint to a (i.e. if $a = (a^{ij})$, then $a^* = (\overline{a^{ji}})$, where $\overline{a}$ denotes the complex conjugation). We note that the Cauchy problem for the adjoint system (2.19) is formulated as follows: we seek for a solution at $(\tau,\xi) \in \Pi_t$ with

$$v|_{\tau=t} = \varphi(\xi)\ , \tag{2.20}$$

Condition α_4 and Theorem 2.2 guarantee that a fundamental solution $\widetilde{Z}(t,x;\tau,\xi)$ to the problem (2.19), (2.20) exists and satisfies (2.16), (2.17). Clearly, τ, ξ are the main and t, x the parametric variables here.

Property. 2.2. If conditions α_1–α_4 are satisfied, then we have

$$\widetilde{Z}(t,x;\tau,\xi) = Z^*(t,x;\tau,\xi)\ . \tag{2.21}$$

Property 2.2 is called the *normality property of fundamental solutions*: $Z(t,x;\tau,\xi)$ as a function of (t,x) is the fundamental solution to the Cauchy problem for the original system (2.9) and the matrix that is Hermitian adjoint to $Z(t,x;\tau,\xi)$ is a fundamental solution to the Cauchy problem (2.19), (2.20) as a function of (τ,ξ). In particular, for a single equation with real coefficients $Z(t,x;\tau,\xi)$ is a fundamental solution to the original Cauchy problem as a function of (t,x) and to the adjoint as a function of (τ,ξ).

We note that in the case of strongly parabolic systems, in particular in the case of a single equation, all assertions remain valid if we replace condition α_1 by a less restrictive:

α_1'. $a_\alpha(t,x)$ are continuous in $\overline{\Pi}_T$, $|\alpha| \le 2b$.

There arises a natural question of what are minimal conditions on coefficients which guarantee existence of classical fundamental solutions satisfying above properties. We shall later discuss to what extent we can eliminate the condition on boundedness of the coefficients and this will turn out a sufficiently serious problem. But now we keep conditions α_1 and α_2, and we try to relax α_3.

Definition 2.1. A function $f(t,x)$ defined in a domain $\overline{Q} \subset \mathbb{R}^{n+1}$ belongs to the class $\widetilde{\mathcal{D}}^1$ (*Dini class*) if its modulus of continuity in $x_1, \dots, x_n$

$$\omega(h) = \sup_{\substack{\{(t,x),(t,y)\}\subset\overline{Q} \\ |x-y|<h}} |\Delta_x^y f(t,x)|$$

satisfies

$$\omega_1(h) = \int_0^h \frac{\omega(z)}{z} dz < \infty$$

for some positive h.

2) A function $f(t,x)$ from $\widetilde{\mathcal{D}}^1$ belongs to the class $\widetilde{\mathcal{D}}^2$ if

$$\int_0^a \frac{\omega_1(h)}{h} dh < \infty \quad \text{for some positive } a\,.$$

Example 2.2. The function $\omega^{(\lambda)}(h) = h^\lambda$, $0 < \lambda \le 1$, is an eigenfunction of the Dini operator: $J_{\widetilde{\mathcal{D}}}(\psi) = \int_0^h (\psi(\beta)/\beta) d\beta$. For we have $J_{\widetilde{\mathcal{D}}}(\omega^{(\lambda)}(h)) = \lambda\omega^{(\lambda)}$. The functions $\omega^{(s)}(h) = (\ln 1/h)^{-s}$ admit the application of the Dini operator for $s > 1$, and twice application for $s > 2$. We note that $\omega(h)$ can be a modulus of continuity if it is non-negative, non-decreasing, bounded, and subadditive, i.e. $\omega(h_1 + h_2) \le \omega(h_1) + \omega(h_2)$.

Theorem 2.3. *Suppose that the coefficients $a_\alpha(t,x)$ of the system* (2.9) *uniformly parabolic in the band $\overline{\Pi}_T$ satisfy α_1, α_2, and*

α_5. *$a_\alpha(t,x)$ belong to the class $\widetilde{\mathcal{D}}^1$ in $\overline{\Pi}_T$;*

α_6. *$a_\alpha(t,x)$, $|\alpha| = 2b$, belong to the class $\widetilde{\mathcal{D}}^2$ in $\overline{\Pi}_T$.*

Then in the class of bounded functions there exists a unique fundamental solution $Z(t,x;\tau,\xi)$ to the system (2.9), *it satisfies* (2.16), *and the increments of the highest order derivatives $D_x^\alpha Z$, $|\alpha| = 2b$, satisfy*

$$|\Delta_x^y D^\alpha Z(t,x;\tau,\xi)|$$

$$\leq C_\alpha \frac{\omega_1(|x-y|)}{\omega_1(t-\tau)}(t-\tau)^{-\frac{n+2b}{2b}} \left[\exp\{-c\rho(t,x;\tau,\xi)\} + \exp\{-c\rho(t,y;\tau,\xi)\}\right] , \tag{2.22}$$

where the constants C_α depend on n, T, δ, M_0, ω, ω_1 and c depends only on δ.

The proof of Theorem 2.3 is given in (Matijchuk and Ejdel'man 1967, 1970). It employes a method originated in the Hopf's paper (1932). As some examples show (Il'in 1962), (Kruzhkov 1967), the rejection of condition α_5 results in general in the lack of a fundamental solution to the Cauchy problem for (2.9) with the above properties. The Dini condition in the classical potential theory (Gyunter 1953), as well as in many problems of the theory of parabolic and elliptic equations, is minimal guaranteeing the existence of solutions with classical properties. We shall later discuss the relevant examples.

1.3. $2\overrightarrow{b}$-parabolic Systems. The results presented above can be extended with natural modifications to $2\overrightarrow{b}$-parabolic systems

$$\mathcal{L}(t,x;D_t,D)u \equiv D_t u - \sum_{\widetilde{\alpha}=1} a_\alpha(t,x) D^\alpha u = 0 , \tag{2.23}$$

where $\widetilde{\alpha} = (\alpha_1/2b_1) + \ldots + (\alpha_n/2b_n)$, $b_1 \geq b_2 \geq, \ldots, \geq b_n$, see (Ivasishen and Ejdel'man 1968), (Ejdel'man 1960).

Theorem 2.4. *The fundamental solution of a model $2\overrightarrow{b}$-parabolic system with constant coefficients*

$$\mathcal{L}_0(D_t,D)u \equiv D_t u - \sum_{\widetilde{\alpha}=1} a_\alpha D^\alpha u = 0$$

considered as a matrix-function of arguments $x_1 t^{-1/2b_1}, \ldots, x_n t^{-1/2b_n}$, $t>0$, is an entire analytic function of the growth order $q_k = 2b_k/(2b_k-1)$ with respect to the k-th argument, $k=1,\ldots,n$, for complex values of the arguments and has the same order of decay for real values. For all its derivatives we have:

$$|D^\alpha \Gamma_0(t,x+iy)| \leq C_\alpha t^{-(\widetilde{n}+\widetilde{\alpha})} \exp\left\{-c_1 \sum_{j=1}^n |x_j|^{q_j} t^{1-q_j} + c_2 \sum_{j=1}^n |y_j|^{q_j} t^{1-q_j}\right\} ,$$

where $\widetilde{n} = (1/2b_1) + \ldots + (1/2b_n)$.

Theorem 2.5. *We assume that the coefficients of the uniformly $2\overrightarrow{b}$-parabolic system (2.23) satisfy $\alpha_1 - \alpha_3$ in the band $\overline{\Pi}_T$. Then (2.23) has a unique bounded fundamental solution $Z(t,x;\tau,\xi)$ which satisfies*

$$|D^\alpha Z(t,x;\tau,\xi)|$$

$$\leq C_\alpha (t-\tau)^{-\widetilde{n}-\widetilde{\alpha}} \exp\left\{-c_1 \sum_{j=1}^{n} |x_j-\xi_j|^{q_j}(t-\tau)^{1-q_j}\right\}, \quad \widetilde{\alpha}\leq 1,$$

$$|\Delta_x^y D^\alpha Z(t,x;\tau,\xi)|$$

$$\leq C_\alpha |x-y|^\lambda (t-\tau)^{-\widetilde{n}-1-\lambda/2b_1}\left[\exp\left\{-c_1 \sum_{j=1}^{n} |x_j-\xi_j|^{q_j}(t-\tau)^{1-q_j}\right\}\right.$$

$$\left.+\exp\left\{-c_1 \sum_{j=1}^{n} |y_j-\xi_j|^{q_j}(t-\tau)^{1-q_j}\right\}\right] \quad \widetilde{\alpha}=1, \tag{2.24}$$

where the constants C_α depend on n, δ, T, M_0, L, λ, $2\overrightarrow{b}$, and c depends only on δ.

Fundamental solutions to $2\overrightarrow{b}$-parabolic systems have properties 1 and 2 and the analogue of Theorem 2.3 is true for them.

1.4. Systems Parabolic in the Sense of Petrovskij with Growing Coefficients. We turn to the analysis of the conditions of existence of fundamental solutions for equations with coefficients growing together with spatial variables. Examples of simple parabolic equations which do not have the uniqueness property of solutions to the Cauchy problem in the class of bounded functions or even in the class $L_2(\mathbb{R}^n)$ are well-known and will be given in the next section. Obviously, in such situations we cannot expect the existence of fundamental solutions with the properties described above. However, it may turn out that a directed growth of coefficients helps us to obtain fundamental solutions with properties in a certain sense better than for the equations with bounded coefficients. This is namely the case we shall begin with and next we shall see what growth conditions allow us to construct fundamental solutions with properties similar to those for the equations with bounded coefficients.

We begin with some simple considerations. If we consider the equation

$$D_t u = \Delta u - q^2(x) u, \tag{2.25}$$

where $q(x)$ is a real function, then it is natural to expect that the quick growth of the positive function $q^2(x)$ together with the growth of $|x|$, which means strong absorption of energy at infinity, should lead to a quick decay of the fundamental solution to this equation. The problem is to define the notion of 'dissipation' for (2.9) and to show that the fundamental solution really decays (and to show the rate of this decay) together with the growth of 'dissipation.

Definition 2.2. We call a function $f(x)$ a *characteristic of dissipation* if it satisfies the following conditions:

1) $f(x)$ is defined and continuous in $\mathbb{R}^n$;
2) $f(x) \geq 1$;
3) $f(x) \to \infty$ as $|x| \to \infty$.

Definition 2.3 (Ejdel'man 1964). We say that a system (2.9) parabolic in the sense of Petrovskij is *dissipative* if there is a characteristic of dissipation $f(x)$ such that the following conditions are satisfied

α_7. the matrix-functions $\widetilde{a}_\alpha(t,x) = a_\alpha(t,x)f(x)^{|\alpha|-2b}$ are bounded in $\overline{\Pi}_T$;

α_8. the polynomial

$$L(t,x;p,\sigma,\sigma_{n+1}) = \det\left(pI - \sum_{|\alpha|\le 2b} \widetilde{a}_\alpha(t,x)\sigma^\alpha\sigma_{n+1}^{2b-|\alpha|}\right)$$

is parabolic in the sense of Petrovskij, i.e. the p-zeros of L satisfy

$$\operatorname{Re} p(t,x;\sigma,\sigma_{n+1}) \le -\delta(|\sigma|^2 + \sigma_{n+1}^2)^b$$

for any $(t,x) \in \overline{\Pi}_T$, $(\sigma,\sigma_{n+1}) \in \mathbb{R}^{n+1}$, where δ is a positive constant.

Example 2.3. We consider equation (2.25). If the function $g(x)$ satisfies the conditions imposed on the characteristic of dissipation, then we put $f(x) = q(x)$ and then $L(p,\sigma,\sigma_{n+1}) = p + |\sigma|^2 + \sigma_{n+1}^2$. Thus (2.25) is a dissipative equation with characteristic of dissipation $q(x)$.

We move to the presentation of theorems on existence of fundamental solutions. To formulate them we need to give conditions on the coefficients $a_\alpha(t,x)$ and on the characteristic of dissipation $f(x)$. We have two sets of these conditions. In the first we impose sufficiently stringent restrictions on smoothness of coefficients but there are no additional conditions on $f(x)$ and in the second, which appeared in the papers of Levitan and Kostyuchenko on spectral theory of differential operators, we impose minimal smoothness conditions but the characteristic of dissipation is assumed to satisfy a certain additional condition.

α_9. a_α have derivatives $D^l a_\alpha(t,x)$, $|l| \le 2b$, which satisfy

$$|D^l a_\alpha(t,x)| \le Cf(x)^{2b-|\alpha|+|l|(1-\varepsilon)}, \quad |l| \le 2b,$$

for some $\varepsilon \in (0,1)$.

α_{10}. $D^l a_\alpha(t,x)$, $|l| \le |\alpha|$, satisfy a local Hölder condition in x, i.e. for every $x \in \mathbb{R}^n$ there is a constant L_x such that

$$|\Delta_x^y D^l a_\alpha(t,x)| \le L_x|x-y|^\lambda$$

for some $\lambda \in (0,1]$ and any y such that $|x-y| \le 1$.

The second set of conditions consists of:

α_{11}. $|\Delta_x^y a_\alpha(t,x)| \le C_x|x-y|^\lambda f(y)^{2b-|\alpha|}$ for $|x-y| \le 1$, $|\alpha| \le 2b$;

and the conditions on the characteristic of dissipation $f(x)$:

α_{12}. for any $x, y \in \mathbb{R}^n$ such that $|x-y| \le 1$ we have $f(x) \le Cf(y)$;

α_{13}. there is a small positive number μ_0 such that for $|x-y| > 1$ we have

$$f(x) \le C \exp\{\mu_0 |x-y| f(y)\} .$$

From the characteristic of dissipation $f(x)$ we construct a smooth function $g(t,x)$ whose derivatives satisfy

$$|D_t g(t,x)| \le Cf(x)^{2b} , \quad |D^\alpha g(t,x)| \le Cf(x)^{|\alpha|} , \quad |\alpha| = 1, \dots, 2b . \quad (2.26)$$

Theorem 2.6. *Suppose that the coefficients of a dissipative system* (2.9) *satisfy conditions* α_9, α_{10} *or condition* α_{11} *with dissipation characteristic* $f(x)$ *satisfying* α_{12} *and* α_{13}. *Then there is a fundamental solution* $Z(t,x;\tau,\xi)$ *to* (2.9) *which satisfies*

$$|D^\alpha Z(t,x;\tau,\xi)|$$

$$\le C_\alpha \sum_{k=1}^{|\alpha|} (t-\tau)^{-\frac{n+|\alpha|-k}{2b}} f(x)^k \exp\{-c\rho(t,x;\tau,\xi)+\eta g(t,x)-\eta g(\tau,\xi)\} , \quad (2.27)$$

$$|\alpha| \le 2b ,$$

where η *is a sufficiently small positive constant.*

The proof of Theorem 2.6 in (Ejdel'man 1964), (Ejdel'man and Malitskaya 1975) employs, depending on the conditions, various variants of the Levy method. Next, a change of unknown function allows us to take into account the characteristic of dissipation in the exponential part of the estimate (2.27).

We move to the case of systems (2.9) with growing coefficients. A number of variant sets of conditions for coefficients can be given and here is one of them:

α_{14}, $a_\alpha(t,x)$, $|\alpha| = 2b$, satisfy α_1–α_3;

α_{15}, $a_\alpha(t,x)$, $|\alpha| < 2b$, satisfy

1) $|a_\alpha(t,x)| \le C(1+|x|)^{\frac{2b-|\alpha|}{2b-1}-\varepsilon}$, $\varepsilon > 0$;
2) for $|x-y| \le 1$ we have $|\Delta_x^y a_\alpha(t,x)| \le C_x |x-y|^\lambda |y|^{\frac{2b-|\alpha|}{2b-1}}$, $\lambda \in (0,1]$.

Theorem 2.7. *Suppose that the coefficients of a uniformly parabolic in the sense of Petrovskij system* (2.9) *satisfy* α_{14} *and* α_{15}. *Then there is a fundamental solution* $Z(t,x;\tau,\xi)$ *to* (2.9) *which satisfies*

$$|D^\alpha Z(t,x;\tau,\xi)|$$

$$\le C_\alpha \sum_{k=1}^{|\alpha|} (t-\tau)^{-\frac{n+|\alpha|-k}{2b}} (1+|x|)^{\frac{k}{2b-1}} \exp\{-c\rho(t,x;\tau,\xi) + \eta|x|^q - \eta|\xi|^q\} ,$$

$$|\alpha| \le 2b . \quad (2.28)$$

Under the basic assumption that $a_\alpha(t,x)$ has order $|x|^{\frac{2b-|\alpha|}{2b-1}}$ (or a little bit smaller) as $|x| \to \infty$, a fundamental solution was constructed in (Zhitomirskij 1959), (Ejdel'man 1964) by various methods. In (Ejdel'man 1964, p.134–135) a simple transformation was presented which reduces an equation with growing coefficients to an equation with dissipation.

Example 2.3. We shall find a fundamental solution to the Cauchy problem for equation (1.3) of Chap. 1

$$\frac{\partial P}{\partial t} = \frac{1}{4}\gamma^2 N_0 \frac{\partial^2 P}{\partial x^2} + \alpha \frac{\partial}{\partial x}(xP) \, ; \quad P(0,x) = P_0(x) \, . \tag{2.29}$$

We note that the coefficients of this equation grow and satisfy α_{15}. For brevity of notation we put $a^2 = \frac{1}{4}\gamma^2 N_0$.

By means of the Fourier transform with respect to x the problem (2.29) is transformed into the following problem for a partial differential equation of the first order

$$\frac{\partial \hat{P}}{\partial t} + \alpha\sigma \frac{\partial \hat{P}}{\partial \sigma} = -a^2\sigma^2 \hat{P} \, , \quad \hat{P}(0,\sigma) = \hat{P}_0(\sigma) \, . \tag{2.30}$$

We write down the characteristics equations

$$\frac{dt}{1} = \frac{d\sigma}{\alpha\sigma} = \frac{d\hat{P}}{-a^2\sigma^2\hat{P}} \, ,$$

whose first integrals have the form

$$\Psi_1 = \sigma \mathrm{e}^{-\alpha t} = c_1 \, , \quad \Psi_2 = \hat{P}\mathrm{e}^{a^2\sigma^2/2\alpha} = c_2 \, .$$

Hence the solution of (2.30) is given by $\hat{P}(t,\sigma) = \mathrm{e}^{-a^2\sigma^2(1-\mathrm{e}^{-2\alpha t})/2\alpha}\hat{P}_0(\sigma\mathrm{e}^{-\alpha t})$, and the solution of (2.29) by

$$\begin{aligned} P(t,x) &= \frac{1}{2\pi}\int_{-\infty}^{\infty} \mathrm{e}^{\mathrm{i}x\sigma - a^2\sigma^2(1-\mathrm{e}^{-2\alpha t})/2\alpha} d\sigma \int_{-\infty}^{\infty} \mathrm{e}^{-\mathrm{i}\xi\sigma\mathrm{e}^{-\alpha t}} P_0(\xi) d\xi \\ &= \int_{-\infty}^{\infty} \left(\frac{1}{2\pi}\int_{-\infty}^{\infty} \mathrm{e}^{\mathrm{i}(x-\xi\mathrm{e}^{-\alpha t})\sigma - a^2\sigma^2(1-\mathrm{e}^{-2\alpha t})/2\alpha} d\sigma \right) P_0(\xi) d\xi \, . \end{aligned}$$

The internal integral in the last formula is the desired fundamental solution $Z(t,x,\xi)$. This integral can be directly computed and we arrive at the formula

$$\begin{aligned} &Z(t-\tau,x,\xi) \\ &= [2\pi a^2(1-\mathrm{e}^{-2\alpha(t-\tau)})]^{-1/2}\alpha^{1/2} \exp\left\{ \frac{-\alpha(x-\xi\mathrm{e}^{-\alpha(t-\tau)})^2}{2a^2(1-\mathrm{e}^{-2\alpha(t-\tau)})} \right\} \, . \end{aligned} \tag{2.31}$$

It is worth noting that in the books on statistical radiotechnics (Tikhonov and Kul'man 1975, pp.86–88) the formula (2.31) is obtained as a result of complicated computations using the Mehler expansion of two-dimensional normal density into a series in orthogonal Chebyshev–Hermite polynomials.

Formula (2.31) shows that the processes considered in Example 1.1 of Chap. 1 are nonstationary, but as $t-\tau\to\infty$, the distribution (2.31) converges to the normal distribution.

1.5. Parabolic Equations of the Second Order. We now consider the most important linear parabolic equation of the second order

$$\frac{\partial u}{\partial t}=\sum_{i,j=1}^{n} a_{ij}(t,x)\frac{\partial^2 u}{\partial x_i\partial x_j}+\sum_{i=1}^{n} b_i(t,x)\frac{\partial u}{\partial x_i}+c(t,x)u\,. \tag{2.32}$$

If the coefficients of (2.32) satisfy conditions α_1', α_2, α_3, then there exists a fundamental solution $Z(t,x;\tau,\xi)$ such that (2.16) and (2.17) hold, $b=1$, and if they satisfy α_1', α_2, α_5, and α_6, then (2.16) and (2.22) hold. We have the following theorem.

Theorem 2.8. *If either α_1', α_2, α_3 or α_1', α_2, α_5, and α_6 are satisfied, then the fundamental solution for* (2.32) *satisfies the two-sided estimates*

$$c_1(t-\tau)^{-n/2}\exp\left\{-\frac{c_1|x-\xi|^2}{(t-\tau)}\right\}\le Z(t,x;\tau,\xi)$$

$$\le c_2(t-\tau)^{-n/2}\exp\left\{-\frac{c_c|x-\xi|^2}{(t-\tau)}\right\}, \tag{2.33}$$

where positive constants c_1 and c_2 depend on n, δ, T, M_0, L, and λ in the former case and on n, δ, T, M_0, ω, and ω_1 in the latter; the constant c_2 depends only on δ and c_1 on δ and M_0.

The estimate of $Z(t,x;\tau,\xi)$ from above was established in Theorems 2.2 and 2.3 respectively, and the estimate from below can be easily deduced from Theorem 1.1 of Krylov and Safonov (1980, p.163) and the analysis of the structure of $Z(t,x;\tau,\xi)$ given by the Levy or Hopf method.

The deepest and most complete results are obtained for *parabolic equations in divergent form*

$$p(x)\frac{\partial u}{\partial t}=\sum_{i,j=1}^{n}\frac{\partial}{\partial x_i}\left(a_{ij}(t,x)\frac{\partial u}{\partial x_j}\right). \tag{2.34}$$

We study (2.34) under the following assumptions:

α_{16}. $p(x)$ and $a_{ij}(t,x)$ are measurable functions in $\mathbb{R}^n$ and $\mathbb{R}^{n+1}_+=\{(t,x): t\ge 0, x\in\mathbb{R}^n\}$.

α_{17}. there is a constant $\mu\ge 1$ such that

$$\mu^{-1}\le p(x)\le\mu\,,\quad \mu^{-1}|\sigma|^2\le\sum_{i,j=1}^{n} a_{ij}(t,x)\sigma_i\sigma_j\le\mu|\sigma|^2\quad\forall\sigma\in\mathbb{R}^n\,.$$

We note that condition α_{17} implies the inequality $|a_{ij}(t,x)| \le \mu$, $i,j = 1,\dots,n$.

The absence of assumptions on regularity of coefficients requires introducing the notion of the generalized solution of (2.34). We define the space $L_{2,\text{loc}}[(0,\infty); H^1_{\text{loc}}(\mathbb{R}^n)]$ of functions $u(t,x)$ defined on $\mathbb{R}^{n+1}_+$. As functions of x they belong to $H^1(\Omega)$, where Ω is any compact set in $\mathbb{R}^n$. Now consider $u(t,x)$ as an element of $H^1(\Omega)$ depending on the parameter t and assume that it belongs to the space $L_2([t_1,t_2])$ for any t_1, t_2, $0 < t_1 < t_2$. By $C^1_0(\mathbb{R}^{n+1}_+)$ we denote the set of functions with compact supports with continuous partial derivatives of the first order, and by $L^0_2(\mathbb{R}^n)$ the set of functions with compact supports belonging to $L_2(\mathbb{R}^n)$.

Definition 2.3. A *weak solution* to (2.34) in $\mathbb{R}^{n+1}_+$ is a function $u(t,x) \in L_{2,\text{loc}}[(0,\infty); H^1_{\text{loc}}(\mathbb{R}^n)]$ satisfying the integral identity

$$\int\!\!\int \left[-pu\frac{\partial \varphi}{\partial t} + \sum_{i,j=1}^{n} a_{ij}\frac{\partial u}{\partial x_j}\frac{\partial \varphi}{\partial x_i} \right] dtdx = 0 \tag{2.35}$$

for any test function $\varphi(t,x) \in C^1_0(\mathbb{R}^{n+1}_+)$.

Definition 2.4. A *weak solution to the Cauchy problem* is a weak solution $u(t,x)$ to (2.34) in $\mathbb{R}^{n+1}_+$ satisfying the initial condition

$$u|_{t=0} = \varphi(x) \tag{2.36}$$

in a weak sense, i.e. for any function $\psi(x) \in C^1_0(\mathbb{R}^{n+1}_+)$

$$\exists \lim_{t\to 0+} \int u(t,x)\psi(x)dx = \int \varphi(x)\psi(x)dx\,.$$

Definition 2.5. A function $\Gamma(t,x;\tau,\xi)$ is a *weak fundamental solution* to the Cauchy problem (2.34), (2.36) if it is defined in the domain $V = \{(t,x;\tau,\xi) : (t,x) \in \mathbb{R}^{n+1}_+, (\tau,\xi) \in \mathbb{R}^{n+1}_+, \tau < t\}$ and for any bounded measurable function $\varphi(x)$ with compact support the function

$$u(t,x) = \int \Gamma(t,x;\tau,\xi)\varphi(\xi)d\xi \tag{2.37}$$

is a weak bounded solution to the Cauchy problem

$$u|_{t=\tau} = \varphi(x)\,. \tag{2.38}$$

One can establish that $\Gamma(t,x;\tau,\xi)$ is a limit of a sequence of classical fundamental solutions to infinitely smooth Cauchy problems. This allows us to prove the assertions in the smooth case first which essentially simplifies the original definitions and proofs.

We move to the presentation of properties of a weak fundamental solution $\Gamma(t,x;\tau,\xi)$. The principal difference from what was presented before is that the estimates we shall obtain will contain constants depending on n and μ and will not depend at all on the smoothness of coefficients nor on the width T of the band. We owe these results to Nach (1958), Moser (1964, 1971), and especially Aronson (1967). A separate survey (Porper and Ejdel'man 1984) is devoted to their presentation. The methods of obtaining the information are based on subtle results of the theory of divergent equations: Harnack inequality, a priori Hölder property of solutions, and special estimates of the type of embedding theorems. If the reader is not familiar with the language of generalized solutions, he can assume that the fundamental solutions are classical with estimates completely independent of the smoothness of coefficients.

In what follows we denote by k (with various indices) positive constants depending only on n and μ. The central result is

Theorem 2.9. *Suppose that conditions α_{16} and α_{17} are fulfilled. The weak fundamental solution $\Gamma(t,x;\tau,\xi)$ to the problem* (2.34), (2.38) *satisfies the two-sided estimate*

$$k_1(t-\tau)^{-n/2}\exp\left\{-\frac{k_1|x-\xi|^2}{(t-\tau)}\right\} \le \Gamma(t,x;\tau,\xi) \le$$

$$\le k_2(t-\tau)^{-n/2}\exp\left\{-\frac{k_2|x-\xi|^2}{(t-\tau)}\right\}. \tag{2.39}$$

Property. 2.3. The function $\Gamma(t,x;\tau,\xi)/p(\xi)$ satisfies the *uniform Hölder condition* with respect to the variables (t,x,τ,ξ). In particular, for all $t>\tau$ and any x, y, ξ we have

$$|\Delta_x^y\Gamma(t,x;\tau,\xi)| \le k(|y-x|(t-\tau)^{-1/2})^{\lambda}(t-\tau)^{-n/2}, \quad \lambda=\lambda(n,\mu)>0.$$

Property. 2.4. The function $\widetilde{\Gamma}(t,x;\tau,\xi) = (p(x)/p(\xi))\Gamma(t,x;\tau,\xi)$ is a unique weak fundamental solution in the class of bounded functions to the Cauchy problem for the equation

$$-p(\xi)\frac{\partial v}{\partial\tau} = \sum_{i,j=1}^{n}\frac{\partial}{\partial\xi_i}\left(a_{ij}(\tau,\xi)\frac{\partial u}{\partial\xi_j}\right), \quad 0\le\tau<t.$$

Property. 2.5. $\Gamma(t,x;\tau,\xi) = \int\Gamma(t,x;\beta,y)\Gamma(\beta,y;\tau,\xi)dy$, $\forall\beta\in(\tau,t)$.

Property. 2.6. $\int\Gamma(t,x;\tau,\xi)d\xi = 1$.

The next property is an integral variant of the estimate of the decay of generalized gradients of a weak fundamental solution $\Gamma(t,x;\tau,\xi)$ with the growth of the difference $|x-\xi|$.

Property. 2.7. For all $A\ge 0$ and all t_1, t_2 such that $\tau<t_1<t_2$, $t_2-t_1\le \min(1,(t_2-\tau)/2)$ we have

$$\int_{t_1}^{t_2} d\tau \int_{|x-\xi|\geq\sqrt{A(t-\tau)}} \sum_{j=1}^{n} \left|\frac{\partial \Gamma}{\partial x_j}(t,x;\tau,\xi)\right|^2 dx \leq K(t_2-\tau)^{-n/2}\exp\{-kA\}\,.$$

Some additional interesting information is obtained for weak fundamental solutions to stationary equations (Porper 1977), (Porper and Ejdel'man 1984), i.e. equations with coefficients independent of time

$$p(x)\frac{\partial u}{\partial t} = \sum_{i,j=1}^{n} \frac{\partial}{\partial x_i}\left(a_{ij}(x)\frac{\partial u}{\partial x_j}\right)\,. \tag{2.40}$$

Theorem 2.10. *A weak fundamental solution $\Gamma(t,x;\xi)$ to the Cauchy problem for* (2.40) *has classical derivatives in t of arbitrary order and for any $t > 0$ they satisfy the estimates*

$$\left|\frac{\partial^\nu}{\partial t^\nu}\Gamma(t,x;\xi)\right| \leq k(\nu,\mu,n)t^{-n/2-\nu}\exp\left\{-k(\mu)\frac{|x-\xi|^2}{t}\right\}\,. \tag{2.41}$$

There arises a natural question of validity of analogous results for equations of second order with lower order terms without assuming their smoothness. Such information is contained in the paper of Aronson (1968) (see also (Porper and Ejdel'man 1984)).

The equation is of the form

$$p(x)\frac{\partial u}{\partial t} = \sum_{i,j=1}^{n} \frac{\partial}{\partial x_i}\left(a_{ij}(t,x)\frac{\partial u}{\partial x_j} + a_i(t,x)u\right)$$
$$+\sum_{i=1}^{n} b_i(t,x)\frac{\partial u}{\partial x_i} + c(t,x)u\,. \tag{2.42}$$

We assume that the coefficients of (2.42) are measurable functions in the band Π_T, conditions α_{16}, α_{17} are fulfilled, and there exist $\theta_0 \in (0,1)$, R_0, M_0 such that

α_{18}. for some $p \geq 2/(1-\theta_0)$ and $\theta_2 = 1/2 - n/2p - 1/q \geq \theta_0/2$

$$a_i(t,x)\,, b_i(t,x) \in L_q([0,T];L_p(\{x : |x| < R_0\}))\,;$$

α_{19}. for $T_0 = \min(1,\sqrt{T})$, for some $p > \frac{1}{1-\theta_x}$ and $\theta_1 = 1 - \frac{n}{2p} - \frac{1}{q} \geq 0$ we have, uniformly in ξ,

$$c(t,x) \in L_q([0,T];L_p(\{x : |x| < T_0\})\,;$$

α_{20}. for almost all x, $|x| > R_0$, $t \in [0,T]$ we have

$$|a_i(t,x)|\,, |b_i(t,x)|\,, c(t,x) \leq M_0\,.$$

One of the reasons for giving these important but not so apparent conditions is to underline again that the conditions for coefficients are related to the structure of the equation.

By C we shall denote various positive constants depending on the structure of (2.42) only, i.e. determined only by the numbers n, μ, θ_0, R_0, T, and the norms $\|a_i\|$, $\|b_i\|$, $\|c\|$ in the corresponding spaces $L_{p,q}$.

Theorem 2.11. *Suppose that the coefficients of the equation* (2.42) *satisfy conditions* α_{16}–α_{20}. *Then this equation has a weak fundamental solution* $\Gamma(t,x;\tau,\xi)$ *to the Cauchy problem which satisfies the two-sided estimates*

$$C_1(t-\tau)^{-n/2}\exp\left\{-\frac{c_1|x-\xi|^2}{(t-\tau)}\right\} \le \Gamma(t,x;\tau,\xi)$$

$$\le C_2(t-\tau)^{-n/2}\exp\left\{-\frac{c_2|x-\xi|^2}{(t-\tau)}\right\}, \tag{2.43}$$

where C_1, C_2, c_1, *and* c_2 *are positive constants depending on the structure of the equation* (2.42).

Despite the similarity in appearance the estimates (2.31) and (2.43) are principally different, in particular (2.31) is valid in $\mathbb{R}^{n+1}_+$ and (2.43) in a finite band Π_T. There arises a question of conditions on $a_i(t,x)$, $b_i(t,x)$, $c(t,x)$ under which one could obtain an estimate of type (2.43) in $\mathbb{R}^{n+1}_+$. Some results in this direction are presented in (Porper and Ejdel'man 1986).

1.6. On Estimates of Fundamental Solutions of Parabolic Systems in $\mathbb{R}^{n+1}_+$ and Fundamental Solutions to Elliptic Systems Generated by Parabolic Ones. We have already noted that the estimates (2.13) of a fundamental solution $\Gamma_0(t,x)$ to a parabolic system with constant coefficients

$$D_t u = \sum_{|\alpha|=2b} a_\alpha D^\alpha u\,, \tag{2.44}$$

and the estimates (2.39) of a weak fundamental solution $\Gamma(t,x;\tau,\xi)$ to (2.34) are valid in $\mathbb{R}^{n+1}_+$. Other instances of validity of analogous estimates are given in (Ejdel'man 1964). On the basis of such estimates we can obtain simple and important formulas which enable the construction of fundamental solutions to elliptic systems, that are right hand sides of parabolic systems (2.9), from fundamental solutions of parabolic systems.

We recall that a system

$$A(x;D)u \equiv \sum_{|\alpha|\le 2b} a_\alpha(x) D^\alpha u = 0\,, \tag{2.45}$$

is called *elliptic* if

$$\det\left(\sum_{|\alpha|=2b} a_\alpha(x)\sigma^\alpha\right) \ne 0 \qquad \forall x\in\mathbb{R}^n\,, \quad \forall \sigma\in\mathbb{R}^n\,,\ \sigma\ne 0\,,$$

and a fundamental solution $\mathcal{E}(x,\xi)$ to such a system is defined as a square matrix of dimension m that provides a formula for the solution $u(x)$ to the elliptic system

$$A(x,D)u = f(x) \tag{2.46}$$

in the form

$$u(x) = \int \mathcal{E}(x,\xi) f(\xi) d\xi \tag{2.47}$$

for any smooth function $f(x)$ with compact support.

Theorem 2.11. *A fundamental solution $\mathcal{E}(x,\xi)$ to the elliptic system*

$$\sum_{|\alpha|=2b} a_\alpha D^\alpha u = 0 \tag{2.48}$$

generated by the parabolic system (2.44) *is given by the formula*

$$\mathcal{E}_0 = \int_0^\infty \left\{ \Gamma_0(t,x) - \sum_{s=0}^{2b-n} \frac{1}{s!} \left(\sum_{\nu=1}^{n} (x_\nu - a_\nu) \frac{\partial}{\partial x_\nu} \right)^s \Gamma_0(t,x)\,|_{x=0} \right\} dt \,, \tag{2.49}$$

where $a = (a_1, \ldots, a_n) \neq 0$.

Thus if $2b < n$, then $\mathcal{E}(x) = \int_0^\infty \Gamma_0(t,x)dt$, and if $2b \geq n$, then $\int_0^\infty \Gamma_0(t,x)dt$ diverges and has to be regularized according to the rule shown in (2.49).

For $x \neq 0$, $\mathcal{E}_0(x)$ has all derivatives satisfying the following estimates:

$$\begin{aligned} \text{for } |x| \leq 1\,, n + |\alpha| < 2b \quad & |D^\alpha \mathcal{E}_0(x)| \leq C\,; \\ \text{for } |x| \leq 1\,, n + |\alpha| = 2b \quad & |D^\alpha \mathcal{E}_0(x)| \leq C \ln(1/|x|) + C_1\,; \\ \text{for } n + |\alpha| < 2b \quad & |D^\alpha \mathcal{E}_0(x)| \leq C_\alpha |x|^{-n-|\alpha|+2b}\,. \end{aligned} \tag{2.50}$$

The estimates (2.50) follow from (2.13) using standard arguments (Ejdel'man 1964, pp.148–154).

Example 2.4. We find a fundamental solution to the *Laplace equation* $\Delta u = 0$ in $\mathbb{R}^3$ and $\mathbb{R}^2$ using (2.1) and (2.49):

$$\begin{aligned} \mathcal{E}_0(x_1,x_2,x_3) &= \frac{1}{(2\sqrt{\pi})^3} \int_0^\infty \exp\left\{-\frac{|x|^2}{4t}\right\} t^{\frac{3}{2}} dt = \left| \begin{matrix} \beta = t^{1/2} \\ -2d\beta = t^{-3/2}dt \end{matrix} \right| \\ &= \frac{1}{4\pi^{3/2}} \int_0^\infty \exp\left\{-\frac{|x|^2}{4}\beta^2\right\} d\beta = \frac{1}{4\pi|x|}\,; \\ \mathcal{E}_0(x_1,x_2) &= \frac{1}{4\pi} \int_0^\infty \left[\exp\left\{-\frac{|x|^2}{4t}\right\} - \exp\left\{-\frac{1}{4t}\right\} \right] dt\,, \end{aligned}$$

(for $a_1 = a_2 = \frac{1}{\sqrt{2}}$).

The last integral can be easily computed by differentiation with respect to the parameter $A = |x|^2$ and we find that $\mathcal{E}_0(x_1,x_2) = \frac{1}{2\pi} \ln \frac{1}{|x|}$.

Definition 2.6. A fundamental solution $\mathcal{E}(x,\xi)$ is called *principal* if

$$\exists \lim_{|x-\xi|\to\infty} \mathcal{E}(x,\xi) = 0$$

and this convergence is uniform in each direction.

Form (2.50) it follows that the fundamental solution to (2.49) is principal if the number n of spatial variables is greater than the order $2b$ of the system.

Inclusion of lower order derivatives into the system allows us, under special conditions, to construct principal fundamental solutions for any n and $2b$. Cases of that sort were considered in (Ejdel'man 1964). We give one of the results in this direction.

Theorem 2.12. *We consider a system parabolic in the sense of Petrovskij with constant coefficients*

$$D_t u = \sum_{|\alpha|\le 2b} a_\alpha D^\alpha u \,. \tag{2.51}$$

Suppose that the real parts of all p-roots of the polynomial

$$\det\left(pI - \sum_{|\alpha|\le 2b} a_\alpha \sigma^\alpha\right)$$

are different from zero for all $\sigma \in \mathbb{R}^n$. Let $\Gamma(t,x)$ be a fundamental solution to the Cauchy problem for the system (2.51). *Then*

$$\mathcal{E}(x) = \int_0^\infty \Gamma(t,x)dt$$

is a principal fundamental solution to the elliptic system

$$\sum_{|\alpha|\le 2b} a_\alpha D^\alpha u = 0$$

and it satisfies the estimate

$$|\mathcal{E}(x)| \le C\mathrm{e}^{-c|x|}\,, \quad |x| > 1\,.$$

We present some corollaries that can be deduced from the results of the study of weak fundamental solutions to (2.34).

In $\mathbb{R}^n$ we consider a uniformly elliptic equation

$$-\sum_{i,j=1}^{n} \frac{\partial}{\partial x_i}\left(a_{ij}(x)\frac{\partial u}{\partial x_j}\right) + \lambda p(x)u = 0 \tag{2.52}$$

assuming that conditions α_{16} and α_{17} are fulfilled. For brevity, we are not giving the definitions of fundamental solution nor that of principal fundamental solution. They are completely analogous to the presented above in similar situations; λ here is a non-negative real parameter.

Theorem 2.13. *If $n \geq 3$ or $n = 2$ and $\lambda > 0$, then the function*

$$\mathcal{E}(x, \xi, \lambda) = \int_0^\infty e^{-\lambda t} \Gamma(t, x, \xi) dt \tag{2.53}$$

is a unique principal fundamental solution to (2.52), *which satisfies the two-sided estimate*

$$\widetilde{g}_1(x - \xi, \lambda) \leq \mathcal{E}(x, \xi, \lambda) \leq \widetilde{g}_2(x - \xi, \lambda) , \tag{2.54}$$

and the function

$$\widetilde{g}(x, \lambda) = k \int_0^\infty t^{-\frac{n}{2}} \exp\left\{-\lambda - \frac{k|x|^2}{t}\right\} dt = k \left(\frac{\sqrt{\lambda}}{|x|}\right)^{n/2-1} K_{n/2-1}(k\sqrt{\lambda}|x|)$$

is a fundamental solution to the equation $-\Delta u + \lambda k u = 0$, where $K_\nu(x)$ is the Macdonald function (Nikiforov and Uvarov 1974). *The functions $\widetilde{g}_1$ and $\widetilde{g}_2$ are the same as $\widetilde{g}$ but with various constants k_1, k_2 depending on n and μ.*

Employing the asymptotics of the Macdonald function or directly deriving the estimates we can show that

1) for $n \geq 3$, $\lambda \geq 0$ and also for $n = 2$, $\sqrt{\lambda}|x - \xi| \geq 1$ we have

$$\mathcal{E}(x, \xi; \lambda) \sim k|x - \xi|^{2-n} e^{-k\sqrt{\lambda}|x-\xi|} ;$$

2) for $n = 2$, $0 < \sqrt{\lambda}|x - \xi| < 1$ we have

$$\mathcal{E}(x, \xi; \lambda) \sim k + k \ln\left(\frac{1}{\sqrt{\lambda}|x - \xi|}\right) .$$

The symbol $\sim$ means here that the function on the right is an upper and lower bound for the function of the left with various k.

1.7. Systems with Constant Coefficients Parabolic in the Sense of Shilov. We consider the Cauchy problem for the system with constant coefficients

$$\mathcal{L}(D_t, D) \equiv D_t u - \sum_{|\alpha| \leq p} a_\alpha D^\alpha u = 0 , \quad u|_{t=0} = \varphi(x) . \tag{2.55}$$

We introduce the polynomial $L(\lambda, s) = \det\left(\lambda I - \sum_{|\alpha| \leq p} a_\alpha s^\alpha\right)$, $s = \sigma + i\tau \in \mathbb{C}^n$, and we define a function $\Lambda(s) = \max_j \mathrm{Re}\lambda_j$, where λ_j are zeros of $L(\lambda, s)$. *Parabolicity in the sense of Shilov* means that there exist positive constants h and C such that

$$\Lambda \leq -C|\sigma|^h + C_1 .$$

The fundamental solution to (2.55) is given by the formula

$$\Gamma(t,x) = \frac{1}{(2\pi)^n} \int e^{ix\cdot\sigma} Q(t,\sigma) d\sigma , \quad Q(t,\sigma) = \exp\left(\sum_{|\alpha|\leq p} a_\alpha \sigma^\alpha \right) .$$

We denote by p_0 the order of growth of the entire analytic function $Q(t,s)$, $p_0 \leq p$.

Lemma 2.1 (Friedman 1963). *For parabolic systems we have* $p_0 > 1$. *For* $Q(t,s)$ *we have the following estimates:*

$$|Q(t,s)| \leq C_2(1 + |s|t^{1/p})^{(m-1)p} e^{c_2 t|s|^{p_0}}$$

$$|Q(t,\sigma)| \leq C_3(1 + |\sigma|t^{1/p})^{(m-1)p} e^{-ct|\sigma|^h} .$$

Definition 2.7. The *genus of a parabolic system* is the greatest number μ such that in the domain $G = \{s : |\tau_j| \leq k(1 + |\sigma_j|)^\mu , j = 1, \ldots, n\}$ the following estimate is valid

$$|Q(t,s)| \leq C_4(1 + |s|t^{1/p})^{(m-1)p} e^{-c_3 t|\sigma|^h} , \quad c_3 > 0 .$$

Theorem 2.14 (Friedman 1963). *The fundamental solution* $\Gamma(t,x)$ *is infinitely differentiable in* t *and* x *for* $t > 0$ *and we have*

$$|D^\alpha \Gamma(t,x)| \leq C_\alpha t^{-\frac{n+|\alpha|+\nu}{h}} \exp\{-c_4 t^{-\mu(h-\mu)} |x|^{h_1}\} \quad \text{for} \quad \mu \leq 0 , \qquad (2.56)$$

$$|D^\alpha \Gamma(t,x)| \leq C_\alpha t^{-\frac{n+|\alpha|+\nu}{h}} \exp\{-c_5 t^{-\mu(p_0-\mu)} |x|^{p_1}\} \quad \text{for} \quad \mu > 0 , \qquad (2.57)$$

where $\nu = (m-1)(p-h)$, $p_1 = p_0/(p_0 - h)$, $h_1 = h/(h-\mu)$, $c_4 > 0$, $c > 0$.

We draw attention to an essential difference between (2.56) and (2.57).

Theorem 2.14 is a direct consequence of special theorems on Fourier transform of entire functions (Gel'fand and Shilov 1958a), (Friedman 1963).

1.8. Diffusion Equation with Inertia. This class of degenerated parabolic equations has, as was first mentioned by Kolmogorov, fundamental solutions with properties quite close to the fundamental solution (2.1) of the heat equation. In the same time, these fundamental solutions have very curious structure properties which is caused by the fact that unlike usual parabolic equation their initial model equations *necessarily* have variable coefficients. We begin with the construction of a fundamental solution for the *model equation of diffusion with inertia*

$$\mathcal{P}u \equiv \frac{\partial u}{\partial t} - \sum_{j=1}^{n} x_j \frac{\partial u}{\partial y_j} - \sum_{j=1}^{n} y_j \frac{\partial u}{\partial z_j} = a^2 \Delta_x u ,$$

$$u|_{t=0} = \varphi(x,y,z) . \tag{2.58}$$

We shall seek for a solution to (2.58), as usually, by means of the Fourier transform:

$$u(t,x,y,z) = \frac{1}{(2\pi)^{3n}} \int_{\mathbb{R}^{3n}} \exp\{\mathrm{i}x \cdot \alpha + \mathrm{i}y \cdot \beta + \mathrm{i}z \cdot \gamma\} \hat{u}(t,\alpha,\beta,\gamma) d\alpha d\beta d\gamma .$$

Using elementary properties of the Fourier transform we obtain from (2.58) the Cauchy problem for a partial differential equation of the first order with respect to $\hat{u}(t,\alpha,\beta,\gamma)$

$$\frac{\partial \hat{u}}{\partial t} + \sum_{j=1}^{n} \beta_j \frac{\partial \hat{u}}{\partial \alpha_j} + \sum_{j=1}^{n} \gamma_j \frac{\partial \hat{u}}{\partial \beta_j} = -a^2|\alpha|^2 \hat{u} ,$$

$$\hat{u}|_{t=0} = \hat{\varphi}(\alpha,\beta,\gamma) . \tag{2.59}$$

The problem (2.59) can be explicitly solved by the method of characteristics (in a simple version this was done in Example 2.3), and next by a direct computation (changing the order of integration) we can arrive at the following formula

$$u(t,x,y,z) = \int_{\mathbb{R}^{3n}} \Gamma_0(t,x,y,z;\xi,\eta,\zeta) \varphi(\xi,\eta,\zeta) d\xi d\eta d\zeta , \tag{2.60}$$

where

$$\Gamma_0(t,x,y,z;\xi,\eta,\zeta) = \frac{(3 \cdot 180)^{n/2} t^{9n/2}}{2^n (a\sqrt{\pi})^{3n}} \exp\left\{ -\frac{1}{4a^2} \left(\frac{|x-\xi|}{t^{1/2}} \right)^2 \right.$$

$$\left. -\frac{3}{a^2} \left| \frac{y-\eta}{t^{3/2}} + \frac{x+\xi}{2t^{1/2}} \right|^2 - \frac{180}{a^2} \left| \frac{z-\zeta}{t^{5/2}} + \frac{y+\eta}{2t^{3/2}} + \frac{x-\xi}{12t^{1/2}} \right|^2 \right\} . \tag{2.61}$$

In an analogous way we can consider the case when the inertia takes place with respect to some variables only (i.e. in (2.57) the summation on the left is taken up to numbers less than n).

A direct generalization of (2.57) is the following equation

$$\mathcal{P}u = \mathcal{L}u = \sum_{k,j=1}^{n} a_{kj}(t,x,y,z) \frac{\partial^2 u}{\partial x_k \partial x_j} + \sum_{j=1}^{n} b_j(t,x) \frac{\partial u}{\partial x_j} + c(t,x)u , \tag{2.62}$$

where $\frac{\partial u}{\partial t} = \mathcal{L}u$ is an equation uniformly parabolic in the sense of Petrovskij.

Under various assumptions the fundamental solutions for this equation were constructed and studied in (Il'in 1964), (Sonin 1967), and (Weber 1951).

We consider the equation (Ejdel'man and Malitskaya 1975)

$$\mathcal{P}u = \mathcal{L}u = \sum_{|\alpha| \le 2b} a_\alpha(t,x,y,z) D_x^\alpha u , \tag{2.63}$$

where $\frac{\partial u}{\partial t} = \mathcal{L}u$ is an equation parabolic in the sense of Petrovskij. We shall call it *equation of diffusion with inertia* (of an arbitrary order).

The model equation for it is of the form

$$\mathcal{P}u = \mathcal{L}_0 u \equiv \sum_{|\alpha|=2b} a_\alpha D_x^\alpha u \,. \tag{2.64}$$

Theorem 2.15. *The model equation* (2.64) *of diffusion with inertia has a fundamental solution $\Gamma_0(t,x,y,z;\xi,\eta,\zeta)$ which is an entire analytic function of order q with respect to the variables*

$$(x_j-\xi_j)t^{-1/2b}\,, \quad (y_j-\eta_j+x_jt)t^{-(1+1/2b)}\,, \quad (z_j-\zeta_j+y_jt+x_jt^2/2)t^{-(2+1/2b)}\,,$$
$$j=1,\dots,n\,.$$

The following estimates for its derivatives are true (for real arguments):

$$|D_x^\alpha D_y^l D_z^s \Gamma_0(t,x,y,z;\xi,\eta,\zeta)|$$
$$\le C_{\alpha ls} t^{-M_{\alpha ls}/2b} \exp\{-c\widetilde{\rho}_{2b}(t,x,y,z;\xi,\eta,\zeta)\}\,, \tag{2.65}$$

where

$$M_{\alpha ls} = (6b+3)n + |\alpha| + (2b+1)|l| + (4b+1)|s|\,,$$
$$\widetilde{\rho}_{2b}(t,x,y,z;\xi,\eta,\zeta)$$
$$= (|x-\xi|t^{-1/2b})^q + (|y-\eta+xt|t^{-(1+1/2b)})^q + (|z-\zeta+yt+xt^2/2|t^{-(2+1/2b)})^q\,.$$

We also give a theorem on existence of fundamental solution to the equation

$$\frac{\partial u}{\partial t} - \sum_{j=1}^{l} x_j \frac{\partial u}{\partial y_j} = \sum_{|\alpha|\le 2b} a_\alpha(t,x,y) D_x^\alpha u\,, \quad l \le n\,. \tag{2.66}$$

Theorem 2.16 (Ejdel'man and Tychinskaya 1979). *Suppose that the coefficients of the equation* (2.66) *are defined in the band $\Pi_T = [0,T]\times\mathbb{R}^{n+l}$ and satisfy the following conditions:*

α_{21}. *$a_\alpha(t,x,y)$, $|\alpha|\le 2b$, are continuous and bounded by M_0;*

α_{22}. *there exist continuous derivatives in $y_1,\dots,y_l$ bounded by M_1;*

α_{23}. *there exist positive constants L and $\lambda\in(0,1]$ such that for any (t,x,y), $(t,\xi,y)\in\Pi_T$ we have*

$$|\Delta_x^\xi a_\alpha(t,x,y)| \le L|x-\xi|^\lambda\,, \quad |\alpha|\le 2b\,.$$

Then for the diffusion with inertia equation (2.66) *there is a fundamental solution $Z(t,x,y;\tau,\xi,\eta)$ which satisfies*

$$|D_x^\alpha Z(t,x,y;\tau,\xi,\eta)|$$
$$\le C_\alpha (t-\tau)^{-\frac{n+l(2b+1)+|\alpha|}{2b}} \Phi(t,x,y;\tau,\xi,\eta)\,, \quad |\alpha|\le 2b\,, \tag{2.67}$$

$$\left|\frac{\partial}{\partial y_j} Z(t,x,y;\tau,\xi,\eta)\right|$$
$$\leq C(t-\tau)^{-\frac{n+l(2b+1)+2b+1}{2b}} \Phi(t,x,y;\tau,\xi,\eta)\,, \tag{2.68}$$

where

$$\Phi(t,x,y;\tau,\xi,\eta) = \sum_{r=1}^{\infty} (C^r/r^r) \exp\{-c_0^r(|y-\eta+x(t-\tau)|(t-\tau)^{-(1+1/2b)})^q$$
$$-c(|x-\xi|(t-\tau)^{-1/2b})^q + c_1|\xi|^q\}\,,$$

and the constants C_α, C, c_0, c, c_1 depend on n, l, $2b$, T, M_0, M_1, L, λ, and the parabolicity constant δ.

We note that the constant c_0 is in general less than one hence the estimates (2.67) and (2.68) preserving the natural order of decay of the fundamental solution guarantee its decay of minimal degree.

For equations (2.62) of the second order we have a similar theorem (Sonin 1967), (Weber 1951) with an essential improvement that instead of Φ in estimates (2.67), (2.68) we take the function

$$\Phi_1(t,x,y,z;\tau,\xi,\eta) = -c\widetilde{\rho}_2(t-\tau,x,y,z;\xi,\eta,\zeta) + c_1|\xi|^2 + c_2|\zeta|^2\,.$$

All these results are established by means of some special modifications of the classical Levy's method.

1.9. Parabolic Pseudodifferential Equations with Nonsmooth Symbols. We seek for a fundamental solution to the *Cauchy problem* for a parabolic equation

$$D_t u + A_0 u + \sum_{k=1}^{s} A_k u = f\,, \quad u|_{t=0} = \varphi(x)\,, \tag{2.69}$$

with pseudodifferential operators described in 1.7 of Chap. 1 (Drin' 1986), (Kochubej 1985), (Ejdel'man and Drin' 1981). We recall that we approach pseudodifferential operators by means of hypersingular integrals.

We formulate a condition for the principal symbol. We fix a natural number $N \geq 2n + 2[\gamma] + 1$.

α_{24}. $a(t,x;\sigma)$ has N continuous derivatives with respect to $\sigma_1,\ldots,\sigma_n$ for all x, y, $\sigma \in \mathbb{R}^n$, $\sigma \neq 0$, $t,\tau \in [0,T]$, $|\alpha| \leq N$ satisfying

$$|D_\sigma^\alpha a(t,x;\sigma)| \leq C_N |\sigma|^{\gamma-|\alpha|}\,,$$
$$|\Delta_{\tau,y}^{t,x} D_\sigma^\alpha a(t,x;\sigma)| \leq C_N\left[|x-y|^\lambda + |t-\tau|^{\lambda/\gamma}\right]|\sigma|^{\gamma-|\alpha|}\,,$$

where $\Delta_{\tau,y}^{t,x} f(t,x) = f(t,x) - f(\tau,y)$, $\lambda \in (0,1)$.

We first consider (2.69) containing pseudodifferential operators of order γ with constant symbol only. In this case the fundamental solution $\Gamma_0(t,x)$ is defined, as usually, by means of the Fourier transform:

$$\Gamma_0(t,x) = \frac{1}{(2\pi)^n} \int \exp\{ix \cdot \sigma - a(\sigma)t\} d\sigma . \tag{2.70}$$

Theorem 2.17. *Suppose that condition* α_{24} *for a constant principal symbol is satisfied. Then* $\Gamma_0(t,x)$ *defined by* (2.70) *is a fundamental solution to* (2.69) *containing a pseudodifferential operator with constant symbol only. The following estimates hold:*

$$|D^\alpha \Gamma_0(t,x)| \le C t^{-\frac{n+|\alpha|}{\gamma}} [1 + |x| t^{-\frac{1}{\gamma}}]^{-n-\gamma-|\alpha|} , \quad |\alpha| \le N - 2n - [\gamma] , \tag{2.71}$$

$$\left| \frac{\partial}{\partial t} \Gamma_0(t,x) \right| \le C t^{-\frac{n+\gamma}{\gamma}} [1 + |x| t^{-\frac{1}{\gamma}}]^{-n-\gamma} . \tag{2.72}$$

Estimates (2.71) and (2.72) for $n = 1$ are obtained in (Ejdel'man and Drin' 1981) and for any n in (Kochubej 1985). We wish to emphasize that unlike in the case of parabolic differential equations, the estimates (2.71), (2.72) are of the power type. Using the asymptotic expansions it was shown in (Fedoryuk 1978) that they are exact.

We shall now consider equation (2.69) with variable coefficients and pseudodifferential operator of lower order.

Theorem 2.18 (Kochubej 1985). *We consider a parabolic pseudodifferential equation* (2.69). *Suppose that conditions* α_{24} *and* α_{25} *are fulfilled.*

α_{25}. *Suppose that the lower order symbols* $a_k(t,x;\sigma)$, $k = 1, \ldots, s$, *have derivatives in* $\sigma_1, \ldots, \sigma_n$ *up to order* N *and the following estimates are satisfied*

$$|D_\sigma^\alpha a_k(t,x;\sigma)| \le C_N |\sigma|^{\gamma_k - |\alpha|} ;$$

$$|\Delta_{t,x}^{\tau,y} D_\sigma^\alpha a_k(t,x;\sigma)| \le C_N \big(|x-y|^\lambda + |t-\tau|^{\lambda/\gamma}\big) |\sigma|^{\gamma_k - |\alpha|}$$

for all x, y, $\sigma \in \mathbb{R}^n$, $\sigma \ne 0$, $t,\tau \in [0,T]$, $|\alpha| \le N$.

Then there exists a fundamental solution to the Cauchy problem for (2.69) *definable by the Levy method*

$$Z(t,x;\tau,\xi) = \Gamma_0(t-\tau, x-\xi; \tau, \xi) + W(t,x;\tau,\xi) ,$$

where the first term is a fundamental solution to the equation with the symbol $a(\tau,\xi,\sigma)$ *and for* $W(t,x;\tau,\xi)$ *we have the estimate*

$$|W(t,x;\tau,\xi)| \le C(t-\tau)^{-n/\gamma} \Big\{ (t-\tau)^{\lambda/\gamma} \big[1 + |x-\xi|(t-\tau)^{-1/\gamma}\big]^{-n-\gamma} +$$

$$+ \sum_{k=1}^{s+1} (t-\tau)^{\gamma - \gamma_k/\gamma} \big[1 + |x-\xi|(t-\tau)^{-1/\gamma}\big]^{-n-\gamma_k} \Big\} , \quad \gamma_{s+1} = \gamma - \lambda .$$

Some additional conditions are necessary in the case when the degree of homogeneity of the symbols is an integer (Kochubej 1985). This is related to specific features of the representation of pseudodifferential operators by means of hypersingular integrals.

The following result is very interesting (Kochubej 1985).

Theorem 2.19. *Suppose that* $\gamma \leq 2$ *and conditions* α_{24}, α_{25} *are satisfied. In addition we assume that the characteristics of all hypersingular integrals are real and non-negative, all symbols with orders not less than one are even, and all differential operators appearing in* (2.69) *have real coefficients. Then the fundamental solution* $Z(t, x; \tau, \xi)$ *to* (2.69) *is non-negative.*

Hence, in particular, it follows that if the pseudodifferential operators are defined for all $t \geq 0$ and conditions α_{24}, α_{25} are fulfilled for any T, then $Z(t, x; \tau, \xi)$ is the transition density of so called right continuous strong Markov process without cutoff and without discontinuities of the second kind (Dynkin 1959).

§2. Cauchy Problem

2.1. Introduction. Fundamental solutions constructed in Sect. 1 and their analytic description and estimates allow us to obtain sufficiently complete information on existence, uniqueness, and continuous dependence of the solution to the Cauchy problem on initial functions and right hand terms. The greatest progress in obtaining precise theorems is achieved by an appropriate combination of the method of fundamental solutions and a priori estimates in various functional spaces. It is essential that the solution to the Cauchy problem can be written, as a rule, by means of the fundamental solution in a form convenient for study and further applications.

We note that from the theorems presented in Chap. 3, concerning *well-posedness* of parabolic boundary-value problems in spaces of Hölder functions and in spaces $\mathcal{H}^l$, it follows, in particular, that the Cauchy problems for systems parabolic in the sense of Petrovskij and the initial problem for the systems parabolic in the sense of Solonnikov are well-posed. These results, unlike the results presented below, do not admit the growth of the solutions, initial data, and right hand side terms as $|x| \to \infty$.

To establish uniqueness or nonuniqueness of solutions to the Cauchy problem we use the method of fundamental solutions and a priori estimates, as well as other strong methods based on the theory of distributions, theorems on quasi-analytic classes of functions, and the theory of positive solutions of partial differential equations.

In conclusion to this section we briefly dwell on the problem of existence of the limit of the solution to the Cauchy problem with unbounded growth of time (stabilization of solutions). This problem, which is exceptionally interesting for applications as well as for the theory of partial differential equations, has been in the focus of attention of researchers during the last twenty years.

2.2. Well-Posedness. We study well-posedness of the Cauchy problem for systems parabolic in the sense of Petrovskij

$$\mathcal{L}(t,x;D_t,D)u \equiv \left(ID_t - \sum_{|\alpha|\le 2b} a_\alpha(t,x)D^\alpha \right) u = f(t,x)\,, \qquad (2.73)$$
$$u|_{t=0} = \varphi(x)\,.$$

From the estimates (2.16) of the fundamental solution $Z(t,x;\tau,\xi)$ it directly follows that the functions $f(t,x)$ and $\varphi(x)$ can be allowed to grow like $\exp\{a\sum_{i=1}^n |x_i|^q\}$ as $|x| \to \infty$ and the class of solutions in question has also to be subject to this condition. We note that even in the class of functions growing like $\exp\{a\sum_{i=1}^n |x_i|^{q+\varepsilon}\}$ for all $\varepsilon > 0$ the uniqueness of solutions to the problem (2.73) defined in Π_T is violated. Simple argument shows that in this class we do not even have the existence of solution defined in the cylinder Π_T for arbitrarily small T.

We consider a simple example

Example 2.4. We find a *solution to the Cauchy problem for the heat equation*

$$\frac{\partial u}{\partial t} = \frac{\partial^2 u}{\partial x^2}\,, \quad u|_{t=0} = \exp\{ax^2\}\,; \qquad (2.74)$$

$$u(t,x) = \int_{-\infty}^{\infty} \frac{1}{2\sqrt{\pi t}} \exp\left\{ -\frac{(x-\xi)^2}{4t} + a\xi^2 \right\} d\xi$$
$$= \frac{1}{\sqrt{1-4at}} \exp\left\{ \frac{ax^2}{1-4at} \right\}\,. \qquad (2.75)$$

From (2.75) it follows that the solution to the problem (2.74) exists in the cylinder $\Pi_{[0,T]}$ where $T = (1/4a) - \varepsilon$ for any $\varepsilon > 0$ and $u(t,x) \to \infty$ as $t \to (1/4a)$ has the growth of order two (for $|x| \to \infty$) and the growth type $k_1(t,a) = a/(1-4at)$. The function $k_1(t,a)$, which plays an essential role in what follows, has the interesting semigroup property $k_1(t-\tau, k_1(\tau,a)) = k_1(t,a)$ which can be directly verified.

In the case of the system (2.73) the role of $k_1(t,a)$ is played by the function $k(t,a) = ca[c^{2b-1} - a^{2b-1}t]^{1/(2b-1)} \equiv k(t)$ where the positive constant c depends on the parabolicity constant δ. The function $k(t,a)$ also has the semigroup property.

The shortage of space does not allow us to discuss various variants of theorems on well-posedness of the problem (2.73), see (Ivasishen and Ejdel'man 1968), (Matijchuk and Ejdel'man 1967), and (Ejdel'man 1964). We restrict ourselves to the formulation and brief discussion of one of them which contains, in our opinion, quite precise information.

We consider the *space* $C^{2b,1}_{\mathrm{K}(t)}(\overline{\Pi}_T)$ of functions $u(t,x)$ with continuous derivatives with respect to $x_1,\ldots,x_n$ up to order $2b$ and continuous derivative with respect to t in $\overline{\Pi}_T$ and such that the following norm is finite

$$\|u;\Pi_T\|^{k(t)}_{2b,1} = \sup_{(t,x)\in\Pi_T} \left\{ \left(\sum_{|\alpha|\le 2b} |D^\alpha u| + |D_t u| \right) E(t,x) \right\}\,,$$

$$E(t,x) = \exp\left\{-k(t)\sum_{i=1}^{n} x_i^q\right\} .$$

Suppose that $\omega(h)$ and $F(h)$ are two non-negative, nondecreasing, subadditive bounded functions (functions of the modulus of continuity type). We define a subspace $\mathcal{D}^{2b,1}_{\omega,F}(\Pi_T)$ of the Banach space $C^{2b,1}_{k(t)}(\overline{\Pi}_T)$ that consists of functions $u(t,x) \in C^{2b,1}_{k(t)}(\overline{\Pi}_T)$ with finite norm

$$\begin{aligned} \|u;\Pi_T\|^{\omega,F}_{2b,1} =& \|u;\Pi_T\|^{k(t)}_{2b,1} + \|u;\Pi_T\|^{k(t),F}_{2b} + \|u(0,x);\mathbb{R}^n\|^{a,\omega}_{2b} \\ &+ \|\mathcal{L}u;\Pi_T\|^{k(t),\omega}_0 , \end{aligned} \tag{2.76}$$

where

$$\|u;\Pi_T\|^{k(t),F}_{2b} = \sum_{|\alpha|=2b} \sup_{\{(t,x),(\tau,y)\}\subset\Pi_T} \left\{\frac{|\Delta^y_x D^\alpha u(t,x)|}{F(|x-y|)}[E(t,x)+E(t,y)]\right\} ,$$

and the remaining norms in (2.76) are defined in a similar way. We denote by $D_F(\Pi_T)$ the subspace of

$$C_{k(t)}(\Pi_T) = \{u(t,x) : \|u;\Pi_T\|^{k(t)} = \sup_{(t,x)\in\Pi_T} (|u(t,x)|E(t,x))\}$$

that consists of functions $u(t,x)$ with finite quasi-norm

$$\|u;\Pi_T\|^{k(t),F} = \sup_{\{(t,x),(\tau,y)\}\subset\Pi_T} \left\{\frac{|\Delta^y_x u(t,x)|}{F(|x-y|)}[E(t,x)+E(t,y)]\right\} ,$$

and by $\mathcal{D}^{2b}_{\omega,a}(\mathbb{R}^n)$ the subspace of $C^{2b}_a(\mathbb{R}^n)$ of functions $u(0,x)$ with finite quasi-norm

$$\|u(0,x);\mathbb{R}^n\|^{a,\omega}_{2b} = \sum_{|\alpha|=2b} \sup_{\{x,y\}\subset\mathbb{R}^n} \left\{\frac{|\Delta^y_x D^\alpha u(0,x)|}{\omega(|x-y|)}[E(0,x)+E(0,y)]\right\} .$$

To study Cauchy problems we need the description of the above functional spaces in the particular case that $\omega(h)$ satisfies the Dini condition and $F(h) = \int_0^h (\omega(z)/z)dz \equiv \omega_1(h)$.

We define Banach spaces $E_1 = \mathcal{D}^{2b,1}_{\omega,\omega_1}(\Pi_T)$ and $E_2 = \mathcal{D}^0_{\omega_1}(\Pi_T) \times \mathcal{D}^{2b}_{\omega,a}(\mathbb{R}^n)$ (we shall use the same letters to denote the corresponding spaces of vector-valued functions), the operator $\mathfrak{A} = (\mathcal{L}u, Iu|_{t=0})$ of the Cauchy problem (2.73), and the vector $\psi = (f(t,x), \varphi(x))$.

Theorem 2.20. *1) Suppose that the coefficients of the system* (2.73) *satisfy conditions* α_1, α_2, *and* α_5. *The operator* $\mathfrak{A}$ *gives a one-to-one correspondence between the spaces* E_1 *and* E_2.

2) If, in addition, condition α_6 *is satisfied, then the unique solution of* (2.73) *in* $\mathcal{D}^{2b,1}_{\omega,\omega_1}(\Pi_T)$ *can be represented by means of the fundamental solution* $Z(t,x;\tau,\xi)$ *by the formula*

$$u(t,x) = \int Z(t,x;0,\xi)\varphi(\xi)d\xi + \int_0^t d\tau \int Z(t,x;\tau,\xi)f(\tau,\xi)d\xi \,. \qquad (2.77)$$

We make some comments on Theorem 2.20. It implies that if the coefficients of the system are continuous, bounded, and satisfy the Dini condition with respect to x, then there is a unique solution to the Cauchy problem (2.73) constructed by means of the initial function $\varphi(x)$ of the same regularity as the solution, and the right hand side term $f(t,x)$ satisfying the Dini condition with respect to x. The solution is in the class of functions that grow like $\exp\{k(t)\sum_{i=1}^{n} x_i^q\}$ and satisfies the following two-sided estimate

$$C_1\|\psi\|_{E_2} \le \|u;\Pi_T\|_{2b}^{\omega,\omega_1} \le C_2\|\psi\|_{E_2} \,,$$

where

$$\|\psi\|_{E_2} = \|\varphi;\mathbb{R}^n\|_{2b}^{a} + \|\varphi;\mathbb{R}^n\|_{2b}^{a,\omega} + \|f;\Pi_T\|_0^{k(t)} + \|f;\Pi_T\|^{k(t)\omega} \,.$$

A proof of Theorem 2.20 is given in (Matijchuk and Ejdel'man 1970). It is based on a priori estimates of the solution to the Cauchy problem in the above Dini spaces. As usually, the properties of various parabolic potentials, sometimes quite subtle, were used.

Definition 2.8. We say that the functions $F(h)$ and $\omega(h)$ of the type of modulus of continuity are *equivalent* if there exist positive constants C_1 and C_2 such that

$$C_1\omega(h) \le F(h) \le C_2\omega(h) \,.$$

In this case the spaces $D_{\omega,F}^{2b,1}$ and $D_{\omega,\omega}^{2b,1}$ coincide.

To make Theorem 2.20 more precise we face an essential question whether the functions $\omega(h)$ and $\omega_1(h) = \int_0^h (\omega(z)/z)dz$ are equivalent (of course, under the assumptions that this integral converges).

Lemma 2.2 (Matijchuk and Ejdel'man 1970). *For $\omega(h)$ and $\omega_1(h)$ to be equivalent it is necessary and sufficient that one of the following two conditions be fulfilled:*

1) $\liminf_{z\to 0}(\omega(2z)/\omega(z)) \ge 1$;

2) $\varphi(2z)/\varphi(z) \le C$ *for* $z \in (0,A)$, *where* $\phi(z)$ *is the function inverse to* $\omega(z)$.

Example 2.5. Examples of $\omega(h)$ equivalent to $\omega_1(h)$:

1) $\omega(h) = h^\lambda$, $0 < \lambda \le 1$;
2) $\omega(h) = h^\lambda(\ln 1/h)^\mu$, $\lambda \in (0,1)$.

It can be directly verified that the assumptions of Lemma 2.1 are satisfied in the variant 1).

Thus, particular cases of Theorem 2.20 are, for example, precise theorems on well-posedness in the classes of Hölder functions and functions with moduli of continuity of the form $\omega(h) = h^\lambda(\ln 1/h)^\mu$.

We can also consider other variants:

1) do not assume regularity of the initial function (for example, assume only that is continuous or impose some other minimal regularity); then in the description of the classes of well-posedness of the Cauchy problem employ the hypotheses on the character of the behavior of the derivatives of the solutions and their moduli of continuity as t converges to zero;
2) to consider in the assumptions the moduli of continuity with respect to the *parabolic distance* $d((t,x),(\tau,\xi)) = (|t-\tau|^{1/b} + |x-\xi|^2)^{1/2}$ instead of the moduli of continuity with respect to the spatial variables;
3) to construct the classes of well-posedness of the Cauchy problem in the whole scale of smooth increasing functions with higher derivatives from the Dini or Hölder classes.

All the variants are considered in the literature (Ivasishen and Ejdel'man 1968), (Ladyzhenskaya et al. 1967), (Matijchuk and Ejdel'man 1967, 1970), and (Ejdel'man 1964).

We give examples showing the accuracy of the results obtained.

Example 2.6. We consider the *volume heat potential in* $\mathbb{R}^3$

$$u(t,x) = \int_0^t d\tau \int_{\mathbb{R}^3} (2\sqrt{\pi(t-\tau)})^3 \exp\left\{-\frac{|x-\xi|^2}{4(t-\tau)}\right\} f(\xi)d\xi\,, \tag{2.78}$$

where $f(x)$ is continuous with compact support and is different from zero in a ball of radius $R<1$ with centre at the origin. We change the order of integration and we introduce a new variable $\beta = |x-\xi|^2(t-\tau)^{-1}$ to the integral with respect to τ. Then $u(t,x)$ can be written in the form

$$u(t,x) = \frac{1}{4\pi}\int_{\mathbb{R}^3} |x-\xi|^{-1} f(\xi)d\xi$$
$$-\frac{1}{(2\sqrt{\pi})^3}\int_{\mathbb{R}^3} |x-\xi|^{-1} f(\xi)\int_0^{|x-\xi|^2/t} \mathrm{e}^{-\beta/4}\beta^{-1/2}d\beta d\xi\,. \tag{2.79}$$

It is easy to show that for any continuous $f(x)$ the second term has two continuous derivatives in x_1, x_2, x_3 and one in t satisfying the Hölder condition in x with exponent $1/2$. We consider the first term in (2.79)

$$J(x) = \frac{1}{4\pi}\int_{\mathbb{R}^3} |x-\xi|^{-1} f(\xi)d\xi \tag{2.80}$$

This is the volume potential for the Laplace equation which is investigated in detail in the classical monograph of Gyunter (1953). Following (Gyunter 1953) we take $f(x) = \{(3x_1^2/|x|) - 1\}\psi(|x|)$ where $\psi(0) = 0$ and $\psi(r)$ is a nondecreasing positive continuous function. In (Gyunter 1953) it is shown that if $\psi_1^{(\varepsilon)}(h) = \int_\varepsilon^h (\psi(r)/r)dr$ does not have a finite limit as $\varepsilon \to 0$, then $J(x)$ does not have second order derivatives at the origin. Thus, the Dini condition in the case in question is necessary for the potential $J(x)$ to have two derivatives. A further analysis shows that if the integral $\psi_1(h) = \int_0^h (\psi(r)/r)dr$ exists but the functions $\psi(h)$ and $\psi_1(h)$ are not equivalent, then the modulus of continuity of the second order derivatives is precisely $\psi_1(h)$, which is in complete agreement with Theorem 2.20.

Example 2.7 (Il'in 1962). We consider the equation

$$\frac{\partial u}{\partial t} = a(t,x)\frac{\partial^2 u}{\partial x^2}\ ; \quad a(t,x) = 1 + b(t,x)\ , \tag{2.81}$$

where

$$b(t,x) = \begin{cases} -(\ln 1/t)^{\gamma_1} & \text{if } \ 0 < (x^2/t) \le 2 - \varepsilon_1(\ln 1/t)^{-\varepsilon\gamma} \\ 0 & \text{if } \ 2 - 2\varepsilon_1(\ln 1/t)^{-\varepsilon\gamma} \le (x^2/t) \le 2\ . \\ (\ln 1/t)^{\gamma_1} & \text{if } \ (x^2/t) \ge 2 + \varepsilon \end{cases}$$

For the remaining (t,x), $b(t,x)$ is extended as an even function with respect to x, nondecreasing in x for $x > 0$, smooth, and bounded ($|b(t,x)| \le 1/2$). The numbers ε, ε_1 are chosen in a special way and $0 < \gamma_1 < 2\gamma < 1$.

In (Il'in 1962) it is established that the modulus of continuity in x of $a(t,x)$ is $\omega(h) = (\ln 1/h)^{\gamma_1}$. For $\gamma_1 \in (0,1]$, $a(t,x)$ does not satisfy the Dini condition. For $\gamma_1 \in (0,1)$ we construct a positive smooth function $v(t,x)$ such that the solution $u_\tau(t,x)$ to the Cauchy problem $u|_{t=\tau} = v(t,x)$ for equation (2.81) has the property

$$\lim_{\tau\to 0}\left(u_\tau(t,x)\Big/\int_{-\infty}^{\infty} v(\tau,\xi)d\xi\right) = \infty \tag{2.82}$$

If equation (2.81) had a fundamental solution with the usual properties (see Theorem 2.3), then we would have

$$u_\tau(t,x) = \int_{-\infty}^{\infty} Z(t,x;\tau,\xi)v(\tau,\xi)d\xi\ . \tag{2.83}$$

From the positivity of $v(t,x)$ and $u_\tau(t,x)$, the estimate of the fundamental solution, and (2.83) we deduce the following inequality

$$u_\tau(t,x) \le (t-\tau)^{-1/2}\int_{-\infty}^{\infty} v(\tau,\xi)d\xi\ ,$$

which contradicts (2.82). Thus we have shown that the rejection of the Dini condition for the coefficients of the equations results in the loss of the existence of a fundamental solution with the usual properties.

Example 2.8 (Kruzhkov 1967). Suppose that $\omega(h)$ is a modulus of continuity that does not satisfy the Dini condition, i.e.

$$\int_0^2 (\omega(h)/h)dh = \infty\ . \tag{2.84}$$

We shall use $\omega(h)$ to construct a convex modulus of continuity $\lambda(h)$ with continuous derivatives of order $N \geq 2b$ on any interval $[\delta, 2] \subset [0, 2]$ and satisfying the following properties:

1) (2.84);
2) $\lambda(h) \leq \omega(h)$;
3) $|\lambda^{(k)}(h)| \leq Mh^{-k}\lambda(h)$, $k = 1, \dots, 2b$;
4) $(h^{-1}\lambda(h))' < 0$.

The function

$$v(t,x) = \varepsilon x_1^b x_2^b \int_0^{\sqrt{r^2+t^2}} (\lambda(h)/h)dh + \frac{x_1^{2b}}{2b!} + (-1)^{b-1}t\,, \quad r^2 = x_1^2 + x_2^2 + x_3^2\,,$$

defined in the cylinder $Q = \{(t,x) : r^2 \leq 1/2, -1/2 \leq t \leq 0\}$ satisfies for sufficiently small ε and for $r^2 + t^2 \neq 0$ the uniformly parabolic equation

$$D_t u = (-1)^{b-1} B(t, x_1, x_2, x_3) \left[D_{x_1}^{2b} + D_{x_2}^{2b} + \sum_{i=1}^{n} D_{x_i}^{2b} \right] u$$

where a positive function B has in Q the modulus of continuity not greater than $\omega(h)$. Obviously,

$$\lim_{t^2+r^2\to 0} \frac{\partial^{2b} v}{\partial x_1^b \partial x_2^b} = -\infty$$

and on the side boundary and on the bottom base of the cylinder Q the function $v(t,x)$ can be assumed to be as smooth as we wish. Therefore, for parabolic equations of arbitrary order the rejection of the Dini condition for the coefficients results in the absence of classical solutions of the Dirichlet problem in this case.

Theorems on the well-posedness of the Cauchy problem with roughly the same content as Theorem 2.20 have been established for $2\vec{b}$-parabolic systems and for the systems parabolic in the sense of Petrovskij of arbitrary order in t. For the shortage of space we do not present them here. We refer to (Ivasishen and Ejdel'man 1968), where the relevant range of problems are exhaustively discussed for $2\vec{b}$-parabolic systems in the classes of Hölder functions.

2.3. Existence. The results on existence and properties of fundamental solutions presented in Sect. 1 allow us to establish in a sufficiently uniform way various variants of existence theorems for the Cauchy problems for systems parabolic in the sense of Petrovskij with growing coefficients, systems parabolic in the sense of Shilov with constant coefficients, diffusion equation with inertia of arbitrary order, and pseudodifferential parabolic equations. In all these cases a precise and complete theory of well-posedness that was discussed in 2.2 has not been achieved yet.

We give, as an example, an existence theorem for the Cauchy problem (2.73) for parabolic systems with growing coefficients.

Theorem 2.21 (Zhitomirskij 1959). *Suppose that the coefficients of the system* (2.73) *satisfy the following conditions*

α_{26}. *for* $|\alpha| = 2b$, $a_\alpha(t,x)$ *has* $2b+1$ *continuous and bounded derivatives in* Π_T;

α_{27}. *for* $|\alpha| < 2b$, $a_\alpha(t,x)$ *are continuous together with their first order derivatives with respect to* $x_1, \ldots, x_n$ *in* Π_T *and satisfy the following inequalities*

$$|D^\beta a_\alpha(t,x)| \le C\left(1 + |x|^{k_{|\alpha|}}\right),$$

$$|\beta| = 0,\,1, \quad \text{where} \quad k_{|\alpha|} < \frac{2b - |\alpha|}{2b-1}\,.$$

Then the Cauchy problem (2.73) *has a classical solution* $u(t,x)$ *constructed by the initial function* $\varphi(x)$ *that is continuous in* $\mathbb{R}^n$ *and satisfies the estimate*

$$|\varphi(x)| \le C \exp\left\{\alpha \sum_{j=1}^{n} x_j^q\right\}.$$

The solution is defined in the cylinder Π_{T_1} *(for some* $T_1 < T$*) and can be written as the* Poisson integral

$$u(t,x) = \int Z(t,x;0,\xi)\varphi(\xi)d\xi\,,$$

where $Z(t,x;\tau,\xi)$ *is a fundamental solution to the problem* (2.73).

Similar results, but more precise with respect to the smoothness, can be obtained by means of a transform that reduces a system with growing coefficients to a system with dissipation (Ejdel'man 1964).

For systems (2.55) parabolic in the sense of Shilov with constant coefficients we can establish the existence of a classical solution to the Cauchy problem using the results presented in Theorem 2.14. The solution is constructed by the initial function whose growth is consistent with the decay of the fundamental solution $\Gamma(t,x)$ (given by the estimates (2.56) and (2.57)) but the initial condition is satisfied in the sense of the convergence in the topology of suitable spaces of type W (Gel'fand and Shilov 1958a,b, 1959), (Friedman 1963).

For diffusion equations with inertia we have to analyse very carefully the dependence of the fundamental solution and the solution to the Cauchy problem on variables y and z with respect to which the equation is degenerate and with respect to these variables we have to impose additional conditions on the smoothness of the coefficients (Il'in 1962), (Sonin 1967), (Ejdel'man and Tychinskaya 1979), (Weber 1951).

We now consider the Cauchy problem for a parabolic equation (2.69) with pseudodifferential operators. By a *solution to the problem* we understand a bounded function $u(t,x)$ continuous in $\overline{\Pi}_T$, satisfying (2.69) with A and A_k replaced by conditionally convergent hypersingular integral operators D^{γ}_{Ω} and $D^{\gamma_k}_{\Omega_k}$, and satisfying the initial condition in the classical sense.

Theorem 2.22 (Kochubej 1985). *Suppose that conditions α_{24} and α_{25} are fulfilled. Then there is a solution $u(t,x)$ to the problem (2.69) constructed by a bounded continuous initial function $\varphi(x)$ and a right hand side term $f(t,x)$ continuous, bounded, and satisfying the uniform Hölder condition in Π_T. The solution is given by the formula*

$$u(t,x) = \int Z(t,x;0,\xi)\varphi(\xi)d\xi + \int_0^t d\tau \int Z(t,x;\tau,\xi)f(\tau,\xi)d\xi \,,$$

where $Z(t,x;\tau,\xi)$ is the fundamental solution described by Theorem 2.18.

2.4. Uniqueness. In the case of a simple parabolic equation

$$\frac{\partial u}{\partial t} = (-1)^{b-1}\frac{\partial^{2b} u}{\partial x^{2b}} \tag{2.85}$$

Täclind (1936) established that the solution to the Cauchy problem in the class of functions

$$|u(t,x)| \le C\exp\{|x|h(|x|)\} \tag{2.86}$$

is unique if and only if the integral

$$\int_1^{\infty} h(r)^{1-2b}dr \tag{2.87}$$

diverges.

Täclind's proof of uniqueness employs the fundamental solution of (2.85), the Green formula, and a special construction of the 'Täclind stairs' that relate the motion in time and space variables. The construction of a solution showing nonuniqueness is based on the *Carleman–Ostrovskij* theorem of the theory of quasi-analytic classes of functions.

We give a typical theorem of *Täclind type.*

Theorem 2.23 (Ejdel'man 1964). *1) If the coefficients of the system* (2.9) *parabolic in the sense of Petrovskij satisfy conditions α_1–α_4, then any classical solution $u(t,x)$ in Π_T to the Cauchy problem with zero initial condition, belonging to the class* (2.86), *where the function $h(r)$ has divergent integral, is identically zero.*

2) In the class of functions (2.86) *with $h(r)$ such that the integral* (2.87) *converges there is a nontrivial solution $v(t,x)$ of a parabolic system $D_t u = \sum_{|\alpha|\le 2b} a_\alpha D^\alpha u$ with constant coefficients, satisfying zero initial condition.*

Recently, theorems of that type are proved for quite various problems and systems of equations. We mention, for example, the paper of Chaus (1968), where the Täclind type theorems are proved for the Cauchy problem in the case of arbitrary systems of the form $D_t u = A(D_x)u$ with constant coefficients.

One should, in our opinion, ponder over the following question. The usual argument is as follows: suppose that $u_1(t,x)$ and $u_2(t,x)$ are two solutions of the same system of equations with the same initial conditions; then $u(t,x) = u_1(t,x) - u_2(t,x)$ is a solution of a linear homogeneous system with zero initial conditions. Thus, the class of uniqueness is determined by the condition that only trivial solutions correspond to zero initial data. A solution for zero initial data always exists in any class. Now, what may be the reason for uniqueness: either in a broader class we do not have solutions constructed by initial functions growing so rapidly, or the solutions exist but from the fact that they coincide at the initial instant it follows that they coincide everywhere in the domain. It is interesting to note that in (Petrushko and Ejdel'man 1967) using the Täclind arguments from (Täclind 1936) it is shown that if the initial function satisfies the inequality $\varphi(x) \geq \exp\{|x|h(|x|)\}$, where $\lim_{r\to\infty}(h(r)/r) = \infty$, then there is no band Π_T where a solution to the Cauchy problem for the heat equation exists. A natural domain of existence of a solution is defined in this case in a special way by the function $\varphi(x)$.

We also have another important class of uniqueness theorems. The first theorem from this class is the classical Widder's result (1944): a non-negative solution of the heat equation in Π_T with zero initial data is identically equal to zero. The assumption of positivity can be the only information available about the solution, for example, in the problems of the probabilistic origin. The theory of positive solutions of partial differential equations (Kondrat'ev and Ejdel'man 1974) allows us, in particular, to establish various generalizations and refinements of this theorem. For simplicity of formulation we consider the case of one equation parabolic in the sense of Petrovskij of the following form

$$D_t u = \sum_{|\alpha|\leq 2b} D^\alpha(a_\alpha(t,x)u) . \tag{2.88}$$

Theorem 2.24 (Kondrat'ev and Ejdel'man 1974). *Suppose that the coefficients of the equation* (2.81) *satisfy the following conditions:*

α_{28}. *$a_\alpha(t,x)$ are functions measurable and bounded in Π_T;*

α_{29}. *for $|\alpha| = 2b$ the functions $a_\alpha(t,x)$ are continuous or real-valued.*

We study real solutions $u(t,x)$ of (2.86), *weak in the sense of integral identities. The following assertions are valid.*

1) If the negative part $u^-(t,x)$ of $u(t,x)$ is such that

$$\int u^-(t,x)\exp\{-|x|h(|x|)\}dx < +\infty , \tag{2.89}$$

where $h(r)$ is such that the integral (2.87) *diverges, then $u(0,x) \equiv 0$ implies $u(t,x) \equiv 0$ in Π_T.*

2) Suppose that the weight in the equation (2.86) *is $2b = 2(2s+1)$. If two solutions $u_1(t,x)$ and $u_2(t,x)$ have negative parts satisfying* (2.89) *with*

$h(r)$ such that the integral (2.87) *diverges, then* $u_1(0,x) \equiv u_2(0,x)$ *implies* $u_1(t,x) \equiv u_2(t,x)$ *in* Π_T.

We note that we do not have any a priori information about the difference of two positive solutions therefore for the proof of more natural, in our opinion, variant 2) of the theorem we need to impose some special hypotheses on the possible growth of positive solutions. It is interesting that in the theorems of that kind there is an essential difference between equations with the parabolic weight $2b = 2(2s+1)$ and $2b = 4s$.

The study of uniqueness in the case when the coefficients of the equations grow is a topic for a separate survey.

We mention, for example, that for the equation

$$\frac{\partial u}{\partial t} = \frac{\partial^2 u}{\partial x^2} + q(x)u$$

where $q(x)$ is a smooth function coinciding with $|x|^{2+\varepsilon}$ for $|x| > 1$ and ε is a positive number, the Cauchy problem has nonunique solution even in $L_2(\mathbb{R}^n)$; if $q(x)$ coincides with $|x|^{2-\varepsilon}$ for $|x| > 1$, then the usual uniqueness theorems hold and if $q(x) \to -\infty$ as $|x| \to \infty$, then the classes of uniqueness can essentially get larger as the rate of the convergence of the dissipation $-q(x)$ to infinity increases.

The equation $\partial u/\partial t = a(x)^2 \partial^2 u/\partial x^2 + a(x)a'(x)\partial u/\partial x$, which reduces to the heat equation by introducing a new independent variable $z = \int_0^x (dy/a(y))$, can, depending on $a(x)$, have quite various classes of uniqueness of solutions to the Cauchy problem.

2.5. Stabilization. We shall very briefly describe some results on stabilization of solution to the Cauchy problem. First theorems on stabilization are presented in (Ejdel'man 1964). In the survey (Denisov and Repnikov 1984) equations with constant coefficients are considered, and the references to the papers on equations with variable coefficients can be found in (Porper and Ejdel'man 1984).

Definition 2.9. A function $\varphi(x)$ defined in $\mathbb{R}^n$ has the *limit mean* at $x^0 \in \mathbb{R}^n$ if

$$\exists \lim_{R\to\infty} \left(\int_{K_R^{x^0}} \varphi(x)dx/|K_R^{x^0}| \right) = m(\varphi; x^0) , \tag{2.90}$$

where $K_R^{x^0}$ is the ball in $\mathbb{R}^n$ with centre at x^0 and radius R, and $|K_R^{x^0}|$ is its Lebesgue measure.

Theorem 2.25. (Repnikov and Ejdel'man 1966). *A bounded solution $u(t,x)$ to the Cauchy problem*

$$D_t u = (-1)^{b-1} \Delta^b u , \qquad u|_{t=0} = \varphi(x) ,$$

constructed by a continuous bounded initial function $\varphi(x)$, stabilizes pointwisely, *i.e.*

$$\exists \lim_{t\to\infty} u(t,x^0) = A(x^0)\,, \tag{2.91}$$

if and only if the initial function $\varphi(x)$ *has the limit mean* $m(\varphi; x^0) = A(x^0)$.

We note that for bounded functions the limits (2.90), (2.91), do not depend on x^0, i.e. the point is that the solution $u(t,x)$ becomes constant as $t \to \infty$.

For the proof of Theorem 2.25 we apply the Wiener Tauberian theorem to the solution $u(t,x)$ represented by the formula (2.15) and the verification of its hypotheses employs the theory of Bessel functions.

We note that it is much simpler to establish the theorem saying that for *uniform* in $x^0 \in \mathbb{R}^n$ *stabilization* it is necessary and sufficient that there exist the uniform limit (2.90). It is also important fact that for the uniform stabilization the form of the family of domains on which the limit means of the initial function exist is quite arbitrary. This is not the case for the pointwise stabilization.

The second topic we shall dwell on is the following one. The only information is that $u(t,x)$ is a positive solution of the heat equation in the half-space $t > 0$. From this piece of information it follows that $u(t,x)$ can be written as the *Poisson–Stieltjes integral*

$$u(t,x) = \frac{1}{(2\sqrt{\pi t})^n} \int \exp\left\{-\frac{|x-\xi|^2}{4t}\right\} \mu(d\xi)\,, \tag{2.92}$$

where $\mu(E)$ is a positive measure defined on the σ-algebra of Borel subsets of $\mathbb{R}^n$ such that

$$\int \exp\{-\varepsilon|x|^2\}\mu(dx) < +\infty \tag{2.93}$$

for any positive ε.

Theorem 2.26 (Valitskij and Ejdel'man 1976). *For the pointwise stabilization of the Poisson–Stieltjes integral* (2.92) *to a constant A it is necessary and sufficient that the measure* $\mu(E)$ *have the* limit mean over balls, *i.e.*

$$\exists \lim_{R\to\infty} \{\mu(K_R^{x^0})/|K_R^{x^0}|\} = A\,.$$

Two proofs of Theorem 2.26 are given in (Valitskij and Ejdel'man 1976). One is based on some elementary considerations and the other on a generalization of the Wiener Tauberian theorem.

The passage to equations with coefficients depending on all spatial variables requires fundamentally different methods. These methods in their essential part use subtle a priori Hölder estimates of solutions. Recently, we have a series of interesting competing approaches to the solution of the stabilization problem for equations with variable coefficients. An important method of proving stabilization theorems based on the ideas of the theory of means is due to Zhikov. We present one of the results obtained in this way.

Theorem 2.27 (Zhikov 1977). *We consider a bounded solution to the Cauchy problem for equation* (2.34). *Suppose that the following conditions are fulfilled:* α_{16}, α_{17},

α_{30}. $a_{ij}(\omega^2 t, \omega x)$ *converge strongly in* $L_2(Q)$ *as* $\omega \to \infty$ *to constants* $\overline{a}_{ij}$, *where* Q *is an arbitrary compact set in* $\mathbb{R}^{n+1}_+$;

α_{31}. $\exists \lim_{R\to\infty} \left(\int_{(Bx,x)\le R^2} p(x)dx/R^n \right)$, *where* B *is the inverse matrix to* $(\overline{a})^n_{i,j=1}$ *and* $(Bx,x) \le R^2$ *is the ellipsoid* $\sum^n_{i,j=1} b_{ij}x_i x_j \le R^2$.

For the stabilization of $u(t,0)$ *at a constant* A *it is necessary and sufficient that for the initial function* $\varphi(x)$ *exist the limit*

$$\lim_{R\to\infty} \left(\int_{(Bx,x)\le R^2} p(x)\varphi(x)dx/R^n \right) \Big/ \lim_{R\to\infty} \int_{(Bx,x)\le R^2} p(x)dx = A\,.$$

Chapter 3
Parabolic Boundary-Value Problems

§1. Shauder Theory

1.1. Introduction. We present the main facts of the theory of parabolic boundary-value problems in Hölder spaces. They compose the foundation of the Schauder theory of parabolic boundary-value problems. The theory is called so to commemorate J.Schauder, who first in 1934 obtained the basic results in the case of Dirichlet problem for elliptic equations of the second order.

We first discuss this theory for parabolic problems of the form

$$\mathcal{L}(t,x;D_t) \equiv D_t u - \sum_{|\alpha|\le 2b} a_\alpha(t,x) D^\alpha u = f\,, \tag{3.1}$$

$$u|_{t=0} = \varphi(x) \tag{3.2}$$

$$B_\beta(t,x;D_t,D)u|_{S_T} = \sum_{2b\alpha_0+|\alpha|\le r_\beta} b^\beta_{\alpha_0\alpha}(t,x) D_t^{\alpha_0} D^\alpha u|_{S_T} = q_\beta\,, \tag{3.3}$$

$$\beta = 1,\dots,bm$$

in the cylinder $Q_T = (0,T] \times \Omega$, $S_T = (0,T] \times \partial\Omega$. We recall that a_α are square matrices of dimension m and $b^\beta_{\alpha_0\alpha}$ are matrix-rows of length m.

Next, we formulate the main result for systems of general form.

1.2. The Theorem on Well-Posedness. We use the notation and information on spaces of Hölder functions presented in 3.2, Chap. 1. By $C^l_0(\overline{Q}_T)$ and

$C_0^l(\overline{S}_T)$ we denote the sets of functions u that belong to the spaces $C_l(\overline{Q}_T)$ and $C_l(\overline{S}_T)$, respectively, and satisfy the initial conditions

$$D_t^{\alpha_0} u|_{t=0} = 0 \,, \qquad \alpha_0 = 1, \ldots, [l/2b] \tag{3.4}$$

In the Schauder theory the precise dependence of differential properties of solutions to boundary-value problems on the right hand side terms is established by showing that the solutions belong to the corresponding Hölder spaces. The restrictions on smoothness of the coefficients, the boundary of the domain, and the functions f and g should be precisely consistent with the choice of space in which we seek the solution. We shall use the following conditions.

β_1. The entries in the matrices $a_\alpha(t,x)$ belong to $C^l(\overline{Q}_T)$.
β_2. The entries in the matrices $b^\beta_{\alpha_0\alpha}(t,x)$ belong to C^{l+2b-r_β}.
β_3. $\partial\Omega \in C^{l+2b}$.

We shall also explain the requirements that should be imposed on the functions f, φ, and g_β, $\beta = 1, \ldots, bm$ for the problem (3.1)–(3.3) to have a solution in $C^{l+2b}(\overline{Q}_T)$. Necessary conditions can be directly found. By the definition of the Hölder spaces we have

$$f \in C^l(\overline{Q}_T) \,, \quad \varphi \in C^{l+2b}(\overline{\Omega}) \,, \quad g_\beta \in C^{l+2b-r_\beta}(\overline{S}_T) \,. \tag{3.5}$$

(We understand (3.5) in the sense that each coordinate of the vectors belongs to the corresponding space or that the vectors belong to the Cartesian product of the corresponding spaces.) Moreover, there exists a constant C independent of u such that

$$\|f; Q_T\|_l + \|\varphi; \Omega\|_{l+2b} + \sum_{\beta=1}^{bm} \|g_\beta; S_T\|_{l+2b-r_\beta} \le C\|u; Q_T\|_{l+2b} \,. \tag{3.6}$$

Moreover, f, φ, and g_β are necessarily linked by the *compatibility conditions* for $t = 0$, $x \in \partial\Omega$. To explain how such conditions appear, we consider a simple parabolic problem for the heat equation:

$$D_t u = a^2 \Delta u + f \,; \quad u|_{t=0} = \varphi \quad u|_{x_n=0} = g \,, \quad Q_T = (0,T] \times \mathbb{R}^n_+ \equiv \Pi^+_T \,.$$

If the solution u belongs to $C^{l+2}(\overline{\Pi^+_T})$, then from the heat equation and the initial condition we can find the derivatives $D_t^{\alpha_0} u|_{t=0}$, $\alpha_0 = 0, \ldots, [(l+2)/2]$. For $x_n = 0$ these derivatives coincide with the derivatives $D_t^{\alpha_0} u|_{t=0}$, $\alpha_0 = 0, \ldots, [(l+2)/2]$, so that we have

$$\varphi|_{x_n=0} = g|_{t=0} \,,$$
$$(a^2\Delta\varphi + f|_{t=0})|_{x_n=0} = D_t g|_{t=0} \,,$$
$$\{a^2\Delta(a^2\Delta\varphi + f|_{t=0}) + D_t f|_{t=0}\} = D_t^2 g|_{t=0}$$

and so forth.

In the case of the problem (3.1)–(3.3) the *compatibility conditions* mean that the derivatives $D_t^{\alpha_0} u|_{t=0}$, $k_0 = 0, \dots, [(l+2b)/2b]$ determined by the system (3.1) and the initial condition (3.2) satisfy a relation obtained from (3.3) in the following way: in the β-th equality and in its derivative with respect to t of order $[(l+2b-r_\beta)/2b]$ we substitute $t=0$. These conditions are called compatibility conditions of order $[(l+2b)/2b]$. If, in particular, $\varphi \equiv 0$ and $D_t^{\alpha_0} f|_{t=0} = 0$ for $\alpha_0 = 0, \dots, [l/2b]$, then $D_t^{\alpha_0} u|_{t=0} = 0$ for $\alpha_0 = 0, \dots, [(l+2b)/2b]$ and the compatibility conditions of order $[(l+2b)/2b]$ reduce to the equalities $D_t^{\alpha_0} g_\beta|_{t=0} = 0$ for $\alpha_0 = 0, \dots, [(l+2b-r_\beta)/2b]$.

Thus, if $u \in C^{l+2b}(\overline{Q}_T)$ is a solution to (3.1)–(3.3), then the functions f, φ, and g_β satisfy conditions (3.6) and the compatibility conditions of order $[(l+2b)/2b]$. It turns out that these conditions also are sufficient for the existence of a solution in $C^l(Q_T)$. This is the central result of the Schauder theory. In the case of general parabolic boundary-value problems this result was established by Solonnikov (1965).

Theorem 3.1. *We consider the parabolic boundary-value problem* (3.1)–(3.3) *in* $\overline{Q}_T$, *satisfying conditions* β_1–β_3. *If the right hand sides of the problem* (3.1)–(3.3) *satisfy* (3.5) *and the compatibility condition of order* $[(l+2b)/2b]$, *then for any noninteger* l, $l > l_0 = \max(0, r_1 - 2b, \dots, r_{bm} - 2b)$ *the problem* (3.1)–(3.3) *has a unique solution* $u(t,x)$ *in the space* $C^{l+2b}(\overline{Q}_T)$ *and the following estimate holds*

$$\|u; Q_T\|_{l+2b} \le C \left(\|f; Q_T\|_l + \|\varphi; \Omega\|_{l+2b} + \sum_{\beta=1}^{bm} \|g_\beta; S_T\|_{l+2b-r_\beta} \right) , \tag{3.7}$$

where C *is a positive constant independent of* u, f, φ, *and* g_β.

1.3. On the Proof of the Main Theorem. The most laborious stage of the proof of Theorem 3.1 is to show that the problem (3.1)–(3.3) is well-posed in the cylinder Q_h of small height $h \le T$ in the case that $\varphi = 0$, $f \in C_0^l(\overline{Q}_h)$, and $g_\beta \in C_0^{l+2b-r_\beta}(\overline{S}_h)$, $\beta = 1, \dots, bm$. This problem is treated as the problem of solving the operator equations

$$Au = F , \tag{3.8}$$

where A is a linear operator acting from the linear space

$$E_1 = \left(C_0^{l+2b}(\overline{Q}_h)\right)^m \quad \text{to} \quad E_2 = \left(C_0^l(\overline{Q}_h)\right)^m \times \prod_{\beta=0}^{bm} C_0^{l+2b-r_\beta}(\overline{S}_T)$$

and assigning to each vector-function $u(t,x) \in E_1$ the vector-function $F = (f, g_1, \dots, g_{bm})$. The spaces E_1 and E_2 are Banach spaces with respect to the norms

$$\|u\|_{E_1} = \|u; Q_h\|_{l+2b} , \qquad \|F\|_{E_2} = \|f; Q_h\|_l + \sum_{\beta=1}^{bm} \|g_\beta; S_T\|_{l+2b-r_\beta} ,$$

respectively.

The fact that the equation (3.8) is well-posed means that the operator A has the bounded inverse operator A^{-1} (acting from the whole E_2 to E_1). The operator A^{-1} is constructed in the following way. *A regularizator of the problem or of the operator* A is an operator $R: E_2 \to E_1$ such that

$$AR = I + V ; \qquad RA = I + W , \tag{3.9}$$

where I is the identity operator in E_2 in the former case and in E_1 in the latter and V, W are bounded operators from E_2 to E_1, respectively, whose norms are less than one. From this and from the contraction principle it follows that the operators $I+V$, $I+W$ have bounded inverse $(I+V)^{-1}$, $(I+W)^{-1}$. Hence from (3.9) it follows that

$$AR(I+V)^{-1} = I , \qquad (I+W)^{-1}RA = I .$$

This means that A has left and right inverse operators equal to $R(I+V)^{-1}$ and $(I+W)^{-1}R$, respectively. These operators, as is well known, coincide, and consequently,

$$A^{-1} = R(I+V)^{-1} = (I+W)^{-1}R .$$

Thus, in this special case the proof of Theorem 3.1 reduces to the construction of a regularizator. It is 'glued together' by means of a suitable partition of unity from the operators that solve model problems of two types. The first type is the Cauchy problem for the parabolic system $\mathcal{L}_0(0, y; D_t, D)u = f$ with coefficients 'frozen' at the point $(0, y)$, $y \in \Omega$. That problem was studied in detail in Chap. 2 and the results are directly applicable to the present situation. The second type of model problems are boundary-value problems in the domains of the type $(0, T] \times \mathbb{R}^n_+$. As usually, these problems are obtained by means of the passage to local coordinates with the origin at $y^0 \in \partial\Omega$ in the differential expressions $\mathcal{L}_0(0, y; D_t, D_x)$ and $B^0_\beta(0, y; D_t, D_x)$, $\beta = 1, \ldots, bm$, involving the operators entering the system of equations and the boundary conditions. The required properties of R are established on the basis of precise estimates of solutions to the model problems. These estimates can be obtained from the convenient formulas for solutions to these problems that will be given below.

1.4. Solution of a Model Parabolic Boundary-Value Problem. The formulas that we obtain are basic for the Schauder theory as well as for an analogous theory for $W^{l/2b,l}_{p,t,x}$ spaces.

We consider a model parabolic problem:

$$\left.\begin{aligned} &\mathcal{L}_0(D_t, D_{x'}, D_{x_n})u = f\,, \\ &u|_{t=0} = \varphi\,, \\ &B^0_\beta(D_t, D_{x'}, D_{x_n})u|_{x_n=0} = g_\beta\,, \quad \beta = 1, \ldots, bm \end{aligned}\right\} \tag{3.10}$$

in $\mathbb{R}^{n+1}_{++} = (0, \infty) \times \mathbb{R}^n_+$.

We begin with the problem

$$\left.\begin{aligned} &\mathcal{L}_0(D_t, D_{x'}, D_{x_n})u = 0\,, \\ &u|_{t=0} = 0\,, \\ &B^0_\beta(D_t, D_{x'}, D_{x_n})u|_{x_n=0} = g_\beta\,, \quad \beta = 1, \ldots, bm\,. \end{aligned}\right\} \tag{3.11}$$

We seek the solution to (3.11) by taking smooth functions $g_\beta(t, x')$ with compact supports and using the Laplace transform in t and the Fourier transform in $x_1, \ldots, x_{n-1}$:

$$\hat{u}(x_n; p, \sigma') = \int_0^\infty \mathrm{e}^{-pt} dt \int_{\mathbb{R}^{n-1}} \mathrm{e}^{-\mathrm{i}x'\cdot\sigma'} u(t, x', x_n) dx'\,.$$

The problem (3.11) is thus converted into a problem for a system of ordinary differential equations

$$\begin{aligned} &\mathcal{L}_0(p, \sigma'; D_{x_n})\hat{u} = 0\,, \\ &B^0_\beta(p, \sigma', D_{x_n})\hat{u}|_{x_n=0} = \hat{g}_\beta(p, \sigma')\,, \\ &|\hat{u}(x_n; p, \sigma')| \to 0 \quad \text{as} \quad x_n \to \infty\,. \end{aligned} \tag{3.12}$$

The problem (3.12) is solved by the residual method briefly presented in Sect. 2 of Chap. 1. So we write the solution in the form

$$\hat{u}_j(x_n; p, \sigma') = \sum_{\beta=1}^{bm} \hat{G}_{j\beta}(x_n; p, \sigma')\hat{g}_\beta(p, \sigma')\,, \quad j = 1, \ldots, m\,, \tag{3.13}$$

where the functions $\hat{G}_{j\beta}(x_n; p, \sigma')$, called the elements of the *Poisson basis*, are written in the forms of contour integrals of functions we are sufficiently familiar with.

Using the inversion formulas of the theory of integral Laplace and Fourier transforms and, in particular, drawing on the fact that these transforms convert products into convolutions, we find important formulas

$$u_j(t, x) = \sum_{\beta=1}^{bm} \int_0^t d\tau \int_{\mathbb{R}^{n-1}} G_{j\beta}(t - \tau, x - \xi')g_\beta(\tau, \xi')d\xi'\,, \tag{3.14}$$

$$G_{j\beta}(t, x) = -\frac{\mathrm{i}}{(2\pi)^n} \int_{\mathbb{R}^{n-1}} \mathrm{e}^{\mathrm{i}x'\cdot\sigma'} d\sigma' \int_{a_0-\mathrm{i}\infty}^{a_0+\mathrm{i}\infty} \mathrm{e}^{pt}\hat{G}_{j\beta}(x_n; p, \sigma')dp\,, \tag{3.15}$$

where a_0 is an arbitrary positive constant.

Functions $G_{j\beta}(t,x)$ are called the *Poisson kernels* for the problem (3.12). To obtain substantial information on Poisson kernels we need a detailed and careful study of the elements of the Poisson basis for complex p and σ' by a special transformation of the contour in the complex plane along which the integral is taken.

Theorem 3.2 (Solonnikov 1965), (Ejdel'man 1964). *The Poisson kernels $G_{j\beta}(t,x)$ are defined and infinitely differentiable in $\mathbb{R}^{n+1}_{++}$. Their derivatives satisfy the inequalities*

$$|D_t^{\alpha_0} D^\alpha G_{j\beta}(t,x)| \le C_{\alpha_0\alpha} t^{-\frac{n+2b-r_\beta+2b\alpha_0+|\alpha|}{2b}} \exp\{-c|x|^q t^{1-q}\} . \tag{3.16}$$

We note that for large t, (3.16) can be essentially improved in some cases. For complete solution of (3.10) it remains to consider the problem

$$\left.\begin{aligned} &\mathcal{L}_0(D_t, D_{x'}, D_{x_n})u = f ,\\ &u|_{t=0} = \varphi ,\\ &B^0_\beta(D_t, D_{x'}, D_{x_n})u|_{x_n=0} = 0 , \end{aligned}\right\} \tag{3.17}$$

where f and φ are smooth functions with compact supports.

To solve (3.17) is a clear form convenient for application we introduce, so called, *homogeneous Green function (matrix-function).*

Definition 3.1. The *homogeneous Green function* $G_0(t,x,\xi)$ of the problem (3.10) is a matrix of dimension m such that the function

$$u(t,x) = \int_0^t d\tau \int G_0(t-\tau, x, \xi) f(\tau,\xi) d\xi$$

that is a solution to (3.10) with zero initial and boundary conditions and smooth function $f(t,x)$ with compact support.

We search for the homogeneous Green function $G_0(t,x,\xi)$ in the form

$$G_0(t,x,\xi) = \varGamma_0(t, x-\xi) - V(t,x,\xi) , \tag{3.18}$$

where $\varGamma_0(t,x)$ is the fundamental solution of the equation $\mathcal{L}_0(D_t, D)u = 0$. Then $V(t,x,\xi)$ should be a solution to the boundary-value problem

$$\left.\begin{aligned} &\mathcal{L}_0(D_t, D)u = 0 ,\\ &u|_{t=0} = 0 ,\\ &B^0_\beta(D_t, D)u|_{x_n=0} = B^0_\beta(D_t, D)\varGamma_0|_{x_n=0} . \end{aligned}\right\} \tag{3.19}$$

The problem (3.19) is solved by means of (3.14) and (3.15). Using the estimates (1.13) of the fundamental solution $\varGamma_0(t,x)$ and the estimates (3.17) of the Poisson kernels we can obtain the following information about $G_0(t,x,\xi)$ (Ivasishen 1987).

Theorem 3.3. *There exists a homogeneous Green function $G_0(t,x,\xi)$ for the problem* (3.10), *given by* (3.18), *infinitely differentiable with respect to all arguments for $t>0$, $\{x,\xi\}\subset\mathbb{R}^n_+$, and such that the following estimates hold*

$$|D^l_\xi D^{\alpha_0}_t D^\alpha_x G_0(t,x,\xi)| \le C_{l\alpha_0\alpha} t^{-\frac{n+|l|+2b\alpha_0+|\alpha|}{2b}} \exp\{-c|x-\xi|^q t^{1-q}\}\,, \quad (3.20)$$

$$|D^l_\xi D^{\alpha_0}_t D^\alpha_x V(t,x,\xi)| \le C_{l\alpha_0\alpha} t^{-\frac{n+|l|+2b\alpha_0+|\alpha|}{2b}} \exp\{-c(|x-\xi|+\xi_n)^q t^{1-q}\}\,. \quad (3.21)$$

The solution $u(t,x)$ to (3.10) *constructed by functions f, φ, g_β, $\beta=1,\dots,bm$ with compact supports is given by*

$$u(t,x) = \int G_0(t,x,\xi)\varphi(\xi)d\xi + \sum_{\beta=1}^{bm}\int_0^t d\tau \int_{\mathbb{R}^{n-1}} G_\beta(t-\tau,x-\xi')g_\beta(\tau,\xi')d\xi'$$
$$+\int_0^t d\tau \int G_0(t-\tau,x,\xi)f(\tau,\xi)d\xi\,, \quad G_\beta=(G_{1\beta},\dots,G_{m\beta})\,. \quad (3.22)$$

We conclude with a remark that to obtain *precise* estimates of higher order derivatives and their Hölder constants, which are necessary to build the Schauder theory, we need a special assumption that some integrals of these derivatives are equal to zero. This is illustrated by the following lemma.

Lemma 3.1 (Ivasishen 1987). *Suppose that the function*

$$\Gamma(t,x) = t^{-1-\frac{(n-1)}{2b}}\,\Omega(xt^{-\frac{1}{2b}})\,, \quad (t,x)\in\mathbb{R}^{n+1}_{++}$$

is a solution of the parabolic system $\mathcal{L}_0(D_t,D)u=0$ and $\Omega(z)$ satisfies the inequality

$$|D^\alpha_z\Omega(z)| \le C_\alpha \exp\{-c|z|^q\}\,.$$

Then we have

$$\int_{\mathbb{R}^{n-1}}\Omega(z',0)dz' = 0\,.$$

Example 3.1. We consider model boundary-value problems for the heat equation

$$\begin{aligned} &D_t u = a^2\Delta u\,,\\ &B^0(D_t,D)u|_{x_n=0} = g(t,x')\,,\\ &u|_{t=0}=0\,, \end{aligned} \quad (3.23)$$

given in Example 1.7.

The *Poisson basis* of (3.23) consists of one function

$$\hat G(x_n;p,\sigma) = (B^0(p,\sigma',\sigma_n^+))^{-1}\exp\{-\sigma_n^+ x_n\}\,, \quad \text{where} \quad \sigma_n^+ = \sqrt{p/a^2+\sigma'^2}\,.$$

In the case of the *Dirichlet problem* we have $B^0 \equiv 1$, in the *Neuman problem* $B^0 = -\sigma_n^+$, and in the *problem with directional derivative* $B_0 = i\sum_{\beta=1}^{n-1} b_\beta\sigma_\beta - b_n\sigma_n^+$. So, using (3.15) and the formulas of the theory of Laplace and Fourier transforms, we find that the *Poisson kernel for the Dirichlet problem* is given by

$$G^{(1)}(t,x) = \frac{x_n}{(2a\sqrt{\pi})^n} t^{-\frac{n+2}{2}} \exp\left\{-\frac{|x|^2}{4a^2t}\right\} ,$$

for the *Neuman problem* by

$$G^{(2)}(t,x) = -2a^2 \frac{1}{(2a\sqrt{\pi t})^n} \exp\left\{-\frac{|x|^2}{4a^2t}\right\} ,$$

and for the *problem with directional derivative*, by

$$G^{(3)}(t,x) = -2a^2 \frac{1}{(2a\sqrt{\pi t})^n} \left(b_n \exp\left\{-\frac{|x|^2}{4a^2t}\right\} \right.$$

$$\left. + \frac{|b|^2 x_n - b_n(b\cdot x)^2}{|b|a\sqrt{t}} \exp\left\{-\frac{|x|^2}{4a^2t} + \frac{b\cdot x}{4a^2|b|^2t}\right\} \int_{(b\cdot x)/2a|b|\sqrt{t}}^{\infty} \exp\{-z^2\}dz \right) ,$$

$$b = (b_1, \dots, b_n) .$$

To construct the homogeneous Green function $G_0(t,x,\xi) = \Gamma_0(t,x-\xi) - V(t,x,\xi)$ we first have to find V which is expressed by means of the Poisson kernel by the formula

$$V(t,x,\xi) = \int_0^t d\tau \int_{\mathbb{R}^{n-1}} G(t-\tau, x-y')B^0\Gamma_0(\tau, y'-\xi)dy' . \tag{3.24}$$

Substituting the concrete expressions for the Poisson kernels and the function Γ_0 to (3.24) we find, for example, that the *homogeneous Green function for the Dirichlet problem* has the form

$$G_0^1(t,x,\xi) = \frac{1}{(2a\sqrt{\pi t})^n} \exp\left\{-\frac{|x'-\xi'|^2}{4a^2t}\right\} \left(\exp\left\{-\frac{(x_n-\xi_n)^2}{4a^2t}\right\} \right.$$

$$\left. - \exp\left\{-\frac{|x_n+\xi_n|^2}{4a^2t}\right\} \right)$$

and for the *Neuman problem*, the form

$$G_0^2(t,x,\xi) = \frac{1}{(2a\sqrt{\pi t})^n} \exp\left\{-\frac{|x'-\xi'|^2}{4a^2t}\right\} \left(\exp\left\{-\frac{|x_n-\xi_n|^2}{4a^2t}\right\} \right.$$

$$\left. + \exp\left\{-\frac{(x_n+\xi_n)^2}{4a^2t}\right\} \right) .$$

Of course, the last formulas can be obtained by a simple argument using, so called, *reflection method*, but here we have presented a general method of finding formulas of this kind.

1.5. Necessity of Parabolicity Conditions. If a boundary-value problem is parabolic, then, by Theorem 3.1, the Schauder theory is valid for it. Here we prove the converse assertion following (Agranovich and Sukhorutchenko 1967) and (Ladyzhenskaya et al. 1967) (see also (Ivasishen 1987)).

We consider a model parabolic boundary-value problem

$$\mathcal{L}_0(D_t, D)u = f\ , \quad (t,x) \in \Pi_T^+\ , \tag{3.25}$$

$$u|_{t=0} = \varphi\ , \quad x \in \mathbb{R}_+^n\ , \tag{3.26}$$

$$B_\beta^0(D_t, D)u|_{x_n=0} = g_\beta\ , \quad \beta = 1, \dots, bm\ ,$$
$$(t, x') \in S_T = (0,T] \times \mathbb{R}^{n-1}\ . \tag{3.27}$$

Theorem 3.4. *For the system* (3.25)–(3.27) *to be parabolic it is necessary that for some noninteger* $l > l_0$ *there exists a constant* $C > 0$ *such that for any* $u \in C^{2b+l}(\Pi_T^+)$ *we have*

$$\|u; \Pi_T^+\|_{2b+l} \le C \left(\|\mathcal{L}_0 u; \Pi_T^+\|_l + \|u|_{t=0}; \mathbb{R}_+^n\|_{l+2b} + \sum_{\beta=1}^{bm} \|B_\beta^0 u|_{x_n=0}; S_T\|_{2b-r_\beta+l} \right) . \tag{3.28}$$

Proof. We shall prove that from the validity of (3.28) for any function $u \in C^{l+2b}(\Pi_T^+)$ it follows that

1) the system (3.25) is parabolic in the sense of Petrovskij,

2) the boundary operator $\mathcal{B}_0 = (B_{\beta j}^0)_{\beta=1,\dots,bm,\, j=1,\dots,m}$ of (3.27) satisfies the complementarity condition relative to the operator $\mathcal{L}_0$.

The proof is by reductio ad absurdam: we assume the contrary and we find an example solution $u_\lambda(t,x)$ depending on a parameter λ such that the estimate (3.28) cannot be valid for large λ.

1) Suppose that the system $\mathcal{L}_0(D_t, D)u = 0$ is not parabolic, that is, for some $\sigma^0 \in \mathbb{R}^n$, $\sigma^0 \neq 0$, the function $\det\mathcal{L}_0(\sigma^0, p)$ has a p-zero p^0 with $\operatorname{Re} p^0 \ge 0$. Let $v_\lambda(t,x)$ be a column-matrix of height m whose all entries are equal to $\exp\{\lambda^{2b} p^0 t + \mathrm{i}x \cdot \sigma^0\}$ and let $\zeta(t,x)$ be an infinitely differentiable function with compact support defined in Π_T^+. We define the functions $u_\lambda(t,x) = [\hat{\mathcal{L}}_0(D_t, D)v_\lambda(t,x)]\zeta(t,x)$, where $\hat{\mathcal{L}}_0$ is the matrix reciprocal to $\mathcal{L}_0$. Since $u_\lambda|_{t=0} = 0$ and $\mathcal{B}u_\lambda|_{x_n=0} = 0$, the estimate (3.28) in this case takes the form

$$\|u_\lambda; \Pi_T^+\|_{l+2b} \le C\|\mathcal{L}_0 u; \Pi_T^+\|_l\ . \tag{3.29}$$

We shall examine the growth as $\lambda \to \infty$ of the functions whose norms are on the left and right in (3.29). The modulus of the leading term, with respect to the growth, on the left hand side has the form $C\lambda^{2bm+l} \exp\{\lambda^{2b} t \operatorname{Re} p^0\}$ and since $\mathcal{L}_0\hat{\mathcal{L}}_0 = \det\mathcal{L}_0 I$, by the choice of σ^0 and p^0 the modulus of the leading term on the right hand side has the form $C\lambda^{2bm+l-1}\exp\{\lambda^{2b}t\,\mathrm{Re}\,p^0\}$ therefore (3.29) is violated for large λ. Thus the system (3.25) is parabolic in the sense of Petrovskij.

2) We prove that the boundary conditions (3.27) satisfy the complementarity conditions relative to the operator $\mathcal{L}_0$. Suppose that this is not the case so there exist $p^0 \in \mathbb{C}$ and $\sigma^0 \in \mathbb{R}^{n-1}$ such that $|p^0| + |\sigma^0| > 0$, $\mathrm{Re}\,p^0 \geq 0$, and the problem

$$\begin{aligned}
&\mathcal{L}_0(p^0, \sigma^{0'}; D_{x_n})v = 0\\
&B^0_\beta(p^0, \sigma^{0'}; D_{x_n})v|_{x_n=0} = 0\,, \quad \beta = 1, \ldots, bm\,,\\
&|v(x_n)| \to 0 \quad \text{as} \quad x_n \to \infty
\end{aligned}$$

has nontrivial solution $v_0(x_n)$.

We consider the function

$$u_\lambda(t,x) = \lambda^{-l-2b}\exp\{\lambda^{2b}p^0(t-T) + \mathrm{i}\lambda x'\cdot\sigma^{0'}\}v_0(\lambda x_n)\,, \quad (t,x) \in \Pi^+_T\,.$$

By the definition of $v_0(x_n)$ from the generalized homogeneity of the operators $\mathcal{L}_0$ and B^0_β it follows that $u_\lambda(t,x)$ is a solution of the system $\mathcal{L}_0 u = 0$ satisfying zero boundary conditions and for $t = 0$ we have $u_\lambda = \varphi_\lambda$, where $\varphi_\lambda = \lambda^{-l-2b}\exp\{-\lambda^{2b}p^0T + \mathrm{i}\lambda x'\cdot\sigma^{0'}\}v_0(\lambda x_n)$. First we consider the case $\mathrm{Re}\,p^0 > 0$. For sufficiently large $\lambda > 0$ we have

$$\|\varphi_\lambda; \mathbb{R}^n_+\|_{2b+l} \leq C\exp\{-\lambda^{2b}\mathrm{Re}\,p^0T\}\,. \tag{3.30}$$

We estimate the norm $\|u_\lambda; \Pi^+_T\|_{2b+l}$ from below (bearing in mind the definition):

$$\|u_\lambda; \Pi^+_T\|_{2b+l} \geq \|u_\lambda(T,x); \mathbb{R}^n_+\|_{2b+l} \geq H(u_\lambda(t,x)) = H(u_1(t,x))\,, \tag{3.31}$$

where

$$H(w) = \sum_{|\alpha|=2b+[l]} \sup\left\{\frac{|\Delta^y_x D^\alpha w(x)|}{|x-y|^{l-[l]}}\right\}\,.$$

From (3.28), (3.30), and (3.32) we have the inequality

$$H(u_1(T,x)) \leq C\exp\{-\lambda^{2b}\mathrm{Re}\,p^0T\}$$

which does not hold for large λ.

Finally, in the case $\mathrm{Re}\,p^0 = 0$ we have to consider the function $w_\lambda(t,x) = \zeta(3t/2T)u_\lambda(t,x)$, where $\zeta(\gamma)$ is infinitely differentiable, equal to zero for $\gamma \leq 0$, equal to one for $\gamma \geq 1$, and such that $0 \leq \zeta(\gamma) \leq 1$. Using a similar argument as before we can contradict (3.28). □

Since the definition of parabolicity of a boundary-value problem has been introduced through the model problems, it follows from Theorem 3.4 that if at least at one point of the cylinder $\overline{Q}_T$ the parabolicity condition is violated (for interior points of Q_T this means that the system is not parabolic and for the points of S_T this means that the complementarity condition does not hold), then for such a problem we do have the Schauder theory described above which, of course, does not exclude good properties of this problem in other functional spaces.

1.6. General Boundary-Value Problems. We turn to the formulation of the main result of the Schauder theory for systems parabolic in the sense of Solonnikov

$$\begin{aligned}
&\sum_{j=1}^{m} L_{kj}u_j = f_k\,, \quad k = 1,\ldots,m\,;\\
&\sum_{j=1}^{m} C_{\gamma j}u_j|_{t=0} = \varphi_\gamma\,, \quad \gamma = 1,\ldots,r\,;\\
&\sum_{j=1}^{m} B_{\beta j}u_j|_{S_T} = g_\beta\,, \quad \beta = 1,\ldots,br\,.
\end{aligned} \tag{3.32}$$

We formulate the conditions on the coefficients of the operators $\mathcal{L}$, C, $\mathcal{B}$ and on the boundary S_T.

β_1'. The coefficients of L_{kj} belong to $C^{l-s_k}(\overline{Q}_T)$.

β_2'. The coefficients of $B_{\beta j}$ belong to $C^{l-\theta_\beta}(\overline{S}_T)$.

β_3'. S_T is a surface of class $C^{l+t_{\max}}$, where $t_{\max} = \max_j t_j$.

β_4. The coefficients of $C_{\gamma j}$ belong to $C^{l-\rho_\gamma}(\overline{\Omega})$.

Theorem 3.5 (Ladyzhenskaya et al. 1967). *Consider an arbitrary $l > l_0 = \max(0, \theta_1, \ldots, \theta_{br})$ and suppose that conditions β_1', β_2', β_3' and β_4 are satisfied. The system* (3.32) *is parabolic if and only if there exists a unique solution $u = (u_1, \ldots, u_m)$, $u_j \in C^{l+t_j}(\overline{Q}_T)$ constructed by arbitrary $f_k \in C^{l-s_k}(\overline{Q}_T)$, $\varphi_\gamma \in C^{l-\rho_\gamma}(\overline{\Omega})$, and $g_\beta \in C^{l-\theta_\beta}(\overline{S}_T)$ satisfying natural compatibility conditions, and this solution satisfies the two-sided estimate*

$$C_1\|F\|_l \le \sum_{j=1}^{m} \|u_j; Q_T\|_{l+t_j} \le C_2\|F\|_l\,, \tag{3.33}$$

where

$$\|F\|_l = \sum_{k=1}^{m} \|f_k; Q_T\|_{l-s_k} + \sum_{\gamma=1}^{r} \|\varphi_\gamma; \Omega\|_{l-\rho_\gamma} + \sum_{\beta=1}^{br} \|g_\beta; S_T\|_{l-\theta_\beta}\,.$$

Analogous results are valid for general parabolic problems of conjugation and nonlocal problems considered in 2.6 of Chap. 1.

The condition on the arrangement of the domains Ω^0, Ω^1, Ω^2 and the diffeomorphism between $\partial\Omega^0$ and $\partial\Omega^1$ are presented in 2.6 in Chap.1. We seek for a function

$$u(t,x) = \begin{cases} u^1(t,x)\,, & (t,x) \in Q^1\,, \\ u^2(t,x)\,, & (t,x) \in Q^2\,, \end{cases} \tag{3.34}$$

that is a solution to the following problem

$$\begin{aligned} &\sum_{j=1}^{m_\mu} L^\mu_{kj} u^\mu_j = f^\mu_k\,, \quad k = 1,\ldots,m_\mu\,, \quad (t,x) \in Q^\mu\,; \\ &\sum_{j=1}^{m_\mu} C^\mu_{\gamma_j} u^\mu_j|_{t=0} = \varphi^\mu_\gamma\,, \quad \gamma = 1,\ldots,r_\mu\,, \quad \mu = 1,2\,; \end{aligned} \tag{3.35}$$

$$\left.\left(\sum_{j=1}^{m_1} B^1_{\beta j} u^1_j + \sum_{j=1}^{m_2} B^2_{\beta j} u^2_j + \sum_{j=1}^{m_3} B^3_{\beta j} u^2_j\right)\right|_{S_1} = g_\beta \qquad \beta = 1,\ldots,b(r_1 + 2r_2)\,. \tag{3.36}$$

The operators (L^μ_{kj}) are parabolic in the sense of Solonnikov with weights s^μ_k, t^μ_j, $\sum_{i=1}^{m_\mu}(s^\mu_i + t^\mu_i) = 2bt_\mu$, the initial operators $(C^\mu_{\gamma j})$ with weights ρ^μ_γ, $\gamma = 1,\ldots,r$, and the boundary operators $(B^\mu_{\beta j})$ with weights θ_β, $\beta = 1,\ldots,b(r_1 + 2r_2)$. We define the number $l_0 = \max_\beta(0, \theta_\beta)$.

We note that the condition of nonlocal joint covering in the general situation is quite similar to the condition of 2.6 in Chap. 1.

Theorem 3.6 (Zhitarashu and Ejdel'man 1970). *Let $l > l_0$ and suppose that the operators L^μ_{kj}, $B^\mu_{\beta j}$, and $C^\mu_{\gamma j}$ satisfy conditions β_1', β_2', and β_4 respectively, and $S^1 \in C^{l+t_{\max}}$, where $t_{\max} = \max(t^1_1,\ldots,t^1_{m_1}, t^2_1,\ldots,t^2_{m_2})$. The parabolicity condition for the problem* (3.35), (3.36) *is necessary and sufficient for the existence of a unique solution $u(t,x)$ with $u^\mu_j \in C^{l+t^\mu_j}(Q^\mu)$, $j = 1,\ldots,m_\mu$, $\mu = 1,2$, constructed by arbitrary functions $f^\mu_k \in C^{l-s^\mu_k}(\overline{Q}^\mu)$, $\varphi^\mu_\gamma \in C^{l-\rho^\mu_\gamma}(\overline{\Omega}^\mu)$, $g_\beta \in C^{l-\theta_\beta}(\overline{S}^1)$ satisfying natural nonlocal compatibility conditions, where the solution satisfies the two-sided estimate*

$$C_1\|F\|_l \le \sum_{\mu=1}^{2}\sum_{j=1}^{m_\mu} \|u^\mu_j; Q^\mu\|_{l+t^\mu_j} \le C_2\|F\|_l\,,$$

where

$$\|F\|_l = \sum_{\mu=1}^{2}\sum_{k=1}^{m_\mu} \|f^\mu_k; Q^\mu\|_{l-s^\mu_k} + \sum_{\mu=1}^{2}\sum_{\gamma=1}^{r_\mu} \|\varphi^\mu_\gamma; \Omega^\mu\|_{l-\rho^\mu_\gamma} + \sum_{\beta=1}^{b(r_1+2r_2)} \|g_\beta; S^1\|_{l-\theta_\beta}\,.$$

In a similar way we formulate the main result of the Schauder theory for parabolic conjugation problems.

We make some remarks.

Remark 1. To obtain from Theorems 3.5 and 3.6 the results for systems in general form, parabolic in the sense of Petrovskij, we have to put $t_i = n_i$, $s_k = 0$, assume that the orders of the boundary operators $B_{\beta j}$ are equal to $\theta_\beta + n_j$, and instead of the initial problem consider the Cauchy one

$$D_t^{\alpha_0} u_i|_{t=0} = \varphi_i^{(\alpha_0)} \quad i = 0, 1, \ldots, n_i - 1 .$$

We note that in Theorems 3.5 and 3.6 we do not require that the domain Ω be bounded but we assume that the conditions described are fulfilled on the boundary.

In particular, from Theorem 3.5 we obtain a theorem on well-posedness of the initial problem for systems parabolic in the sense of Solonnikov and the Cauchy problem for the systems parabolic in the sense of Petrovskij in spaces $C^l(\Pi_T)$.

Remark 2. From Theorems 3.5 and 3.6 we can deduce the theorems on improved smoothness of solutions up to the boundary depending on improved smoothness of the data.

Remark 3. The results remain valid for noncylindrical domains that can be locally rectifiable into cylinders with elements parallel to the t axis.

§2. L_2-Theory

2.1. Introduction. Here we briefly present the main results of the theory of parabolic boundary-value problems in Hilbert spaces $\mathcal{H}^l(Q_T)$ of smooth as well as generalized functions. A detailed exposition of the theory in the smooth case is contained in (Agranovich and Vishik 1964), (Solonnikov 1965), and in the spaces of generalized functions in (Zhitarashu 1985, 1987). The construction of this theory is based on successive applications of the Fourier transform of smooth and generalized functions, the theory of isotropic and anisotropic spaces of functions of many real variables, and theorems on traces of functions on manifolds. As in the Schauder theory we essentially employ the method of regularizator and detailed investigation of model parabolic boundary-value problems. In the L_2-theory the methods we apply are based on the use of fundamental solutions and Poisson kernels as well as the well developed theory of integral transforms.

In the case of spaces of generalized functions it is necessary to give a new definition of solution and also of parabolic boundary-value problem which, in particular, is connected with the fact that a new model problem emerges, a problem on the edge $t = 0$, $x \in \partial\Omega$. A properly formulated parabolic boundary-value problem remains well-posed on the corresponding spaces. It is interesting to note that the simplest examples taken from applications show that such an understanding of the boundary-value problem is natural.

2.2. Spaces of Smooth Functions. We proceed to the presentation of the theory of parabolic boundary-value problems in the spaces of smooth functions. We formulate the central result of this theory

We consider the general parabolic problem (3.32). We list the conditions.

β_5. $l = 2b\nu_0$, ν_0 is an integer, $l > \max_\beta \theta_\beta$, $s_k = 2bs_k'$, $t_j = 2bt_j'$, s_k' and t_j' are integers, and the numbers $\{(l-\theta_\beta)/2b - (2b+1)/4b\}$ are noninteger for $\beta = 1, \dots, br$.

β_6. The coefficients L_{kj}^0 have continuous derivatives $D_t^\lambda D_x^\mu$ in $\overline{Q}_T$ for $2b\lambda + |\mu| \le l - s_k$.

β_1''. The coefficients L_{kj} have bounded derivatives $D_t^\lambda D_x^\mu$ in $\overline{Q}_T$ for $2b\lambda + |\mu| \le l - s_k$.

β_2''. The coefficients $B_{\beta j}$ belong to $C^{l-\theta_\beta-1/2+\varepsilon}(\overline{S}_T)$, $\varepsilon > 0$.

β_4''. The coefficients $C_{\gamma j}$ belong to $C^{l-\rho_\gamma-1/2+\varepsilon}(\overline{\Omega})$, $\varepsilon > 0$.

Theorem 3.7. *Suppose that conditions β_1'', β_2'', β_3', β_4'', β_5, and β_6 are satisfied. The problem (3.32) is parabolic if and only if there exists a unique solution $u(t,x) = \big(u_1(t,x), \dots, u_m(t,x)\big)$, $u_j(t,x) \in \mathcal{H}^{l+t_j}(Q_T)$, $j = 1, \dots, m$ constructed by $f_k \in \mathcal{H}^{l-s_k}(Q_T)$, $k = 1, \dots, m$, $g_\beta \in \mathcal{H}^{l-\theta_\beta-1/2}(S_T)$, $\beta = 1, \dots, br$, $\varphi_\gamma \in H^{l-\rho_\gamma-b}(\Omega)$, $\gamma = 1, \dots, r$, and satisfying the two-sided estimates*

$$C_1\|F\|_l \le \sum_{j=1}^m \|u_j; Q_T\|_{l+t_j} \le C_2\|F\|_l \,,$$

where

$$\|F\|_l = \sum_{k=1}^m \|f_k; Q_T\|_{l-s_k} + \sum_{\beta=1}^{br} \|g_\beta; S_T\|_{l-\theta_\beta-1/2} + \sum_{\gamma=1}^r \|\varphi_\gamma; \Omega\|_{l-\rho_\gamma-b} \,.$$

One of the methods of proof of Theorem 3.7 (Agranovich and Vishik 1964) is that first we consider the stationary problem, i.e. the problem where the coefficients of the equations and of the boundary conditions do not depend on time t, the initial condition is zero, and a solution is sought for in the cylinder $Q^+ = (0,\infty) \times \Omega$ infinite in t. By means of the Laplace transform in t the problem (3.32) is converted into an elliptic problem with complex parameter p

$$\sum_{j=1}^m L_{kj}(x; p, D_x)\hat{u}_j = \hat{f}_k(p,x) \,, \qquad k = 1, \dots, m \,, \quad x \in \Omega \,,$$

$$\sum_{j=1}^m B_{\beta j}(x; p, D_x)\hat{u}_j|_{\partial\Omega} = g_\beta(p, x') \,, \qquad \beta = 1, \dots, br \,.$$

It turns out that parabolicity of the problem (3.32) implies well-posedness of this elliptic boundary-value problem in the scale of spaces dual (in the sense of the Laplace transform) to the spaces of type $\mathcal{H}^l$ of smooth functions if the parameter p varies in a sector of the complex p-plane and is sufficiently large in modulus.

Precise theorems on isomorphisms of spaces of functions $u(t,x)$ and $u(p,x)$ permit us to establish that the parabolic boundary-value problem is also well-posed in the stationary case. The passage to the general situation is done by a standard argument.

2.3. Spaces $\mathcal{H}^l$ and H^l of Generalized Functions. We have to supplement the information on the spaces $\mathcal{H}^l$ and H^l given in 3.3 in Chap. 1.

We begin with the definition of anisotropic spaces $\mathcal{H}^l(\mathbb{R}^{n+1})$ for any real l. We denote by $\hat{u}(\sigma_0,\sigma)$ the Fourier transform of $u(t,x)$. The space $\mathcal{H}^l(\mathbb{R}^{n+1})$ is defined as the completion of $C_0^\infty(\mathbb{R}^{n+1})$ in the norm

$$\|u;\mathbb{R}^{n+1}\|_l = \left(\iint_{\mathbb{R}^{n+1}} (1+|\sigma_0|^{l/b}+|\sigma|^{2l})|\hat{u}(\sigma_0,\sigma)|^2 d\sigma_0 d\sigma\right)^{1/2} .$$

For $l \geq 0$ these are Hilbert spaces of ordinary functions and for $l < 0$ they contain generalized functions of finite order (Volevich and Paneah 1965). And so, for example, the Dirac delta-function $\delta(t,x)$ belongs to all $\mathcal{H}^l(\mathbb{R}^{n+1})$ for $l < -(b+n/2)$. It is easy to find out to which spaces $\mathcal{H}^l(\mathbb{R}^{n+1})$ belong its derivatives and also delta-functions concentrated on manifolds of various dimensions in $\mathbb{R}^{n+1}$.

In a similar way we define the isotropic *space* $H^l(\mathbb{R}^n)$ for real l as the completion of $C_0^\infty(\mathbb{R}^n)$ in the norm

$$\|u;\mathbb{R}^n\|_l = \left(\int_{\mathbb{R}^n} (1+|\sigma|^{2l})|\hat{u}(\sigma)|^2 d\sigma\right)^{1/2} .$$

We now define the *spaces* $\mathcal{H}^l(Q_T)$ for any real l. For $l \geq 0$ this definition was given in 3.3, Chap. 1. Now for $l > 0$ we consider the subspace $\mathcal{H}^{\circ l}(Q_T)$ of the space $\mathcal{H}^l(Q_T)$ that is the closure in the norm of $\mathcal{H}^l(Q_T)$ of the set of infinitely differentiable functions in Q_T which are zero with all their derivatives for $t = T$.

We define $\mathcal{H}^l(Q_T)$ for any negative l as the space dual to $\mathcal{H}^{-l}(Q_T)$ with respect to the extension of the scalar product of the space $\mathcal{H}^0(Q_T) = L_2(Q_T)$ with a natural norm of the space of linear continuous functionals.

We give yet another description of $\mathcal{H}^l(Q_T)$. Let us consider all generalized functions $V \in \mathcal{H}^l(\mathbb{R}^{n+1})$ with supports in $\overline{Q}_T$. With an element of $\mathcal{H}^l(Q_T)$ we associate all generalized functions V_u which coincide with u in $[0,T]\times\overline{\Omega}$. By $\mathcal{H}^l(Q_T)$ we understand here the set of restrictions u of generalized functions V_u to $[0,T]\times\overline{\Omega}$ with the norm

$$\|u;Q_T\|_l = \inf_{V_u} \|V_u;\mathbb{R}^{n+1}\|_l .$$

The anisotropic spaces $\mathcal{H}^l(S_T) = \left(\mathcal{H}^{-l}(Q_T)\right)^*$, $l < 0$, and the isotropic spaces $H^l(\Omega) = \left(H^{-l}(\Omega)\right)^*$, $H^l(\partial\Omega) = \left(H^{-l}(\partial\Omega)\right)^*$, are defined analogously. We recall that the surfaces on which the functions are defined are infinitely smooth. They admit finite coverings by domains that can be locally rectified. The norms are then defined by a suitable partition of unity.

2.4. Green Formulas. We move to the question of how to transform in a natural way a parabolic boundary-value problem in the case when it can be solved in suitable Hilbert spaces of generalized functions of finite order. We do this obeying the following requirements:

1) the problem remains well-posed in the full scale of spaces of smooth as well as generalized functions,
2) in the spaces of smooth functions the problem coincides with the usual parabolic boundary-value problem.

For clarity, we perform all construction in the case of the following equation parabolic in the sense of Petrovskij

$$L(t,x;D_t,D_x)u \equiv \sum_{2b\alpha_0+|\alpha|\le 2br} a_{\alpha_0\alpha}(t,x)D_t^{\alpha_0}D_x^{\alpha}u(t,x) = f(t,x)\,, \quad (t,x)\in Q_T\,, \tag{3.37}$$

$$D_t^{\alpha_0-1}u|_{t=0} = \varphi_{\alpha_0}(x)\,, \qquad \alpha_0 = 1,\dots,r\,; \quad x\in\Omega\,, \tag{3.38}$$

$$B_\beta(t,x;D_t,D_x)u|_{S_T} = \sum_{2b\alpha_0+|\alpha|\le r_\beta} b^\beta_{\alpha_0\alpha}(t,x)D_t^{\alpha_0}D_x^{\alpha}u(t,x)|_{S_T} = g_\beta\,, \quad \beta = 1,\dots,br\,. \tag{3.39}$$

Here $D_t = \frac{\partial}{\partial t}$ and $D_{x_k} = \frac{\partial}{\partial x_k}$. To define a generalized solution we need a special form of the *Green formula*. We introduce the scalar products

$$(u,v) = \iint_{Q_T} u\bar{v}dtdx'\,, \quad <u,v> = \int_{S_T} u\bar{v}dtdx'\,,$$

$$<u,v>' = \int_{\partial\Omega} u\bar{v}dx'\,, \quad [u,v] = \int_{\Omega} u\bar{v}dx\,.$$

We write $L(\cdot;p,\sigma)$ and $B_\beta(\cdot;p,\sigma)$ in the following form:

$$L(\cdot,p,\sigma',\sigma_n) = \sum_{2b\alpha_0+\alpha_n\le 2br} l_{\alpha_0\alpha_n}(\cdot;p,\sigma')p^{\alpha_0}\sigma_n^{\alpha_n}\,,$$

$$B_\beta(\cdot,p,\sigma',\sigma_n) = \sum_{\alpha_n=0}^{r_\beta} b_{\beta\alpha_n}(\cdot;p,\sigma')\sigma_n^{\alpha_n}\,,$$

where $b_{\beta\alpha_n} = \sum_{\alpha_0} b_{\beta\alpha_0\alpha_n}(\cdot,\sigma')p^{\alpha_0}$. Accordingly, we define the polynomials

$$L_{\mu\lambda}(\cdot;\cdot,\sigma',\sigma_n) = \sum_{\alpha_0\geq\mu,\alpha_n\geq\lambda} l_{\alpha_0\alpha_n}(\sigma')p^{\alpha_0-\mu}\sigma_n^{\alpha_n-\lambda} ,$$

$$b_{\beta\alpha_n\mu} = \sum_{\alpha_0\geq\mu} b_{\beta\alpha_0\alpha_n}(\sigma')p^{\alpha_0-\mu} .$$

To write the Green formula we have write the operators L and B_β in local coordinates in a neighborhood of the boundary of the cylinder Q_T. The coordinate x_n is taken along the inward normal ν, the derivative D_{x_n} corresponds to the normal derivative D_ν, and the remaining coordinates lie in the plane tangent to $\partial\Omega$.

For functions $u(t,x)$ and $v(t,x)$ smooth in $\overline{Q}_T$, where $v(t,x)$ vanishes together with all its derivatives at $t=T$, we obtain by integration by parts the *Green formula:*

$$\begin{aligned}(u,L^*v)-(Lu,v) &= \sum_{\lambda=1}^{2br} <D_\nu^{\lambda-1}u, L_{0\lambda}^*v> + \sum_{\mu=1}^{r}[D_t^{\mu-1}u, L_{\mu 0}^*v] \\ &\quad - \sum_{\mu,\lambda\in\mathcal{P}_0} <D_t^{\mu-1}D_\nu^{\lambda-1}u, L_{\mu\nu}^*v>' ,\end{aligned} \tag{3.40}$$

$$<B_\beta u,v> = \sum_{\lambda=1}^{r_\beta+1} <D_\nu^{\lambda-1}u, b_{\beta\lambda-1}^*v> - \sum_{\mu,\lambda\in\mathcal{P}_\beta} <D_t^{\mu-1}D_\nu^{\lambda-1}u, b_{\beta\lambda-1\mu}^*v>' \tag{3.41}$$

$\beta=1,\dots,br$, where $\mathcal{P}_0$ and $\mathcal{P}_\beta$ are the sets of pairs of natural numbers (μ,λ) such that $L_{\mu\lambda}$ and $b_{\beta\lambda-1\mu}$ are different from zero, respectively, and we put $\mathcal{P}=\bigcup_{\alpha=0}^{br}\mathcal{P}_\alpha$. We have

$$[D_t^{\mu-1}u,v] = [\varphi_\mu,v] , \quad \mu=1,\dots,r . \tag{3.42}$$

Let $u(t,x)$ be a smooth solution to the problem (3.37)–(3.39). We introduce the following notations: $u_0 = u(t,x)$, $(t,x)\in Q_T$; $u_\lambda = D_\nu^{\lambda-1}u$, $(t,x)\in S_T$, $\lambda=1,\dots,2br$; $v_\mu = D_t^{\mu-1}u|_{t=0}$, $\mu=1,\dots,r$, $x\in\Omega$; and $\omega_{\mu\lambda} = D_t^{\mu-1}D_\nu^{\lambda-1}u|_\Gamma$, $\Gamma=\{(0,x) : x\in\partial\Omega\}$, $(\mu,\lambda)\in\mathcal{P}$. Let U be a vector with coordinates $(u_0,u_\lambda,v_\mu,\omega_{\mu\lambda})$. Then we have the following integral identities for coordinates of this vector.

$$\begin{aligned}(u_0,L^*v) - \sum_{\lambda=1}^{2br} <u_\lambda, L_{0\lambda}^*v> - \sum_{\mu=1}^{r}[v_\mu, L_{\mu 0}^*v] + \sum_{(\mu,\lambda)\in\mathcal{P}_0} <\omega_{\mu\lambda}, L_{\mu\lambda}^*v>' \\ = (f,v) ,\end{aligned} \tag{3.43}$$

$$\begin{aligned}\sum_{\lambda=1}^{r_\beta+1} <u_\lambda, b_{\beta\lambda-1}^*v> - \sum_{(\mu,\lambda)\in\mathcal{P}_\beta} <\omega_{\mu\lambda}, b_{\beta\lambda-1\mu}^*v>' = <g_\beta,v> , \\ \beta=1,\dots,br ,\end{aligned} \tag{3.44}$$

$$[v_\mu, v] = [\varphi_\mu, v]\,, \quad \mu = 1, \ldots, r\,, \tag{3.45}$$

where v is an arbitrary *test function*, i.e. a function infinitely differentiable in Q_T equal to zero together with all its derivatives for $t = T$.

If $u_0 \in \mathcal{H}^l$ for $l > l_0 = 1/2 + \max(0, r_\beta - 2br)$, then u_0 uniquely determines the remaining coordinates of the vector U as the corresponding traces of derivatives of u_0 on S_T, on the hyperplane $t = 0$, $x \in \Omega$, and on the edge Γ. And precisely in this sense we understand the solution to (3.37)–(3.39).

For $l \le l_0$ this problem makes no longer sense and a question arises of whether in this situation we can find a vector U such that its coordinates satisfy in a reasonable sense the equalities (3.43)–(3.45). If u_0 is a generalized function, then it is natural to consider the scalar product (u_0, L^*v) as the value of u_0 on the test function L^*v and we can analogously treat the other terms in (3.43)–(3.45). We have to take into account the fact that the number of unknown functions is greater than the number of identities (3.43)–(3.45) and therefore to find a well-posed problem we have to learn how to write additional identities which together with the existing ones compose a uniquely jointly solvable system. We proceed to a realization of this idea.

2.5. Parabolic Boundary-Value Problems in the Space $\widetilde{\mathcal{H}}^l$. We now define the spaces in which we shall solve the problem. We put $\lambda_0 = \max_\beta(2br, r_\beta + 1)$

The equalities (3.43)–(3.45) for $l > l_0$ will be considered as a system of integral identities with respect to the coordinates of the vector $U = (u_0, u_\lambda, v_\mu, \omega_{\mu\lambda})$. From the trace theorem it follows that

$$U \in K^l = \mathcal{H}^l(Q_T) \times \prod_{\lambda=1}^{\lambda_0} \mathcal{H}^{l-\lambda+1/2}(S_T) \times \prod_{\mu=1}^{r} H^{l-2b\mu+b}(\Omega)$$
$$\times \prod_{(\mu,\lambda)\in\mathcal{P}} H^{l-\lambda-2b\mu+b+1/2}(\Gamma)\,. \tag{3.46}$$

In the case when $l \le l_0$ we treat the scalar products in (3.43)–(3.45) as definitions of the action of a generalized function on the corresponding test functions as prescribed by (3.46). Thus, we face a problem of defining unknown generalized functions whose coordinates compose the vector U.

We introduce the *space* $\widetilde{\mathcal{H}}^l_{\lambda_0,r,\mathcal{P}}(Q_T) \equiv \widetilde{\mathcal{H}}^l(Q_T)$ which occupies the central position in the theory we present.

Definition 3.1. For l such that l and $l/2b$ are not multiplicities of $1/2$ the space $\widetilde{\mathcal{H}}^l(Q_T)$ is the completion of $C^\infty(\overline{Q}_T)$ in the norm

$$|||u; Q_T|||_l^2 = \|u_0; Q_T\|_l^2 + \sum_{\lambda=1}^{\lambda_0} \|u_\lambda; S_T\|'^2_{l-\lambda+1/2} + \sum_{\mu=1}^{r} \|v_\mu; \Omega\|^2_{l-2b\mu+b}$$
$$+ \sum_{(\mu,\lambda)\in\mathcal{P}} \|\omega_{\mu\lambda}; \partial\Omega\|^2_{l-\lambda-2b\mu+b+1/2}\,. \tag{3.47}$$

For the excluded values of l we define $\widetilde{\mathcal{H}}^l(Q_T)$ by interpolation (Triebel 1978).

For $u \in C^\infty(\overline{Q}_T)$ we consider the map $\chi : u \to U$.

Lemma 3.2. *The closure of the map χ in the norm (3.47) gives an isometry between the space $\mathcal{H}^l(Q_T)$ and the subspace $\widetilde{K}^l$ of the space K^l that consists of the vectors with coordinates satisfying the compatibility conditions which follow from the trace theorems:*

$$u_\lambda = D_\nu^{\lambda-1} u_0|_{S_T} \quad \forall \lambda < l + \frac{1}{2}\,;$$
$$v_\mu = D_t^{\mu-1} u_0|_{t=0} \quad \forall \mu < l + b,$$
$$\omega_{\lambda\mu} = D_\nu^{\lambda-1} D_t^{\mu-1} u_0|_\Gamma\,, \qquad (\mu, \lambda) \in \mathcal{P} \quad 2b\mu + \lambda < l + b + \frac{1}{2}\,.$$

From the lemma it follows that $u \in \widetilde{\mathcal{H}}^l$ and $U \in \widetilde{K}^l$ can be identified (up to isometry) and we do this in what follows.

Definition 3.2. We call $u \in \widetilde{\mathcal{H}}^l(Q_T)$ a *generalized solution to the problem* (3.37)–(3.39) in $\widetilde{\mathcal{H}}^l(Q_T)$ if the coordinates of the corresponding vector $U \in \widetilde{K}^l$ satisfy (3.43)–(3.45).

The choice of the space $\widetilde{\mathcal{H}}^s(Q_T)$ in which we seek a solution is determined by the number of dependent and independent coordinates of U. We have already said that for $l > l_0$ the first coordinate u_0 allows us to reconstruct the whole vector U. If l is a negative number sufficiently large in modulus, then all coordinates of U are quite arbitrary generalized functions which belong to spaces determined by the space K^l to which U belongs. In this case, if $1/2 < l < l_0$, then it is necessary to supplement the integral identities (3.43)–(3.45) with the compatibility conditions of the lemma written in a weak form.

Definition 3.3. The problem (3.37)–(3.39) is called a *parabolic boundary-value problem in the spaces* $\widetilde{\mathcal{H}}^l(Q_T)$, $-\infty < l < \infty$, if

1) (3.37)–(3.39) is a parabolic boundary-value problem with infinitely differentiable coefficients in $\widetilde{Q}_T$, $\partial\Omega \in C^\infty$;
2) the solution $u \in \widetilde{\mathcal{H}}^l(Q_T)$ is such that U satisfies (3.43) and (3.45);
3) we have $\omega_{\mu\lambda} = \varphi_{\mu\lambda}$ where μ and λ satisfy $l - 2b\mu - \lambda + b + 1/2 \le 0$ if the corresponding coordinates enter U.

We denote $\mathcal{P}^{(l)} = \{(\mu, \lambda) : (\mu, \lambda) \in \mathcal{P}, l - 2b\mu - \lambda + b + 1/2 \le 0\}$. We define the space

$$\mathcal{F}^l = \mathcal{H}^{l-2br}(Q_T) \times \prod_{\beta=1}^{br} \mathcal{H}^{l-r_\beta-1/2}(S_T) \times \prod_{\mu=1}^{r} H^{l-2b\mu+b}(\Omega)$$
$$\times \prod_{(\mu,\lambda)\in\mathcal{P}^l} H^{l-\lambda-2b\mu+b-1/2}(\Gamma)$$

with the norm $\|\cdot\|_{\mathcal{F}^l}$ of the direct product of Hilbert spaces. The elements of $\mathcal{F}^l$ will be denoted by $F = (f, g_1, \ldots, g_{br}, \varphi_1, \ldots, \varphi_r, \varphi_{\mu\lambda})$. We introduce the subspaces $\widetilde{\mathcal{F}}^l$ of $\mathcal{F}^l$ consisting of those F that satisfy natural compatibility conditions.

Theorem 3.8. *The boundary-value problem* (3.37)–(3.39) *is parabolic in* $\widetilde{\mathcal{H}}^l(Q_T)$, $-\infty < l < \infty$, *if and only if there is a unique solution* u *in* $\widetilde{\mathcal{H}}^s(Q_T)$ *constructed by an arbitrary vector* $F \in \widetilde{\mathcal{F}}^l$ *and satisfying the two-sided estimate*

$$C_1\|F\|_{\mathcal{F}^l} \leq |||u; Q_T|||_l \leq C_2\|F\|_{\mathcal{F}^l}$$

Analogous results are true for general parabolic problems and for parabolic nonlocal problems.

Example 3.2. We consider the problem

$$\left.\begin{aligned} &\frac{\partial u}{\partial t} = \frac{\partial^2 u}{\partial x^2} \quad (t,x) \in (0,T) \times \mathbb{R}^1_+ \equiv Q_T\,; \\ &u|_{t=0} = 0 \quad x > 0\,; \\ &\left(\frac{\partial u}{\partial t} - a\frac{\partial u}{\partial x}\right)_{x=0} = 0 \quad a > 0\,. \end{aligned}\right\} \tag{3.48}$$

The problem (3.48) is parabolic. We write down the formulas (3.43) and (3.44) in this case.

$$\left(u, -\left(\frac{\partial}{\partial t} + \frac{\partial^2}{\partial x^2}\right)v\right) - <u_1, \frac{\partial v}{\partial x}> + <u_2, v> = 0\,. \tag{3.43'}$$

The operators $B \equiv \frac{\partial}{\partial t} - a\frac{\partial}{\partial x}$ corresponds to the polynomial $B(p,\sigma) = p - a\sigma$, i.e. $b_0 = -a$, $b_1 = p$, $b_{01} = 0$, and $b_{11} = 1$, therefore (3.44) will take the form, which, of course, can be also found directly,

$$<u_1, -\frac{\partial v}{\partial t}> + <u_2, av> + \omega_{11}v(0,0) = 0\,. \tag{3.44'}$$

Should we need the compatibility condition at (0,0), then the theory of parabolic boundary-value problems in spaces of smooth functions would apply to the problem (3.48) and this theory would imply that there are no solutions to (3.48) in these classes except the trivial one. If we apply Theorem 1.8, then for the problem to be well-posed in $\widetilde{\mathcal{H}}^l(Q_T)$ with $l < 1/2$ we have to define ω_{11} additionally.

In fact, the problem (3.48), as can be directly verified, has a nontrivial solution (Ditkin and Prudnikov 1966)

$$u(t,x) = C\exp\{ax + a^2t\}\int_{(x/2\sqrt{t})+a\sqrt{t}}^{\infty} \exp\{-z^2\}dz\,,$$

where C is an arbitrary constant.

If ω_{11} is given, then we can define $C = 2\omega_{11}/\sqrt{\pi}$. It is interesting to note that

$$\frac{\partial u}{\partial t} - a\frac{\partial u}{\partial x} = \omega_{11}\frac{2}{\sqrt{\pi}} \cdot \frac{x}{4t^{3/2}} \exp\left\{-\frac{x^2}{4t}\right\} = \omega_{11}\frac{x}{2\sqrt{\pi}t^{3/2}} \exp\left\{-\frac{x^2}{4t}\right\}$$
$$= \omega_{11} G_1(t,x) ,$$

where G_1 is the Poisson kernel of the Dirichlet problem for the heat equation. That is, for $x \to 0$ we have $\frac{\partial u}{\partial t} - a\frac{\partial u}{\partial x} \to \omega_{11}\delta(t)$.

§3. Green Function

3.1. Introduction. Theorems 3.6–3.8 guarantee well-posedness of parabolic boundary-value problems in spaces of smooth and generalized functions. From this it follows that the operator of the problem has the inverse defined on its whole range. General theorems of functional analysis guarantee that in this situation the inverse operator is an integral one whose kernel is a generalized function of finite order. This kernel is called the *Green function of the parabolic boundary-value problem.* It is interesting to find out when the Green function is an ordinary function, to obtain its precise estimates as well as the estimates of its derivatives with respect to the principal as well as parametrical variables, and to find various variants of its asymptotic expansions. The following papers are devoted to the Green functions for systems parabolic in the sense of Petrovskij: (Drin' and Ivasishen 1977), (Ivasishen 1981), (Solonnikov 1969), and (Ejdel'man and Ivasishen 1970). They present three methods of studying the Green functions of the parabolic boundary-value problems. The first is to analyse the structure of the inverse operator constructed by the regularizator method (see 1.3). In the second the Green function is found as the limit of solutions to parabolic boundary-value problems constructed by a delta-like sequence of the data of the problem. The third develops the method of integral operators with singularities that are not summable, presented by Krasovskij in the investigation of Green's function of elliptic boundary-value problems. And this is the method that allowed Ivasishen to obtain the most complete information. However, when applicable, the first two approaches give more precise results with respect to the smoothness conditions. We shall call these methods, respectively, the method of regularizator, delta-like sequences, and integral operators. In our opinion all three methods are competitive but their realization is technically not so simple. Regretfully, the formulation of precise results of this study is quite complicated so we restrict ourselves to a description only.

3.2. Green Functions. Homogeneous Green functions. We consider the parabolic boundary-value problem (3.37)–(3.39). By the *Green function* of this problem we understand a matrix function $G = (G_0, G_1, G_2)$ such that any solution $u(t,x)$ from a suitable function space can be written by the formula

$$u(t,x) = \int_0^t d\tau \int_\Omega G_0(t,x;\tau,\xi) f(t,\xi) d\xi + \int_0^t d\tau \int_{\partial\Omega} G_1(t,x;\tau,\xi') g(t,\xi') d\xi' + \int_\Omega G_2(t,x;0,\xi)\varphi(\xi) d\xi \,, \tag{3.49}$$

where $g = (g_1, \dots, g_{br})$ and $\varphi = (\varphi_1, \dots, \varphi_r)$. In this case G_0 is a function and $G_1 = (G_{11}, \dots, G_{1br})$, $G_2 = (G_{21}, \dots, G_{2r})$ are vectors.

If the initial and the boundary conditions are equal to zero, then the problem (3.37)–(3.39) is called the homogeneous boundary value problem and the corresponding Green function the *homogeneous Green function.* In the case of model parabolic equations and systems of the first order in t we have discussed the homogeneous Green function in Theorem 3.3. By estimates (3.20) and (3.21) we can judge the properties of the homogeneous Green function and its derivatives. However the assumption that $f(t,x)$ and $\varphi(x)$ have compact supports, which was imposed in 1.4, does not allow us to analyse completely all possible situations. To this end we do the following. First we assume that the orders of all boundary operators are less than the order of the equation and by the limit passage in (3.22) we remove the assumption on compact supports of the data (leaving, of course, the compatibility condition). And now we consider an example.

Example 3.3 (Ivasishen 1981, 1987). We consider the following problem

$$\begin{aligned} D_t u &= \frac{\partial^2 u}{\partial x^2} + f(t,x) \,, \\ u|_{t=0} &= \varphi(x) \,, \\ D_t u|_{x=0} &= g(t) \,. \end{aligned} \tag{3.50}$$

In the discussion of Example 1.7 we have shown that the problem (3.50) is parabolic. It is equivalent to the following one

$$\begin{aligned} D_t u &= \frac{\partial^2 u}{\partial x^2} \,, \\ u|_{t=0} &= \varphi(x) \,, \\ u|_{x=0} &= \varphi(0) + \int_0^t g(\tau) d\tau \,. \end{aligned} \tag{3.51}$$

We write the solution to the problem (3.51) ((3.50)) using (3.22)

$$\begin{aligned} u(t,x) &= \int_0^t d\tau \int_0^\infty G_0^{(1)}(t-\tau,x,\xi)f(\tau,\xi)d\xi \\ &+ \int_0^t G^{(1)}(t-\tau,x)\left\{\varphi(0)+\int_0^\tau g(\beta)d\beta\right\}d\tau \\ &+ \int_0^\infty G_0^{(1)}(t,x,\xi)\varphi(\xi)d\xi\,, \end{aligned} \tag{3.52}$$

where $G_0^{(1)}$ is the Green function and $G^{(1)}$ is the Poisson kernel of the Dirichlet problem.

We write (3.52) in the form (3.49):

$$\begin{aligned} u(t,x) &= \int_0^t d\tau \int_0^\infty G_0^{(1)}(t-\tau,x,\xi)f(\tau,\xi)d\xi \\ &+ \int_0^t \widetilde{G}(t-\tau,x)g(\tau)d\tau \\ &+ \int_0^\infty \left\{G_0^{(1)}(t,x,\xi)+2\int_0^t G^{(1)}(t-\tau,x)d\tau\delta(\xi)\right\}\varphi(\xi)d\xi\,, \end{aligned}$$

where $\widetilde{G}(t,x)=(2/\sqrt{\pi})\int_{x/2\sqrt{t}}^\infty \exp\{-z^2\}dz$.

Thus, the Green function G of the problem (3.50) is given by

$$\left(G_0^{(1)}(t-\tau,x,\xi),\widetilde{G}(t-\tau,x),G_0^{(1)}(t,x,\xi)+2\int_0^t G^{(1)}(t-\tau,x)d\tau\delta(\xi)\right).$$

So the third component G_3 of the Green function of (3.50) is a generalized function.

We return to the problem (3.37)–(3.39). We write the boundary operator B_β in local coordinates with centre at $(\gamma,y)\in S_T$. We assume that at this point the order of the normal derivative is maximal and we denote it by $n_\beta(\gamma,y)$. We also put $n_\beta=\max_{(\gamma,y)\in S_T} n_\beta(\gamma,y)$, $\rho_\beta=n_\beta-2br$, $\rho_0=\max_\beta\rho_\beta$, and we denote by m_β the maximal order of derivative with respect to t in the operator B_β.

Lemma 3.3 (Ivasishen 1981). *1) If $n_\beta<2br$ and $m_\beta\le r$, $\beta=1,\dots,b$, then G_0 is an ordinary function.*

2) If $m_\beta=0$ and $2b(n-1)<-\rho_0$, then $G_{2\beta}$, $\beta=1,\dots,r$, are ordinary functions.

In the remaining cases G_0 and $G_{2\beta}$ in general contain terms that are finite linear combinations of derivatives of the delta-functions concentrated on S_T or on the hyperplane $t=0$, $x\in\Omega$. In such a situation, the integrals in (3.49), whose kernels are generalized functions mentioned above, denote the results of application of generalized functions to test functions.

We now dwell on the case most frequently met in applications where (3.37) is of the first order in t and the orders r_β of boundary operators are less than the order $2b$ of the equation. Then all components of the Green function are ordinary functions.

In the general situation the problem adjoint to a parabolic differential problem is no more differential so for complete investigation of adjoint problems we have to extend the frames of the theory of boundary-value problems including equations with pseudodifferential operators. However, under special conditions of, so called, *normality of the boundary-value problem* the adjoint problem is also differential and parabolic (if we change the orientation of the time axis) and the complete Green function is defined by the homogeneous Green function (Ejdel'man and Ivasishen 1970). We shall briefly dwell on that.

In the case of normal boundary-value problems Green's formula can be given the form

$$\begin{aligned}(Lu, v) + \sum_{\beta=1}^{b} < B_\beta u, C_\beta^* v> + [u(0,x), v(0,x)] \\ = (u, L^* v) + \sum_{\beta=1}^{b} < C_\beta u, B_\beta^* v> + [u(T,x), v(T,x)] \ ,\end{aligned} \tag{3.53}$$

where B_β^*, C_β, and C_β^* are differential operators given on S_T.

And so, the *adjoint differential parabolic problem* arises

$$\begin{aligned}&L^*(\tau, \xi; D_\tau, D_\xi) v = F \ ; \\ &v|_{t=T} = \Phi \ , \\ &B_\beta^*(\tau, \xi; D_\xi) v|_{S_T} = \Psi_\beta \ .\end{aligned} \tag{3.54}$$

From (3.53) it directly follows that the function $G_0^*(t, x; \tau, \xi)$ as a function of (τ, ξ) is the homogeneous Green function of the problem (3.54) and any solution $u(t, x)$ of (3.37)–(3.39) is determined by the given functions by

$$\begin{aligned}u(t,x) = &\int_0^t d\tau \int_\Omega G_0(t,x;\tau,\xi) f(t,\xi) d\xi \\ &+ \sum_{\beta=1}^{b} \int_0^t d\tau \int_{\partial\Omega} \{C_\beta^*(\tau,\xi'; D_\xi) G_0^*(t,x;\tau,\xi')\}^* g(\tau,\xi') d\xi' \\ &+ \int_\Omega G_0(t,x;0,\xi)\varphi(\xi) d\xi \ .\end{aligned} \tag{3.55}$$

Comparing (3.49) and (3.55) we can define all components of $G(t, x; \tau, \xi)$ by means of the homogeneous Green function $G_0(t, x; \tau, \xi)$.

We conclude by noting that classical problems of mathematical physics for parabolic equations have, as a rule, the normality property, therefore their solutions can be represented by (3.55) by means of the homogeneous Green function.

3.3. Green Functions of Conjugation Problems. Here we note that complete results on Green's functions are established in (Drin' and Ivasishen 1984) for parabolic conjugation problems (1.76). The analogue of the formula (3.49) takes the form

$$\begin{aligned} u^\mu(t,x) = \sum_{\lambda=1}^{2} \int_0^t d\tau \int_{\Omega^\lambda} G_0^{\mu\lambda}(t,x;\tau,\xi) f_\lambda(\tau,\xi) d\xi \\ + \int_0^t d\tau \int_{\partial\Omega^1} G_1^{\mu 1}(t,x;\tau,\xi') g^1(\tau,\xi') d\xi' \\ + \int_0^t d\tau \int_{\partial\Omega^0} G_1^{\mu 2}(t,x;\tau,\xi') g^2(\tau,\xi') d\xi' \\ + \sum_{\lambda=1}^{2} \int_{\Omega^\lambda} G_2^{\mu\lambda}(t,x;0,\xi) \varphi^\lambda(\xi) d\xi , \quad \mu = 1,2 . \end{aligned}$$

Thus the Green function $G = (G_0, G_1, G_2)$ is in this case a matrix-function

$$G = \begin{pmatrix} G_0^{11} & G_0^{12} & G_1^{11} & G_1^{12} & G_2^{11} & G_2^{12} \\ G_0^{21} & G_0^{22} & G_1^{21} & G_1^{22} & G_2^{21} & G_2^{22} \end{pmatrix} .$$

§4. Stabilization

4.1. Introduction A natural problem arises to find precise, if possible, conditions of *stabilization of solutions of parabolic boundary-value problems.* The methods of solution use, as a rule, exact estimates of solutions in infinite in t cylinder $Q = (0,\infty) \times \Omega$ as well as asymptotic as $t \to \infty$ representation of solution.

Here we give two precise results in the case of parabolic boundary-value problems with constant coefficients. We draw the attention to a fundamental difference in the formulation of the problems considered: in the first we find necessary and sufficient conditions on the boundary functions under which we have stabilization of solutions of a given parabolic boundary-value problem and in the second for a given parabolic equation we seek for conditions for the boundary-value problem so that bounded solutions stabilize for any boundary functions that stabilize. This was the formulation of the problem in the pioneer paper of Tikhonov (1950).

4.2. Model Problems. In $\mathbb{R}^{n+1}_{++}$ we consider a model parabolic boundary-value problem where the coefficients of the equation and the coefficients of the boundary conditions are constant.:

$$\begin{aligned} \mathcal{L}_0(D_t, D_x) u \equiv \sum_{\substack{2b\alpha_0+|\alpha|=2br \\ \alpha_0<r}} a_{\alpha_0\alpha} D_t^{\alpha_0} D_x^\alpha u = 0 ; \\ D_t^{\alpha_0} u|_{t=0} = 0 , \quad \alpha_0 = 0, \dots, r-1 , \end{aligned} \tag{3.56}$$

$$B^0_\beta(D_t, D_x)u \equiv \sum_{2b\alpha_0+|\alpha|=r_\beta} b^\beta_{\alpha_0\alpha} D_t^{\alpha_0} D_x^\alpha u|_{x_n=0} = g_\beta(t, x') , \quad \beta = 1, \dots, br . \tag{3.57}$$

We study the stabilization of the solutions to the problem (3.56), (3.57) that are represented by the formula

$$u(t, x) = \sum_{\beta=1}^{br} \int_0^t d\tau \int_{\mathbb{R}^{n-1}} G_\beta(t-\tau, x-\xi') g_\beta(\tau, \xi') d\xi' . \tag{3.58}$$

To this end we consider the Laplace transform $\hat{G}_\beta(p, x)$ of the Poisson kernel $G_\beta(t, x)$. It turns out that $\hat{G}_\beta(p, x)$ is a function analytic with respect to p for $(p, x) \in \mathfrak{A}_\chi = \{(p, x) : p \in \mathbb{C}^1, \operatorname{Re} p > -\chi \operatorname{Im} p, \chi > 0, x \in \mathbb{R}^n_+\}$. We have the following asymptotic (as $|p| \to 0$) representations (Zhitarashu and Ejdel'man 1970, 1979)

$$\begin{aligned} \hat{G}(p, x) = {} & \sum_{s=\tau_\beta}^{r_\beta - n} \mathcal{P}^s_\beta(x) p^{-\frac{r_\beta - n+1-s}{2b}} - F_\beta(x) \left(\frac{1}{2b} \ln p + C_\beta \right) \\ & + \mathcal{E}_\beta(x) + \mathcal{P}_\beta(x) + \mathcal{R}_\beta(p, x) , \end{aligned} \tag{3.59}$$

where τ_β is the so called *reduced order* of the boundary operator B_β and it carries the responsibility for the asymptotic behavior of the Poisson kernel $G_\beta(t, x)$ as $t \to \infty$, $F_\beta(x)$ and $\mathcal{P}_\beta(x)$ are polynomials of degrees $r_\beta - n + 1$ whose coefficients can be explicitly computed

$$\mathcal{P}^s_\beta(x) = \sum_{|k|=s} D^k_x \hat{G}_\beta(1, 0) x^k / k! ,$$

$\mathcal{E}_\beta(x)$ are the Poisson kernels of the elliptic boundary-value problem

$$\mathcal{L}_0(0, D_x)v = 0 \quad B^0_\beta(0, D_x)v|_{x_n=0} = g^0_\beta(x') , \quad \beta = 1, \dots, br ,$$

and the functions $\mathcal{R}_\beta(p, x)$ satisfy the estimates

$$|\mathcal{R}_\beta(p, x)| \le C_\beta(x) |p|^{b/2} .$$

We now define the class of boundary-value problems we are going to consider. We put $\alpha_\beta = (r_\beta - n + 1 - \tau_\beta)/2b$.

Definition 3.4. We say that $g_\beta(t, x')$ belongs to the class $S_{\alpha\beta}$ if

1) for all $x' \in \mathbb{R}^{n-1}$ the Laplace transforms $\hat{g}_\beta(p, x')$ of $g_\beta(t, x')$ are functions analytic with respect to p in the right half-plane of the complex p-plane;
2) for each $\varepsilon > 0$ there is a $C_\varepsilon > 0$ such that $|\hat{g}_\beta(p, x')| \le C_\varepsilon \exp\{\varepsilon |p|^{1/2b}\}$;
3) if $p \to 0$, then we have the following asymptotic expansion

$$\hat{g}_\beta(p,x') = \sum_{q=0}^{r_\beta} a_{\beta q}(x') p^{\gamma_{\beta q}} + \Phi_\beta(p,x') ,$$

$$-1 \le \gamma_{\beta 0} \le \gamma_{\beta 1} \le \ldots \le -1 + \alpha_\beta , \tag{3.60}$$

and moreover

$$\int_{\mathbb{R}^{n-1}} |x'|^{r_\beta - n+2} |a_{\beta q}(x')| dx' < +\infty ,$$

$$\int_{\mathbb{R}^{n-1}} |\Phi(p,x')| |x'|^{r_\beta-n+2} |dx' \le C|p|^{-1+\alpha_\beta+\delta} , \quad \delta > 0 .$$

Example 3.4. The function $g_\beta(t,x')$ belongs to $S_{\alpha\beta}$ if there exists the limit average

$$\lim_{T\to\infty} \left(\int_0^T g_\beta(t,x') dt/T \right) = g_\beta^0(x')$$

and the following integral converges

$$a_{\beta q}(x') = \frac{(-1)^{q-1}}{(q-1)!} \int_0^\infty t^{q-1} [g_\beta(t,x') - g_\beta^0(x')] dt ,$$

$q = 1, \ldots, \alpha_\beta$. In this case $\gamma_{\beta q} = q-1$, $a_{\beta 0}(x') = g_\beta^0(x')$. From the asymptotic representations (3.59) and (3.60) as $p \to 0$ we can derive the asymptotic representation as $t \to \infty$ of the solution of (3.56), (3.57) given by (3.58):

$$u_\beta(t,x) = \sum_{\alpha_{\beta s} - \gamma_{\beta q} > 1} M_{\beta s q}(x) t^{\alpha_{\beta s} - \gamma_{\beta q} - 1} / \Gamma(\alpha_{\beta s} - \gamma_{\beta q})$$
$$+ M_\beta(x)(\ln t/2b + C'_\beta) + v_\beta(x) + Q_\beta(x) + O_\beta(t,x) , \tag{3.61}$$

where

$$M_{\beta s q}(x) = \mathcal{P}_\beta^s(x) * a_{\beta q}(x') , \quad M_\beta(x) = F_\beta * a_{\beta 0}(x') ,$$

$$v_\beta = \mathcal{E}_\beta(x) * a_{\beta 0}(x') , \quad Q_\beta(x) = \mathcal{P}_\beta(x) * a_{\beta 0}(x') + \sum_{\gamma_{\beta q} - \alpha_{\beta s} = -1} M_{\beta s q} ,$$

where $*$ denotes the convolution with respect to x' and $Q_\beta(t,x) \to 0$ as $t \to \infty$. If $\alpha_\beta = 0$, then $M_{\beta s q} \equiv 0$ for any s, q; if $r_\beta - n + 1 < 0$ or $\tau_\beta - n + 1 < 0$, then $M_{\beta s q} \equiv M_\beta(x) \equiv Q_\beta(x) \equiv 0$.

From (3.61) we directly have

Theorem 3.9 (Zhitarashu and Ejdel'man 1979). *Let $n \ge 2$ and suppose that $g_\beta(t,x') \in S_{\alpha\beta}$, $\beta = 1, \ldots, br$. For stabilization of the solution to* (3.56), (3.57) *that is given by (3.58) it is necessary and sufficient that*

$$M_\beta(x) \equiv 0 \quad Q_\beta(x) \equiv \sum_{\alpha_{\beta s} - \gamma_{\beta q} = h} M_{\beta s q}(x) \equiv 0 \tag{3.62}$$

for any h such that for some $\alpha_{\beta s}$ and $\gamma_{\beta q}$ we have $\alpha_{\beta s} - \gamma_{\beta q} = h$.

In addition $u_\beta(t,x)$ stabilizes at $v_\beta(x) + Q_\beta(x)$, $\beta = 1, \dots, br$, uniformly on any compact set of the form $\Omega = \{x : |x| \le R, x_n \ge \delta > 0\}$.

We recall that $v_\beta(x)$ is a solution to the limit elliptic problem and $Q_\beta(x)$ is a polynomial of degree $r_\beta - n + 1$, $\beta = 1, \dots, br$.

The identities (3.62) are equivalent to the fact that all coefficients of the polynomials $M_\beta(x)$ and $Q_\beta^s(x)$ are equal to zero which means that the linear combinations of the moments of $a_{\beta q}(x')$ which appear in the representation (3.60) are equal to zero.

In particular, condition (3.60) is always fulfilled for the Dirichlet problem and for the Neuman problem in the case of the heat equation orthogonality of $g^0(x')$ to one is necessary and sufficient for stabilization of solutions.

4.3. The Tikhonov Theorem and Its Generalizations. We consider the problem of finding necessary and sufficient conditions on the boundary operator $B\left(\frac{\partial}{\partial x}, \frac{\partial}{\partial x_n}\right)$ to guarantee stabilization of solutions to the problem

$$\begin{aligned} &D_t u = \Delta u \,, \\ &u|_{t=0} = 0 \,; \\ &B\left(\frac{\partial}{\partial x}, \frac{\partial}{\partial x_n}\right) u|_{x_n=0} = \sum_{|\alpha'|+\alpha_n \le r} b_{\alpha'\alpha_n} \frac{\partial^{\alpha'}}{\partial x'^{\alpha'}} \frac{\partial^{\alpha_n}}{\partial x_n{}^{\alpha_n}} u|_{x_n=0} = g(t,x') \,, \end{aligned} \tag{3.63}$$

that are given by the formula

$$u(t,x) = \int_0^t d\tau \int_{\mathbb{R}^{n-1}} G(t-\tau, x-\xi') g(\tau,\xi') d\xi' \,, \tag{3.64}$$

where $g(t,x')$ is any boundary function that stabilizes and $G(t,x)$ is the Poisson kernel of the problem (3.63).

Definition 3.5. A function $g(t,x')$ defined in $\mathbb{R}^n_+$ belongs to class S if

1) $g(t,x')$ is continuous and bounded in $\mathbb{R}^n_+$;
2) $g(t,x') \to g^0(x')$ as $t \to \infty$ uniformly in $x' \in \mathbb{R}^{n-1}$.

We denote by $P^-_{\sigma'}$ the half-plane of the complex p-plane where $\operatorname{Re} p < -|\sigma'|^2$ and we put $\bar{P}^-_{\sigma'} = P^-_{\sigma'} \cup \{p : \operatorname{Re} p = -|\sigma'|^2\}$. We denote by $\sqrt{p + |\sigma'|^2}$ the branch with positive real part of the two-valued function involved.

We formulate the following conditions

β_7. p-zeros of $B_0(\sigma', -\sqrt{p+|\sigma'|^2})$ belong to $\bar{P}^-_{\sigma'}$ for all $\sigma' \in \mathbb{R}^{n-1}$.

β_8. p-zeros of $B(\sigma', -\sqrt{p+|\sigma'|^2})$ belong to $P^-_{\sigma'}$ for all $\sigma' \in \mathbb{R}^{n-1}$.

Condition β_7 guarantees parabolicity of the problem and β_8 its stability.

Theorem 3.10 (Kushitskij and Ejdel'man 1988). *Suppose that β_7 and β_8 are fulfilled. Then the Poisson kernel $G(t,x)$ satisfies the following assertions.*

1) We have the representation $G(t,x) = G^{(1)}(t,x) + G^{(2)}(t,x)$, *where*

$$G^{(1)}(t,x) = B(0',0)^{-1}(x_n/t)(4\pi t)^{-\frac{n}{2}} \exp\left\{-\frac{|x|^2}{4t}\right\}$$

is the Poisson kernel for the Dirichlet problem and $G^{(2)}(t,x)$ *satisfies for* $t \geq 1$ *the estimate*

$$|G^{(2)}(t,x)| \leq Ct^{-\frac{n+2}{2}} \exp\left\{-\frac{c|x|^2}{t}\right\}.$$

2) The integral $\int_0^\infty dt \int_{\mathbb{R}^{n-1}} |G(t,x)|dx'$ *is uniformly bounded in* x_n.

3) The function $\mathcal{E}(x) = \int_0^\infty G(t,x)dt$ *is the Poisson kernel of the problem* $B\left(\frac{\partial}{\partial x'}, \frac{\partial}{\partial x_n}\right) v|_{x_n=0} = g(x')$ *for the Laplace equation.*

We note that the main result of Theorem 3.10 is the representation of the Poisson kernel of the problem (3.63) by means of the Poisson kernel of the Dirichlet problem with a good estimate (as $t \to \infty$) of the difference $G^{(2)}(t,x)$.

Theorem 3.11 (Kushitskij and Ejdel'man 1988). *1) Suppose that conditions* β_7 *and* β_8 *are fulfilled. Then the solution* $u(t,x)$ *of* (3.63) *given by* (3.64) *and constructed by an arbitrary boundary function* $g(t,x') \in S$ *stabilizes uniformly in* $x \in \mathbb{R}^n_+$ *at the function*

$$u_0(x) = \int_{\mathbb{R}^{n-1}} \mathcal{E}(x'-\xi', x_n) g^0(\xi') d\xi',$$

which is a solution to the problem $B\left(\frac{\partial}{\partial x'}, \frac{\partial}{\partial x_n}\right) v|_{x_n=0} = g^0(x')$ *for the Laplace equation.*

2) Let us consider the problem (3.63) *with the operator*

$$B\left(\frac{\partial}{\partial x_n}\right) \equiv \sum_{\alpha_n=0}^{r} b_{0\alpha_n} \frac{\partial^{\alpha_n}}{\partial x_n{}^{\alpha_n}}$$

If bounded solutions $u(t,x)$ *to* (3.63) *constructed by arbitrary* $g(t,x') \in S$ *stabilize, then the operator* $B\left(\frac{\partial}{\partial x_n}\right)$ *satisfies condition* β_8.

In the case of one spatial variable Theorem 3.11 was established by Tikhonov (1950).

Example 3.5. We consider the heat exchange problem

$$B\left(\frac{\partial}{\partial x'}, \frac{\partial}{\partial x_n}\right) \equiv \sum_{s=1}^{n} b_s \frac{\partial}{\partial x_s} + b_0.$$

If $b_n \neq 0$, then condition β_7 is fulfilled. Condition β_8 in this case means that $b_0 b_n < 0$. It is necessary and sufficient for the stabilization of solution of the heat exchange problem constructed by any boundary function $g(t,x') \in S$.

References*

The list of references has to be short and, up to some extent, subjective. In addition to the papers cited in the text of the article we have included monographs and survey papers which permit us to create a more complete picture of the theory of parabolic equations.

A presentation of intensively developing branches of the theory of quasilinear parabolic equations can be found in (Maslov et al.1987), (Samarskij et al. 1987), (Henry 1981), and the comments to (Petrovskij 1986, 1987).

We have included to the references the monographs (Klovskij et al. 1984), (Sosulin 1978), (Tikhonov and Kul'man 1975), and (Tikhonov and Mironov 1977), where we find interesting applications of the theory of parabolic equations to the problems of statistical radiotechnics, perhaps less popular among mathematicians. The theory of parabolic equations has deep connections with functional analysis, especially with the theory of evolution equations with unbounded operators in Banach spaces and the semigroup theory. These connections are reflected, perhaps not completely, in (Krein 1967), (Krein and Khazan 1983), and (Ladas and Lakshmikantham 1978).

The theory of stochastic processes, especially the theory of Markov processes and stochastic differential equations, closely interacts with the theory of parabolic equations, differential as well as pseudodifferential. This is reflected in (Gikhman and Skorokhod 1968), (Kochubej 1984), and (Portenko 1982).

A picture of the recent state of the theory of parabolic boundary-value problems in domains with unsmooth boundaries can be obtained from the survey (Kondrat'ev and Olejnik 1983). Important methods of the study of qualitative behaviour of solutions to broad classes of systems of equations are contained in the survey article (Olejnik and Radkevich 1978); in particular they permit us to obtain theorems on uniqueness of solutions of general parabolic boundary-value problems in unbounded domains.

The theory of linear parabolic problems is partially presented in the monographs (Il'in et al. 1962), (Ladyzhenskaya et al. 1967), (Ejdel'man 1964), and (Friedman 1964).

Agranovich, M.S., Sukhorutchenko, V.V. (1967): Necessity for algebraic parabolicity conditions for nonstationary problems. Mat. Zametki **2**, No. 6, 615–625. English transl.: Math. Notes **2**, No. 6, 864–869 (1967). Zbl.153,421

Agranovich, M.S., Vishik, M.I. (1964): Elliptic problems with a parameter and parabolic problems of general form. Usp. Mat. Nauk **19**, No. 3, 53–161. English transl.: Russ. Math. Surv. **19**, No. 3, 53–157 (1964). Zbl.137,296

Aronson, D.G. (1967): Bounds for the fundamental solution of a parabolic equation. Bull. Am. Math. Soc. **73**, No. 6, 890–896. Zbl.153,420

Aronson, D.G. (1968): Non-negative solutions of linear parabolic equations. Ann. Sc. Norm. Super. Pisa., Sci. Fis. Math., III. Ser. **22**, 607–694. Zbl.182,138

Chaus, N.N. (1968): Classes of uniqueness of solutions to the Cauchy problem and representations of positive definite kernels. Tr. Semin. Funkts. Anal. **1**. 176–270 (Russian). Zbl.252.35015

Denisov, V.N., Repnikov, V.D. (1984): Stabilization of a solution of a Cauchy problem for parabolic equations. Differ. Uravn. **20**, No. 1, 20–41. English transl.: Differ. Equations **20**, No. 1, 16–33 (1984). Zbl.589.35054

* For the convenience of the reader, references to reviews in *Zentralblatt für Mathematik* (Zbl.), compiled using the MATH database, have, as far as possible, been included in this bibliography.

Ditkin, V.A., Prudnikov, A.P. (1966): Operator Calculus. Vysshaya Shkola: Moscow. French transl.: Mir: Moscow 1979. Zbl.148,367

Drin', M.M., Ivasishen, S.D. (1984): Green's matrix for the general boundary-value problem for a system parabolic in the sense of Petrovskij with discontinuous coefficients. Dokl. Akad. Nauk Ukr. SSR, Ser. A 1984, No. **11**, 7–10 (Russian). Zbl.565.35052

Drin', Ya. M. (1977): Fundamental solution of the Cauchy problem for a class of parabolic pseudodifferential equations. Dokl. Akad. Nauk Ukr. SSR, Ser. A 1977, No. **3**, 198–203 (Russian). Zbl.347.35080

Dynkin, E.B. (1959): Foundations of the Theory of Markov Processes. Fizmatgiz: Moscow. English transl.: Pergamon Press: Oxford etc. 1961. 91,136

Dzharbashyan, M.M., Nersesyan, A.B. (1958): Application of certain integro-differential operators. Dokl. Akad. Nauk SSSR **121**, 210–213 (Russian). Zbl.95,85

Dzharbashyan, M.M., Nersesyan, A.B. (1968): Fractional derivatives and the Cauchy problem for differential equations of fractional order. Izv. Akad. Nauk Arm. SSR, Mat. **3**, No. 1, 3–29 (Russian). Zbl.165,408

Egorov, Yu.V., Shubin, M.A. (1988): Linear partial differential equations. Foundations of the classical theory. Itogi Nauki Tekh., Ser. Sovrem. Probl. Mat., Fundam. Napravleniya **30**. English transl.: Encyclopaedia of Mathematical Sciences **30**, Springer–Verlag: Berlin–Heidelberg–New York 1991. Zbl.657.35002

Ejdel'man, S.D. (1960): On a class of parabolic systems. Dokl. Akad. Nauk SSSR **133**, No. 1, 40–43. English transl.: Sov. Math. Dokl. **1** 815–818 (1960). Zbl.98,299

Ejdel'man, S.D. (1964): Parabolic Systems. Nauka: Moscow. English transl.: North–Holland: Amsterdam 1969. Zbl.121,319

Ejdel'man, S.D., Drin', Ya.M. (1981): Construction and investigation of classical fundamental solutions of the Cauchy problem for uniformly parabolic pseudodifferential equations. Mat. Issled. **63**, 18–33 (Russian). Zbl.461.35087

Ejdel'man, S.D., Ivasishen, S.D. (1970): Investigation of Green's matrix for a homogeneous parabolic boundary-value problem. Tr. Mosk. Mat. O.-va **23**, 179–234. English transl.: Trans. Mosc. Math. Soc. **23**, 179–242 (1972). Zbl.224.35046

Ejdel'man, S.D., Malitskaya, A.P. (1975): Fundamental solutions and stabilization of solutions of Cauchy's problem for a class of degenerate parabolic equations. Differ. Uravn. **11**, No. 7, 1316–1330. English transl.: Differ. Equations **11**, No. 7, 988–998 (1976). Zbl.322.35045

Ejdel'man, S.D., Porper, F.O. (1971): Properties of solutions of parabolic equations of the second order with dissipation. Differ. Uravn. **7**, No. 9, 1684–1695. English transl.: Differ. Equations **7**, No. 9, 1280–1288 (1974). Zbl.223.35047

Ejdel'man, S.D., Tychinskaya, L.M. (1979): Construction of fundamental solutions of some degenerate parabolic equations of arbitrary order. Dokl. Akad. Nauk Ukr. SSR, Ser. A 1979, No. **11**, 896–899 (Russian). Zbl.426.35054

Fedoryuk, M.V. (1978): Asymptotics of Green's function for a parabolic pseudodifferential equation. Differ. Uravn. **14**, No. 7, 1296–1301. English transl.: Differ. Equations **14**, No. 7, 923–927 (1978). Zbl.406.35070

Fedoryuk, M.V. (1987): Asymptotics. Integrals and Series. Nauka: Moscow (Russian). Zbl.641.41001

Friedman, A. (1963): Generalized Functions and Partial Differential Equations. Prentice-Hall: Englewood Cliffs, New York. Zbl.116,70

Friedman, A. (1964): Partial Differential Equations of Parabolic Type. Prentice-Hall: Englewood Cliffs, New York. Zbl.144,349

Gel'fand, I.M., Shilov, G.E. (1959, 1958a,b): Generalized Functions. Vols. 1, 2, 3. Fizmatgiz: Moscow. English transl.: Academic Press: New York 1964, 1968, 1967. Zbl.91,111

Gikhman, I.I., Skorokhod, A.V. (1968): Stochastic Differential Equations. Naukova Dumka: Kiev. English transl.: Springer–Verlag: Berlin–Heidelberg–New York 1972. Zbl.169,487

Gikhman, I.I., Skorokhod, A.V. (1982): Stochastic Differential Equations and Their Applications. Naukova Dumka: Kiev (Russian). Zbl.557.60041

Gyul'misaryan, A.G. (1965): General boundary problems for parabolic equations with discontinuous coefficients. Izv. Akad. Nauk Arm. SSR **18**, No. 1, 14–33 (Russian). Zbl.158,166

Gyunter, N.M. (1953): Potential Theory and its Application to Main Problems of Mathematical Physics. Gos. Izd. Tekh.-Teor. Lit.: Moscow (Russian). Transl. from the French edition, Gautier: Paris 1934. Zbl.9,113

Henry, D. (1981): Geometric Theory of Semilinear Parabolic Equations. Lecture Notes Math. **840**. Springer–Verlag: Berlin–Heidelberg–New York. Zbl.456.35001

Hopf E. (1931): Über den funktionalen, insbesondere den analytischen Charakter der Lösungen elliptischer Differentialgleichungen zweiter Ordnung. Math.Z. **34**, 194–233. Zbl. 2,340.

Hörmander L. (1983 ab, 1985 a,b): The Analysis of Linear Partial Differential Operators, I, II, III, IV. Springer–Verlag: Berlin–Heidelberg–New York. Zbl.521.35001; Zbl.521.35002; Zbl.601.35001; Zbl.612.35001

Il'in, A.M. (1962): On the fundamental solution for a parabolic equation. Dokl. Akad. Nauk SSSR **147**, No. 2, 768–771. English transl.: Sov. Math., Dokl. **3**, No. 6, 1697–1700 (1962). Zbl.168,82

Il'in, A.M. (1964): On a class of ultraparabolic equations. Dokl. Akad. Nauk SSSR **159**, No. 6, 1214–1217. English transl.: Sov. Math., Dokl. **5**, No. 6, 1673–1676 (1965). Zbl.173,377

Il'in, A.M. (1967): On parabolic equations whose coefficients do not satisfy the Dini condition. Mat. Zametki **1**, No. 1, 71–79. English transl.: Math. Notes **1**, No. 1, 46–51 (1967). Zbl.146,339

Il'in, A.M., Kalashnikov, A.S., Olejnik, O.A. (1962): Linear second order equations of parabolic type. Usp. Mat. Nauk **17**, No. 3, 3–141. English transl.: Russ. Math. Surv. **17**, No. 3, 1–146 (1962). Zbl.108,284

Ivasishen, S.D. (1981a,b): Green's matrices of boundary-value problems for Petrovskij parabolic systems of general form. Mat. Sb., Nov. Ser. **114**, No. 1, 110–166; No. 4, 523–565. English transl.: Math. USSR, Sb. **42**, No. 1, 93–144; No. 4, 461–498 (1982). Zbl.468.35048; Zbl.468.35049

Ivasishen, S.D. (1987): Linear Parabolic Boundary-Value Problems. Vishcha Shkola: Kiev (Russian). Zbl.704.35059

Ivasishen, S.D., Ejdel'man, S.D. (1968): $\overrightarrow{2b}$-parabolic systems. Tr. Semin. Funkts. Anal. **1**, 3–175, 271–273 (Russian). Zbl.245.35043

Ivasishen, S.D., Ejdel'man, S.D. (1987): Parabolic equations: examples, Cauchy problems, property of solutions. In: Contemporary Mathematics. Vishcha Shkola: Kiev, 74–108 (Russian). Zbl.688.35033

Ivasishen, S.D., Lavrenchuk, V.P. (1978); Correct solvability of general boundary-value problems for parabolic systems with increasing coefficients. Ukr. Mat. Zh. **30**, No. 1, 100–106. English transl.: Ukr. Math. J. **30**, No. 1, 75–79 (1978). Zbl.371.35025

Kalashnikov, A.S. (1987): Some problems of the qualitative theory of nonlinear degenerate parabolic equations of second order. Usp. Mat. Nauk **42**, No. 2, 135–176. English transl.: Russ. Math. Surv. **42**, No. 2, 169–222 (1987). Zbl.642.35047

Klovskij, D.D., Kontorovich, V.Ya., Shirokov, S.M. (1984): Models of Continuous Communication Channels on the Basis of Stochastic Differential Equations. Radio i Svyaz': Moscow (Russian). Zbl.609.94001

Kochubej, A. N. (1984): Singular parabolic equations and Markov processes. Izv. Akad. Nauk SSSR, Ser. Mat. **48**, No. 1, 77–103. English transl.: Math. USSR, Izv. **24**, No. 1, 73–97 (1985). Zbl.545.35039

Kochubej, A.N. (1985): Parabolic pseudodifferential equations with homogeneous symbols. Dokl. Akad. Nauk SSSR **297**, No. 3, 534–538. English transl.: Sov. Math., Dokl. **36**, No. 3, 493–497 (1988). Zbl.661.35091

Kolmogorov, A.N. (1934): Zufällige Bewegungen. Ann. Math., II. Ser. **35**, 116–117. Zbl.8,399

Komatsu, T. (1973): Markov processes associated with certain integro-differential operators. Osaka J. Math. **10**, No. 2, 271–303. Zbl.284.60066

Komatsu, T. (1984): Pseudodifferential operators and Markov processes. J. Math. Soc. Japan **36**, No. 3, 387–418. Zbl.539.60081

Kondrat'ev, V.A., Ejdel'man, S.D. (1974): Positive solutions of linear partial differential equations. Tr. Mosk. Mat. O-va **31**, 85–146. English transl.: Trans. Mosc. Math. Soc. **31**, 81–148 (1976). Zbl.325.35010

Kondrat'ev, V.A., Olejnik, O.A. (1983): Boundary problems for partial differential equations in non-smooth domains. Usp. Mat. Nauk. **38**, No. 2, 3–76. English transl.: Russ. Math. Surv. **38**, No. 2, 1–86 (1983). Zbl.523.35010

Krein, S.G. (=Krejn, S.G.) (1967): Linear Differential Equations in Banach Spaces. Nauka: Moscow. English transl.: Am. Math. Soc.: Providence 1972. Zbl.172,419

Krein, S.G., Khazan, M.I. (1983): Differential equations in Banach spaces. Itogi Nauki Tekh., Ser. Mat. Anal. **21**, 130–264. English transl.: J. Sov. Math. **30**, 2154–2239 (1985). Zbl.611.34059

Kruzhkov, S.N. (1967): Estimates of the highest derivatives of solutions of elliptic and parabolic equations with continuous coefficients. Mat. Zametki **2**, No. 5, 549–560. English transl.: Math. Notes **2**, No. 5, 824–830 (1967). Zbl.153,424

Krylov, N.V., Safonov, M.V. (1980): A certain property of parabolic equations with measurable coefficients. Izv. Akad. Nauk SSSR, Ser. Mat. **44**, No. 1, 161–175. English transl.: Math. USSR, Izv. **16**, No. 1, 151–164 (1981). Zbl.439.35023

Kushitskij, Ya.S., Ejdel'man, S.D. (1988): A multidimensional version of Tikhonov's problem. Dokl. Akad. Nauk SSSR **299**, No. 5, 1056–1059. English transl.: Sov. Math., Dokl. **37**, No. 2, 518–521 (1988). Zbl.699.35019

Ladas, G.E., Lakshmikantham, V. (1978): Differential Equations in Abstract Spaces. Academic Press: New York–London. Zbl.257.34002

Ladyzhenskaya, O.A., Solonnikov, V.A., Uraltseva, N.N. (1967): Linear and Quasilinear Equations of Parabolic Type. Nauka: Moscow. English transl.: Am. Math. Soc.: Providence 1968. Zbl.164,123

Lepeltier, J.P., Marchal, B. (1976): Probleme des martingales et equations differentiales stochastiques associées a un operateur integro-differentiel. Ann. Inst. Henri Poincaré, Sect.B **12**, No. 1, 43–103. Zbl.345.60029

Lopatinskij, Ya.B. (1953): A method of reduction of boundary-value problems for systems of differential equations of elliptic type to a system of regular integral equations. Ukr. Mat. Zh. **5**, No. 2, 123–151 (Russian). Zbl.52,102

Marinov, M.L. (1977). Existence of a solution of a boundary problem for general parabolic systems in unbounded domains. Vestn. Mosk. Univ., Ser.I 1977, No. 6, 56–63. English transl.: Mosc. Univ. Math. Bull. **32**, No. 6, 46–52 (1977). Zbl.388.35036

Maslov, V.P., Danilov, V.G., Volosov, K.A. (1987): Mathematical Modelling of Heat and Mass Transfer Processes. Evolution of Dissipative Structures. Nauka: Moscow (Russian). Zbl.645.73049

Matijchuk, M.I., Ejdel'man, S.D. (1967): Fundamental solutions and the Cauchy problem for parabolic systems whose coefficients satisfy the Dini conditions. Tr. Semin. Funkts. Anal. **9**, 54–83 (Russian).

Matijchuk, M.I., Ejdel'man, S.D. (1970): Cauchy problem for parabolic systems whose coefficients have slight smoothness. Ukr. Mat. Zh. **22**, No. 1, 22–36. English transl.: Ukr. Math. J. **22**, No. 1, 18–30 (1970). Zbl.208,366

Maz'ya, V.G. (1985). Sobolev Spaces. Leningrad State University. English transl.: Springer–Verlag: Berlin–Heidelberg–New York 1985. Zbl.692.46023

Moser, J. (1964): A Harnack inequality for parabolic differential equations. Commun. Pure Appl. Math. **17**, No. 1, 101–134; a correction ibid. **20**, No. 1, 231–236 (1967). Zbl.149,69; Zbl.149,70

Moser, J. (1971): On a pointwise estimate for parabolic differential equations. Commun. Pure Appl. Math. **24**, No. 5, 727–740. Zbl.227.35016

Nash, J. (1958): Continuity of solutions of parabolic and elliptic equations. Am. J. Math. **80**, No. 4, 931–954. Zbl.96,69

Nigmatullin, R.R. (1986): The realisation of the generalized transfer equations in a medium with fractal geometry. Phys. Status Solidi B **133**, No. 1, 425–430.

Nikiforov, A.F., Uvarov, V.B. (1974): Foundations of the Theory of Special Functions. Nauka: Moscow. English transl. of the 2nd ed.: Birkhäuser: Basel/Boston 1988. Zbl.567.33001, Zbl.378.33001

Olejnik, O.A., Radkevich, E.V. (1978): The method of introducing a parameter in the study of evolution equations. Usp. Mat. Nauk **33**, No. 5, 7–76. English transl.: Russ. Math. Surv. **33**, No. 5, 7–84 (1978). Zbl.397.35033

Petrovskij, I.G. (1986): Selected Works. Systems of Partial Differential Equations. Algebraic Geometry. Nauka: Moscow (Russian). Zbl.603.01018

Petrovskij, I.G. (1987): Selected Works. Differential Equations. Probability Theory. Nauka: Moscow (Russian). Zbl.621.01020

Petrushko, I.M., Ejdel'man, S.D. (1967): Solvability of the Cauchy problem for parabolic equations of the second order in the class of arbitrarily rising functions. Ukr. Mat. Zh. **19**, No. 1, 108–112. English transl.: Ukr. Math. J. **19**, No. 1, 93–97 (1967). Zbl.154,121

Porper, F.O. (1977): Estimates for the derivatives of a fundamental solution of a stationary divergent parabolic equation in terms of constants that do not depend on the smoothness of the coefficients. Dokl. Akad. Nauk SSSR **235**, No. 5, 1022–1025. English transl.: Sov. Math., Dokl. **18**, No. 4, 1092–1096 (1978). Zbl.376.35034

Porper, F.O., Ejdel'man, S.D. (1984): Two-sided estimates of fundamental solutions of parabolic equations of second order and applications. Usp. Mat. Nauk **39**, No. 4, 107–156. English transl.: Russ. Math. Surv. **39**, No. 3, 119–178 (1984). Zbl.582.35052

Porper, F.O., Ejdel'man, S.D. (1986): Properties of weak fundamental solutions of parabolic equations with lower derivatives. Dokl. Akad. Nauk SSSR **288**, No. 4, 827–831. English transl.: Sov. Math., Dokl. **33**, No. 3, 789–793 (1986). Zbl.682.35005

Portenko, N.I. (1982): Generalized Diffusion Processes. Naukova Dumka: Kiev. English transl.: Transl. Math. Monogr. **83**: Providence 1990. Zbl.727.60089

Repnikov, V.D., Ejdel'man, S.D. (1966): Necessary and sufficient conditions for the establishment of a solution of the Cauchy problem. Dokl. Akad. Nauk SSSR **167**, No. 2, 298–301. English transl.: Sov. Math., Dokl. **7**, No. 2, 388–391 (1966). Zbl.145,357

Romanovskij, Yu.M., Stepanova, N.V., Chernavskij, D.S. (1984): Mathematical Biophysics. Nauka: Moscow (Russian).

Samarskij, A.A., Galaktionov, V.A., Kurdyumov, S.P., Mikhajlov, A.P. (1987): Tightened Conditions in the Problems for Quasi-Linear Parabolic Equations. Nauka: Moscow (Russian). Zbl.631.35002

Samko, S.G. (1984): Hypersingular Integrals and Their Applications. Rostov State University (Russian). Zbl.577.42016

Samko, S.G., Kilbas, A.A., Marichev, O.G. (1987): Integrals and Derivatives of Fractional Order and some of Their Applications. Nauka i Tekhnika: Minsk (Russian). Zbl.617.26004

Shirota, T.O. (1957): Cauchy problem for linear partial differential equations with variable coefficients. Osaka Math. J. **9**, No. 1, 43–60. Zbl.84,87

Skorokhod, A.V. (1986): Stochastic Processes with Independent Increments. Nauka: Moscow (Russian). Zbl.622.60082

Slobodetskij, L.N. (1958): Generalized Sobolev spaces and their applications to boundary problems for partial differential equations. Uch. Zap. Leningr. Gos. Ped. Inst. **197**, 54–112. English transl.: Am. Math. Soc., Transl., II. Ser. **57**, 207–275 (1966). Zbl.192,228

Sneddon, J. (1951): Fourier Transforms. New York–Toronto–London. Zbl.38,268

Solonnikov, V.A. (1965): Boundary value problems for linear parabolic systems of differential equations of general form. Tr. Mat. Inst. Steklova **83**. English transl.: Proc. Steklov Inst. Math. **83** (1965). Zbl.161,64

Solonnikov, V.A. (1967): Estimates in L_p of solutions of elliptic and parabolic systems. Tr. Mat. Inst. Steklova **102**, 137–160. English transl.: Proc. Steklov Inst. Math **102**, 157–185 (1970). Zbl.204,421

Solonnikov, V.A. (1969): Green matrices for parabolic boundary value problems. Zap. Nauch. Sem. Leningr. Otd. Mat. Inst. Steklova **14**, 256–287. English transl.: Semin. Math., Steklov Math. Inst. Leningr. **14**, 109–121 (1971). Zbl.202,376

Sonin, I.M. (1967): On a class of degenerate diffusion processes. Teor. Veroyatn. Primen. **12**, No. 3, 540–547. English transl.: Theor. Probab. Appl. **12**, 490–496 (1967). Zbl.183,199

Sosulin, Yu.G. (1978): Theory of Detection and Estimation of Stochastic Signals. Sov. Radio: Moscow (Russian).

Täcklind, S. (1936): Sur les classes quasianalytiques des solutions des equations aux derivée partielles du type parabolique. Nova. Acta Soc. Sci. Upsal., IV. Ser. **10**, No. 3, 1–57. Zbl.14,22

Tikhonov, A.N. (1950): Boundary problems containing derivatives of order higher than the order of the equation. Mat. Sb., Nov. Ser. **26**, No. 1, 35–56 (Russian). Zbl.41,66

Tikhonov, V.I., Kul'man, N.K. (1975): Nonlinear Filtration and Quasi-Coherent Reception of Signals. Radio i Svyaz': Moscow (Russian).

Tikhonov, V.I., Mironov, M.A. (1977): Markov Processes. Sov. Radio: Moscow (Russian). Zbl.478.60078

Triebel, H. (1978): Interpolation. Function Spaces. Differential Operators. Dtsch. Verlag der Wissenschaften: Berlin. Zbl.387.46033

Valitskij, Yu.N., Ejdel'man, S.D. (1976): Necessary and sufficient condition for stabilization of positive solutions of the heat equation. Sib. Mat. Zh. **17**, No. 4, 744–756. English transl.: Sib. Math. J. **17**, No. 4, 564–572 (1976). Zbl.362.35036

Volevich, L.R., Paneyakh, B.P. (=Paneah, B.P.)(1965): Certain spaces of generalized functions and embedding theorems. Usp. Mat. Nauk **20**, No. 1, 3–74. English transl.: Russ. Math. Surv. **20**, No. 1, 1–73 (1965). Zbl.135,165

Weber, M. (1951): The fundamental solution of a degenerate partial differential equation of parabolic type. Trans. Am. Math. Soc. **71**, 24–37. Zbl.43,99

Widder, D.V. (1944): Positive temperatures on an infinite rod. Trans. Am. Math. Soc. **55**, 85–95. Zbl.61,223

Zagorskij, T.Ya. (1961): A mixed problem for systems of partial differential equations of parabolic type. L'vov State University: L'vov (Russian).

Zhikov, V.V. (1977): On the stabilization of solutions of parabolic equations. Mat. Sb., Nov. Ser. **104**, No. 4, 597–616. English transl.: Math. USSR, Sb. **33**, No. 4 519–537 (1977). Zbl.374.35025

Zhitarashu, N.V. (1966): Schauder estimates and solvability of general boundary problems for general parabolic systems with discontinuous coefficients. Dokl. Akad. Nauk SSSR **169**, No. 5, 511–514. English transl.: Sov. Math., Dokl. **7**, No. 4, 952–956 (1966). Zbl.168,83

Zhitarashu, N.V. (1985): Theorems on the complete set of isomorphisms in L_2-theory of generalized solutions of boundary-value problems for a Petrovskij parabolic equation. Mat. Sb., Nov. Ser. **128**, No. 4, 451–473. English transl.: Math. USSR, Sb. **56**, No. 2, 447–471 (1987). Zbl.609.35045

Zhitarashu, N.V. (1987): On well-posedness of general model parabolic boundary-value problems in spaces $\mathcal{H}^s$, $-\infty < s < \infty$. Izv. Akad. Nauk SSSR, Ser. Mat. **51**, No. 5, 962–993. English transl.: Math. USSR, Izv. **31**, No. 2, 273–305 (1988). Zbl.658.35045

Zhitarashu, N.V., Ejdel'man, S.D. (1970): A nonlocal parabolic boundary-value problem. Mat. Issled. 5, No. **3**(17), 83–100 (Russian). Zbl.232.35057

Zhitarashu, N.V., Ejdel'man, S.D. (1979): Necessary and sufficient conditions of stabilization of solutions of model parabolic boundary-value problems. Dokl. Akad. Nauk SSSR **244**, No. 4, 809–813. English transl.: Sov. Math., Dokl. **20**, No. 1, 110–114 (1979). Zbl.432.35041

Zhitomirskij, Ya. I. (1959): The Cauchy problem for parabolic systems of linear partial differential equations with growing coefficients. Izv. Vyssh. Uchebn. Zaved. Mat. **1**, 55–74 (Russian). Zbl.90,74

Author Index

Subject Index

MIX
Papier aus verantwortungsvollen Quellen
Paper from responsible sources
FSC® C105338

If you have any concerns about our products,
you can contact us on
ProductSafety@springernature.com

In case Publisher is established outside the EU,
the EU authorized representative is:
Springer Nature Customer Service Center GmbH
Europaplatz 3, 69115 Heidelberg, Germany

Printed by Libri Plureos GmbH
in Hamburg, Germany